Lecture Notes in Bioinformatics   8542

Subseries of Lecture Notes in Computer Science

Adrian-Horia Dediu Carlos Martín-Vide
Bianca Truthe (Eds.)

Algorithms for Computational Biology

First International Conference, AlCoB 2014
Tarragona, Spain, July 1-3, 2014
Proceedings

 Springer

Volume Editors

Adrian-Horia Dediu
Rovira i Virgili University, Research Group on Mathematical Linguistics
Avinguda Catalunya, 35, 43002 Tarragona, Spain
E-mail: adrian.dediu@urv.cat

Carlos Martín-Vide
Rovira i Virgili University, Research Group on Mathematical Linguistics
Avinguda Catalunya, 35, 43002 Tarragona, Spain
E-mail: carlos.martin@urv.cat

Bianca Truthe
Justus-Liebig-Universität, Fachbereich 07, Institut für Informatik
Arndtstraße 2, 35392 Gießen, Germany
E-mail: bianca.truthe@informatik.uni-giessen.de

ISSN 0302-9743   e-ISSN 1611-3349
ISBN 978-3-319-07952-3   e-ISBN 978-3-319-07953-0
DOI 10.1007/978-3-319-07953-0
Springer Cham Heidelberg New York Dordrecht London

Library of Congress Control Number: 2014940380

LNCS Sublibrary: SL 8 – Bioinformatics

Typesetting: Camera-ready by author, data conversion by Scientific Publishing Services, Chennai, India

Printed on acid-free paper

Springer is part of Springer Science+Business Media (www.springer.com)

Preface

These proceedings contain the papers that were presented at the First International Conference on Algorithms for Computational Biology (AlCoB 2014), held in Tarragona, Spain, during July 1–3, 2014.

The scope of AlCoB includes topics of either theoretical or applied interest, namely:

- Exact sequence analysis
- Approximate sequence analysis
- Pairwise sequence alignment
- Multiple sequence alignment
- Sequence assembly
- Genome rearrangement
- Regulatory motif finding
- Phylogeny reconstruction
- Phylogeny comparison
- Structure prediction
- Proteomics: molecular pathways, interaction networks, etc.
- Transcriptomics: splicing variants, isoform inference and quantification, differential analysis, etc.
- Next-generation sequencing: population genomics, metagenomics, metatranscriptomics, etc.
- Microbiome analysis
- Systems biology

AlCoB 2014 received 39 submissions. Most papers were reviewed by three and some by two Program Committee members. There were also several external referees consulted; we acknowledge all the reviewers in the next section. After a thorough and vivid discussion phase, the committee decided to accept 20 papers (which represents an acceptance rate of 51.28%). The conference program also included two invited talks and one invited tutorial. Part of the success in the management of this number of submissions is due to the excellent facilities provided by the EasyChair conference management system.

We would like to thank all invited speakers and authors for their contributions, the Program Committee and the reviewers for their cooperation, and Springer for its very professional publishing work.

April 2014

Adrian-Horia Dediu
Carlos Martín-Vide
Bianca Truthe

Organization

AlCoB 2014 was organized by the Research Group on Mathematical Linguistics –
GRLMC, from Rovira i Virgili University, Tarragona.

Program Committee

Tatsuya Akutsu	Kyoto University, Japan
Amihood Amir	Bar-Ilan University, Ramat-Gan, Israel
Alberto Apostolico	Georgia Institute of Technology, Atlanta, USA
Joel Bader	Johns Hopkins University, Baltimore, USA
Pierre Baldi	University of California, Irvine, USA
Serafim Batzoglou	Stanford University, USA
Bonnie Berger	Massachusetts Institute of Technology, Cambridge, USA
Francis Y.L. Chin	University of Hong Kong, Hong Kong
Benny Chor	Tel Aviv University, Israel
Keith A. Crandall	George Washington University, Washington DC, USA
Bhaskar DasGupta	University of Illinois, Chicago, USA
Joaquín Dopazo	Príncipe Felipe Research Center, Valencia, Spain
Liliana Florea	Johns Hopkins University, Baltimore, USA
Olivier Gascuel	LIRMM-CNRS, Montpellier, France
David Gilbert	Brunel University, Uxbridge, UK
Gaston H. Gonnet	ETH Zürich, Switzerland
Roderic Guigó	Center for Genomic Regulation, Barcelona, Spain
Dan Gusfield	University of California, Davis, USA
Vasant Honavar	Pennsylvania State University, University Park, USA
Sorin Istrail	Brown University, Providence, USA
Tao Jiang	University of California, Riverside, USA
Inge Jonassen	University of Bergen, Norway
Anders Krogh	University of Copenhagen, Denmark
Giovanni Manzini	University of Eastern Piedmont, Alessandria, Italy
Carlos Martín-Vide (Chair)	Rovira i Virgili University, Tarragona, Spain
Satoru Miyano	University of Tokyo, Japan
Burkhard Morgenstern	University of Göttingen, Germany

Shinichi Morishita	University of Tokyo, Japan
Cédric Notredame	Center for Genomic Regulation, Barcelona, Spain
Graziano Pesole	National Research Council, Bari, Italy
Mark Ragan	University of Queensland, Brisbane, Australia
Timothy Ravasi	King Abdullah University of Science and Technology, Thuwal, Saudi Arabia
Allen G. Rodrigo	Duke University, Durham, USA
Steven Salzberg	Johns Hopkins University, Baltimore, USA
David Sankoff	University of Ottawa, Canada
Thomas Schiex	INRA Toulouse, France
João Carlos Setubal	University of São Paulo, Brazil
Steven Skiena	Stony Brook University, USA
Peter F. Stadler	University of Leipzig, Germany
Wing-Kin Sung	National University of Singapore, Singapore
Alfonso Valencia	Spanish National Cancer Research Centre, Madrid, Spain
Jacques van Helden	University of Aix-Marseille, France
Arndt von Haeseler	Center for Integrative Bioinformatics Vienna, Austria
Lusheng Wang	City University of Hong Kong, Hong Kong
Limsoon Wong	National University of Singapore, Singapore
Xiaohui Xie	University of California, Irvine, USA
Dong Xu	University of Missouri, Columbia, USA
Zohar Yakhini	Agilent Laboratories, Santa Clara, USA
Alex Zelikovsky	Georgia State University, Atlanta, USA
Michael Q. Zhang	University of Texas, Dallas, USA

External Reviewers

Artyomenko, Alexander	Leibovich, Limor
Chateau, Annie	Leung, Henry
De Givry, Simon	Mandric, Igor
Doi, Koichiro	Park, Hee-Won
Ehsani, Sepehr	Puglisi, Simon J.
Gonnet, Pedro	Sheridan, Paul
Hamelryck, Thomas	Srihari, Sriganesh
Hermelin, Danny	Swenson, Krister M.
Katsirelos, George	Wang, Yi
Kifer, Ilona	Wood, Derrick
Kim, Daehwan	Zagrovic, Bojan
Kurowski, Krzysztof	Zheng, Chunfang

Organizing Committee

Adrian-Horia Dediu, Tarragona
Carlos Martín-Vide, Tarragona (Chair)
Bianca Truthe, Gießen
Lilica Voicu, Tarragona

Table of Contents

Invited Talks

Comparative Genomics Approaches to Identifying Functionally Related
Genes 1
  Michael Y. Galperin and Eugene V. Koonin

Regular Papers

A Greedy Algorithm for Hierarchical Complete Linkage Clustering 25
  *Ernst Althaus, Andreas Hildebrandt, and
  Anna Katharina Hildebrandt*

Vester's Sensitivity Model for Genetic Networks with Time-Discrete
Dynamics 35
  *Liana Amaya Moreno, Ozlem Defterli, Armin Fügenschuh, and
  Gerhard-Wilhelm Weber*

Complexity and Polynomial-Time Approximation Algorithms around
the Scaffolding Problem. 47
  Annie Chateau and Rodolphe Giroudeau

Heuristics for the Sorting by Length-Weighted Inversions Problem on
Signed Permutations 59
  Thiago da Silva Arruda, Ulisses Dias, and Zanoni Dias

On Low Treewidth Graphs and Supertrees 71
  Alexander Grigoriev, Steven Kelk, and Nela Lekić

On Optimal Read Trimming in Next Generation Sequencing and Its
Complexity 83
  *Ivo Hedtke, Ioana Lemnian, Matthias Müller-Hannemann, and
  Ivo Grosse*

On the Implementation of Quantitative Model Refinement 95
  Bogdan Iancu, Diana-Elena Gratie, Sepinoud Azimi, and Ion Petre

HapMonster: A Statistically Unified Approach for Variant Calling and
Haplotyping Based on Phase-Informative Reads. 107
  *Kaname Kojima, Naoki Nariai, Takahiro Mimori,
  Yumi Yamaguchi-Kabata, Yukuto Sato, Yosuke Kawai, and
  Masao Nagasaki*

Mapping-Free and Assembly-Free Discovery of Inversion Breakpoints
from Raw NGS Reads 119
    Claire Lemaitre, Liviu Ciortuz, and Pierre Peterlongo

Modeling the Geometry of the Endoplasmic Reticulum Network 131
    *Laurent Lemarchand, Reinhardt Euler, Congping Lin, and
    Imogen Sparkes*

On Sorting of Signed Permutations by Prefix and Suffix Reversals and
Transpositions 146
    Carla Negri Lintzmayer and Zanoni Dias

On the Diameter of Rearrangement Problems........................ 158
    Carla Negri Lintzmayer and Zanoni Dias

Efficiently Enumerating All Connected Induced Subgraphs of a Large
Molecular Network 171
    Sean Maxwell, Mark R. Chance, and Mehmet Koyutürk

On Algorithmic Complexity of Biomolecular Sequence Assembly
Problem 183
    Giuseppe Narzisi, Bud Mishra, and Michael C. Schatz

A Closed-Form Solution for Transcription Factor Activity Estimation
Using Network Component Analysis 196
    *Amina Noor, Aitzaz Ahmad, Bilal Wajid, Erchin Serpedin,
    Mohamed Nounou, and Hazem Nounou*

SVEM: A Structural Variant Estimation Method Using Multi-mapped
Reads on Breakpoints 208
    *Tomohiko Ohtsuki, Naoki Nariai, Kaname Kojima,
    Takahiro Mimori, Yukuto Sato, Yosuke Kawai,
    Yumi Yamaguchi-Kabata, Testuo Shibuya, and
    Masao Nagasaki*

Analysis and Classification of Constrained DNA Elements with N-gram
Graphs and Genomic Signatures 220
    *Dimitris Polychronopoulos, Anastasia Krithara,
    Christoforos Nikolaou, Giorgos Paliouras, Yannis Almirantis, and
    George Giannakopoulos*

Inference of Boolean Networks from Gene Interaction Graphs Using a
SAT Solver... 235
    *David A. Rosenblueth, Stalin Muñoz, Miguel Carrillo, and
    Eugenio Azpeitia*

RRCA: Ultra-Fast Multiple In-species Genome Alignments   247
    Sebastian Wandelt and Ulf Leser

Exact Protein Structure Classification Using the Maximum Contact
Map Overlap Metric.   262
    Inken Wohlers, Mathilde Le Boudic-Jamin, Hristo Djidjev,
    Gunnar W. Klau, and Rumen Andonov

Author Index   275

Comparative Genomics Approaches to Identifying Functionally Related Genes*

Michael Y. Galperin and Eugene V. Koonin

National Center for Biotechnology Information, National Library of Medicine
National Institutes of Health, Bethesda, Maryland, USA
{galperin,koonin}@ncbi.nlm.nih.gov

Abstract. The rapid progress in genome sequencing makes it possible to address fundamental problems of biology and achieve critical insights into the functioning of the live cells and entire organisms. However, the widening gap between the rapidly accumulating sequence data and our ability to properly annotate these data constitutes a major problem that slows down the progress of genome biology. This paper discusses the notion of "function" as it relates to computational biology, lists the most common ways of assigning function to the new genes, particularly those that specifically rely on comparative genome analysis, and briefly reviews the drawbacks of the current algorithms for semi-automated high-throughput functional annotation of genomes.

Keywords: genome annotation, genomic context, gene neighborhood, operon, functional genomics, orthology databases.

1   Introduction

Next year will mark the 20th anniversary of the sequencing of the first complete genome of a cellular organism, the bacterium *Haemophilus influenzae* [1]. Many bacterial and eukaryotic genomes followed shortly after that, including the first human genome in 2001 [2]. These events led to a revolution in the genome sequencing technologies, which sharply decreased the sequencing costs and dramatically changed the way we do science. It is now often cheaper to isolate the DNA from some obscure environmental sample and do the sequencing than to perform a standard biochemical or biophysical experiment.

The rapid progress in technology has led to a largely unexpected conundrum where the sequencing data are being accumulated at such a fast pace that the ability of the biologists to perform any sensible data analysis inevitably falls behind. As a result, most published research typically addresses only a relatively small number of specific problems that prompted generation of the respective data set, and most sequence data remain underutilized by the researchers. The growing schism between data generation

---

A.-H. Dediu, C. Martín-Vide, and B. Truthe (Eds.): AlCoB 2014, LNBI 8542, pp. 1–24, 2014.
© Springer International Publishing Switzerland 2014

and the use of these data makes post-genomic sequence analysis a particularly promising avenue of research, offering computational biologists ample amounts of raw sequence data that could be used to answer a variety of important questions. The onus therefore shifts to the researcher's ability to ask the right questions and to extract from the databases the right data sets to answer these questions.

One of the most common stumbling blocks in converting the raw sequence data to scientific - or biotechnological - findings is the insufficient level of understanding the functions of numerous genes even in the best-studied genomes, such as the bacteria Escherichia coli and Bacillus subtilis, or the yeast Saccharomyces cerevisiae. Even for Escherichia coli K-12, the workhorse of molecular biology and arguably the best-studied organism in the world, the EcoGene database[1] shows that 1336 genes out of the current list of 4141 still have the 'y' designation, indicating that their functions remain uncharacterized [3]. Further, for products of many other genes, only a general function (e.g., 'cell division protein', stress-induced protein') is known at this time. For less-studied organisms, the fraction of uncharacterized genes can be much higher, with virtually all of their genes are being assigned their functions solely based on the sequence similarity to the genes in other organisms. Thus, comparing different genomes and transferring functional annotation of genes (proteins) from better studied organisms to their orthologs from lesser studied organisms has become the key process in the efforts to provide functional annotation of newly sequenced genomes and use this information to achieve a better understanding of the physiology of the respective organisms.

The goal of this presentation is to a) define the notion of "biological function" as it relates to computational biology, b) describe the most popular ways of assigning function to predicted genes (open reading frames), particularly those that specifically rely on comparative genome analysis, and c) discuss the challenges and drawbacks of the current algorithms for semi-automated high-throughput functional annotation of genomes.

2   What Is the Gene "Function"?

While it is only natural to think of the live cell as a perfectly designed system where every part has its own well-defined role (the "function"), in reality, cell components participate in a complex network of interactions and often have more than one role. Most enzymes can work with a group of related substrates instead of a single one (have group specificity) and catalyze various side reactions. The function of the gene is typically defined as the role that its protein product plays in situ, i.e. the live cell. As a result, a protein that hydrolyzes a natural substrate, e.g. a phosphorylated sugar into sugar and phosphate moieties, will be usually called a phosphatase, even if this protein is more active with a non-natural artificial substrate, such as a sugar phosphonate. Sometimes, however, the name is derived from an easily measurable side activity whereas the genuine native function might not even be known. Thus, the enzymes that catalyzed reduction of certain dyes - and whose activity could be easily measured

---

[1] http://www.ecogene.org/

by changes in color - has been referred to as diaphorase for more 20 years before its activity as NAD(P)H:acceptor oxidoreductase has been established and it became clear that there exists a whole family of such enzymes.

In biology, gene (protein) function is usually defined historically, based on the first description of the properties of the respective mutant or the biochemical activity of the purified protein. For essential genes, where mutations are lethal or conditionally lethal, the function can be defined as something that the gene product needs to do to sustain the cell growth. Operationally, for lethal mutations, the cause of cell death is assumed to be the "function" of the gene in question. For non-essential genes, mutation phenotypes can be quite complicated and, accordingly, the descriptions of "function" may be quite long and fuzzy, and not necessarily physiologically relevant, i.e. reflecting their core functions. For example, studies of the sporulation process in the hay bacterium Bacillus subtilis, a popular model organism, have been used to define functions for hundreds of genes. As a result, certain bacterial genes are being referred to as "sporulation" genes, even though the respective organisms, e.g., cyanobacteria, are unable to sporulate [4,5].

This problem becomes particularly severe for high-throughput enzyme assays, which can be used to define general biochemical activities of the products of previously uncharacterized genes, but are often unable to identify the natural substrates for the respective enzymes or the biochemical pathway involving these enzymes [6,7]. A proper definition of the protein function should probably combine characterization of its biochemical activity, if any (i.e. the nature of the catalyzed reaction and the range of utilized substrates and products) with the description of the biological process (e.g. a metabolic or signaling pathway) that involves this protein. For poorly studied organisms, such information is obviously unavailable and every overly specific assignment should be taken with a grain of salt. We have previously discussed certain functional assignments that, despite being supported by reasonably high similarity scores, do not pass even the cursory "sanity check". Examples include bacterial and archaeal "head morphogenesis protein", "mitochondrial benzodiazepine receptor", "centromere protein", and many others [8,9].

In the course of evolution, homologous genes may adopt new functions, sometimes quite distinct from their 'original' ones. There are several excellent databases that collect such data. The FunShift database[2] at the Stockholm University [10] documents functional shifts between different subfamilies within a single protein domain family. The PANTHER[3] database at SRI International in Menlo Park, California, shows such functional shifts on the phylogenetic trees [11], whereas the Structure-Function Linkage Database[4] at the University of California, San Francisco, analyzes structural and functional details for functionally diverse enzymes that belong to the same superfamilies [12].

A further complication is the phenomenon of so-called "moonlighting proteins" that perform one function in one environment, such as cytoplasm, and an entirely

---

[2] http://funshift.sbc.su.se/
[3] http://www.pantherdb.org/
[4] http://sfld.rbvi.ucsf.edu/

different function in a different environment, such as, for example, when secreted outside the cell [13]. Some of such cases are captured in MultitaskProtDB, a database of multitasking proteins[5] at the Universitat Autònoma de Barcelona in Barcelona, Spain [14]. While the number of such moonlighting proteins appears to be relatively small, that might be due to the fact that such cases are not easy to recognize.

To summarize, the biological notion of 'function' is rather fuzzy, which usually leave sufficient wiggle room for functional annotations to be reasonably close to the reality. However, finding proper balance between overly generic (non-specific) and overly specific functional annotation is a complex task that does not have easy algorithmic solutions. Simply copying the functional annotation of the closest homolog in the database or the closest characterized homolog is hardly an appropriate solution, as it leads to numerous problems, from propagation of errors to generation of annotations that cannot pass the sanity check.

3   Homology-Based Functional Assignments

3.1   Annotation by Similarity

The simplest and the most straightforward way to assign function to a newly sequence gene (protein) is to find a similar gene (protein) with an experimentally characterized function. Every day, numerous researchers use the BLAST program on the NCBI web site to perform sequence comparisons and use them to annotate new genes (proteins) based on the functional information from previously characterized genes. There are also other sequence comparisons algorithms; some of them will be mentioned below.

It is important to remember, however, BLAST and other sequence comparisons algorithms measure the degree of sequences similarity, not functional similarity. In other words, such algorithms evaluate the probability that the given sequences are related solely by chance, i.e. the probability that the given sequences are evolutionarily unrelated. When that value is sufficiently low, e.g. less than one per million, this result can be interpreted as evidence of an evolutionary relationship of those sequences, i.e. their common descent from the same ancestral gene. However, at lower similarity levels, i.e. higher E (expectation) values, the probability that the respective proteins have the same function and, therefore, that transfer of functional information from already known genes (proteins) to the new one is justified, becomes progressively lower. Furthermore, because of the intrinsic diversity of biological sequences, there can be no a priori estimate as to which E-value still allows transfer of functional information and which E-value does not.

A potential way out of this conundrum lies in the development of databases of orthologous proteins or, more precisely, orthologous groups of proteins [15]. In its original implementation in the COG database, the algorithm for identification of orthologs across diverse bacteria and archaea relied on the triangles of genome-specific bidirectional best hits with no cut-off by E-value [15]. Subsequent algorithms preserved the need for bidirectional best hits but included certain cut-offs to eliminate spurious hits.

---

[5] http://wallace.uab.es/multitask/

There is now a wide variety of ortholog databases that use various tool to infer orthology and are geared towards various uses, including functional annotation of genomes [15-23].

3.2   Family/Superfamily Annotation

Despite the best efforts on sequence analysis, a substantial fraction of proteins show only a limited similarity to their experimentally characterized counterparts. In many cases, the similarity is limited to the common sequence motifs and/or to the predicted structural features. In such cases, direct transfer of functional information from is hardly justified. Instead, a much more productive way would be replacing a specific - and most likely inaccurate - annotation of the new protein with a family-based annotation, stressing the general conserved features of the family members but avoiding unnecessary specifics (or, rather, leaving them for the future). We have previously discussed the inherent fuzziness of the functional annotation for the members of the ATP-grasp, alkaline phosphatase, all-alpha NTP-PPase, and other superfamilies [24-27], as well as for transcriptional regulators and membrane transporters [8].

Finally, there are numerous protein families whose functions remain totally enigmatic. Such proteins have been referred to as "hypothetical", "conserved hypothetical", "uncharacterized" or even "putative uncharacterized" [28]). Families of such proteins include Domains of Unknown Function (DUFs) in Pfam, and Uncharacterized Protein Families (UPFs) in UniProt [28,29]. These lists are quite valuable for genome annotation, because clarification of the functions of any of their members immediately allows functional assignments for all other members of that family. From the computational standpoint, the software should allow sufficient flexibility in protein names, so that an amended functional assignment could be quickly propagated to the members of a given protein family without the need for any major revamp of the system. In fact, the continuing process of biological research means that changes in gene (protein) functional annotation are bound to be a constant factor in genomic databases for the foreseeable future.

4   Using Genome Comparisons for Predicting Protein Functions

While sequence similarity searches remain by far the most popular tool for identifying the functions of unknown proteins and RNA, in many cases such searches do not yield satisfactory functional annotation, as no functional assignment can be made with any degree of confidence. For such cases, there are several computational approaches that go beyond sequence comparison. Instead, such methods rely on "genomic context", i.e. common properties that are shared by unrelated (non-homologous) proteins that perform the same or related functions. Examples of such proteins include different subunits of the same complex enzyme, components of the same signaling pathway, alternative enzymes that catalyze the same biochemical reaction, and many others. In order for such non-homologous but functionally related protein pairs to work in concert, they need to be present in the same organism at the same time, they might also physically interact. Accordingly, identification of functionally associated pairs of proteins relies on their joint presence and absence in a certain set of genomes (phylogenetic co-occurrence) and their co-expression, as judged by the presence of common regulatory sites, conservation of their location next to each other in multiple genomes, and/or gene fusions [8,30,31].

These approaches have two important traits: they take advantage of the availability of multiple complete genomes and they treat them as genomes rather than just sets of individual genes. Accordingly, these approaches rely on the same basic premise - that organization of the genetic information in each particular genome is meaningful, in the sense that it reflects a long history of mutations, gene duplications, gene re-arrangements, gene function divergence, gene acquisition and loss that has produced organisms that are uniquely adapted to their environment and are capable of regulating their metabolism in accordance with the environmental conditions. Further, some of these approaches, as the analysis of gene co-expression, gene neighborhoods and protein domain fusions, do not require knowledge of complete genome sequences and therefore can benefit from the enormous amount of sequence data available in the unfinished genomes and metagenomes. This dramatically increases sensitivity and robustness of these approaches, making them indispensable tools in the functional analysis of uncharacterized genes.

The principles and methods of genome context-based functional annotation have been described in detail in numerous publications [8,30-43]. Here we briefly describe the general principles of these approaches and discuss their principal caveats. We also discuss the limitations of applying these tools to infer sensible functional association. It is important to note that all these approaches critically depend on the number of available genome sequences and their diversity. Therefore, recent progress in genome sequencing that leads to the constantly growing number of available genomes, even if incomplete, gradually increases the specificity of all these methods, effectively improving the signal-to-noise ratio. In addition, functional links can be deduced from the results of several high-throughput experimental techniques, such as gene co-expression obtained using microarrays or deep RNA sequencing and various protein-protein interaction data. All this makes genomic context-based methods increasingly powerful in providing valuable clues to inferring gene (protein) function.

4.1   Phylogenetic Profiling

General Approach. The number of genes that are encoded in all known genomes is extremely small, less than a hundred, and functions of all of them are already known. Most of these genes encode ribosomal proteins or subunits of several key enzymes of DNA replication, tRNA aminoacylation, and central metabolism [44]. All other genes are present in some genomes and absent in the others. When comparing the distribution of two genes across multiple genomes, one can come with the following general patterns. First, the genes typically co-occur, i.e. certain genomes carry both these genes while other genomes do not have either of them. In such cases, functional association of the two genes becomes very likely, which makes this method a potentially powerful tool for inferring protein function [15,34,38,45]. However, as mentioned above, this functional association is quite fuzzy in biological terms and may be used only for a very general functional annotation. In other cases, the genes are rarely found together, most genome carry either one or the other, resulting in complementary phylogenetic patterns. Such cases may arise from a specific kind of functional association, the one where the respective gene actually have the same (or closely related) functions, such that the organism only needs either of them. Such cases, referred to as non-orthologous gene displacement [46], are not very common but, when found, could be used for very specific functional annotation [31].

Algorithmic Aspects. The overall approach is quite straightforward: compile a matrix of presence (1) or absence (0) of the given genes in as many genomes as possible and calculate the numbers of (1,1), (1,0), (0,1) and (0,0) combinations. Then compare the fraction of (1,1) cases [as well as the combined fraction of (1,1) and (0,0) cases] with the fraction of other two and evaluate the probability that the difference, if any, arises simply by chance. If that probability is sufficiently low, the pair can be marked as likely to have a functional interaction. For non-orthologous gene displacement, vice versa, the (0,1) and (1,0) cases should be far more common than (1,1) ones.

Unfortunately, this approach has several important caveats. First of all, it relies on recognition of the "same gene" in many distinct genomes, i.e. runs into all the problems described above. Different genes evolve with different rates, and even functionally related genes may accumulate mutations, insertions and deletions at dramatically different pace. As a result, two homologous genes in two different genomes might be very similar (e.g. with E-value of 1x10-10), whereas their partners in the same genomes would show only borderline similarity (e.g. E-value of 1x10-3). Selecting an overly strict cut-off for similarity scores would throw away distant homologs of the given gene and might artificially inflate the fraction of (1,0) cases. On the other hand, selecting an overly permissive cut-off would result in an inflated fraction of (1,1) cases, which would decrease the specificity of the method, highlighting spurious gene pairs as functionally related. To avoid this conundrum, one could specifically look for pairs of orthologs in diverse genomes, which would alleviate most of the problems arising from differences in evolutionary rates. However, this would mean either adding an entirely new layer of computation or relying on the external sources of orthology data, which might have their own problems. For example, some orthology databases, like OMA browser[6] emphasize one-to-one correspondence between orthologous genes and are therefore might be sensitive to lineage-specific gene duplication events [17]. We believe that by defining orthologous groups, as opposed to single orthologs, the COG approach offers the best balance of specificity and sensitivity. However, the COG database covers only 63 genomes and has not been updated since 2003.

Another potential problem of phylogenetic profiling is taxonomic depth. With hundreds of Escherichia coli genomes already in the database, most E. coli gene pairs are already found in hundreds of genomes and are missing in numerous other

---

[6] http://omabrowser.org

genomes. While this ensures predominance of (1,1)+(0,0) cases, that does not mean that such genes necessarily interact. Thus gene pairs that are found in phylogenetically distant organisms (e.g. in members of different phyla) should score much higher than those found only at very short phylogenetic distances. It also makes sense to ignore closely related genomes, e.g. by collapsing at the level of genus or even a family. On the other hand, horizontal gene transfer between organisms that inhabit the same environment can result in groups of unrelated genes being co-transferred across large phylogenetic distances, e.g. from hyperthermophilic bacteria to hyperthermophilic archaea or vice versa. As a result, assigning too much value to the rare sightings of the same genes in phylogenetically distinct organisms might be dangerous and counterproductive.

For the rare genes that are found in relatively few genomes, the above factors combine making phylogenetic profiling particularly unreliable. Thus, when the number of (1,1) is small and the number of (0,0) is large, there is a decent chance that the (1,1) cases are not indicative of a functional relationship.

One more potential problem of phylogenetic profiling is the reliance of the method on the correct identification of all the ORFs in the genome. In practice, automatically annotated genomes often miss short ORFs, those with less than 70-80 codons, and sometimes even longer ones [45]. In addition, ORFs with frameshifts typically get omitted from the protein set, even when these frameshifts s result from sequencing errors, so that the genome encodes a fully functional protein. In some cases, supposed frameshifts create stop codons between separate protein domains and therefore do not result in the loss of function but such proteins still get removed from the respective proteomes. We have previously described how deviations from conserved phylogenetic patterns could be used for improving genome annotation [45], but that required manual intervention. When used semi-automatically on a genome scale, phylogenetic profiling, particularly for short ORFs could be very sensitive to the annotation errors.

Practical Aspects. At this time, there is no universally accepted way to score the results of phylogenetic profiling. As a result, this approach is still widely used but typically on an ad hoc basis: biologists typically use co-occurrence of certain genes as additional evidence of their involvement in the same process or a pathway. There are databases that could be used to extract phylogenetic profiles from the genome data, the best and most widely used being the STRING database[7], maintained by Peer Bork and coworkers at the European Molecular biology Laboratory in Heidelberg, Germany [43]. STRING allows the user to select a gene from a variety of complete genomes and search for genes with the same or similar phylogenetic profiles. This tool is very useful for genome annotation, particularly if combined with other options offered by the same database (see below). FunCoup database[8] at the Stockholm University specifically targets eukaryotic genes and, like STRING, presents various kinds of functional coupling information, including phylogenetic profiles [47].

---

[7] http://string.embl.de
[8] http://funcoup.sbc.su.se/

4.2   Genomic Neighborhood

General Approach. Co-expression of proteins belonging to the same metabolic or signaling pathway is typically achieved thorough co-regulation of the transcription of the respective genes by the same transcriptional regulators. This could be detected by identifying common regulatory sites, although the specificity of such prediction is typically limited and they need to be verified by direct experimentation. In bacteria, co-expressed genes are often located next to each other, forming operons that are transcribed as a single multigenic mRNA. On the other hand, due to the constant events of gene translocation within the genome, as well as gene acquisition through horizontal gene transfer and gene loss, the overall gene order is not conserved even among relatively close relatives that belong to the same genus, and is typically wiped out at the level of the bacterial family. Thus, conserved gene neighborhoods in phylogenetically distinct organisms are relatively rare [48] and analysis of gene may provide important functional clues [36,37]. Therefore, bacterial genome analysis offers an easy way of inferring functional connections by simply looking at the genes that are consistently adjacent to the studied gene in multiple genomes. This approach could even be used for analyzing eukaryotic genes through finding bacterial orthologs of the given eukaryotic gene, followed by an analysis of their genome neighborhoods [49].

Algorithmic Aspects. The general approach to the identification of functionally linked genes through the analysis of their genomic context includes the following steps. First, for a given gene from the given organism, one needs to identify the 'same gene' (or, more precisely, orthologs of this gene) in all available genomes and, at the next step, define other genes that belong to the same operons and therefore are co-expressed. However, genetic studies have revealed co-regulated divergent operons (running in both directions from a common regulatory site), as well as convergent ones. That is why, in practice, the direction of the genes is usually ignored and the algorithm simply selects a certain number of their neighbors (just the nearest neighbors or two, three, or more adjacent genes) on one or both sides in all these genomes. These neighboring genes then need be classified into conserved groups of the same function and ranked by the frequency of their occurrence in these neighborhoods. The genes that show a statistically significant association with the orthologs of the given gene may be expected to have a functional connection to this gene.

Obviously, this approach is subject to the same caveats as phylogenetic profiling, and also additional ones. First, again, the definition of the 'same gene' in various genomes has to rely on sequence comparisons and is subject to all the limitations discussed above. The availability of predefined clusters of orthologs helps but, again, means either extra computation or reliance on an external source of information that the user cannot control. This method, however, requires identification of orthologs not just for the initial query gene but also for the genes that abut its orthologs in all studied genomes. This calls for a far more complex computation and/or far more extensive use of orthology databases.

The other two problems of phylogenetic profiling, the taxonomic depth and the potential effect of horizontal gene transfer, also apply to the analysis of the genomic neighborhood. The high incidence of the same genome neighborhood in numerous closely related genomes is likely to make it difficult to find relatively rare cases where the neighbors might be different. On the other hand, such rare associations could reflect cases of horizontal gene transfer and assigning too much weight to them might be misleading.

One more problem complicating the analysis of the genomic neighborhoods is a rapid increase in the amount of the necessary computation with the expansion of the search field. The chance of finding non-trivial gene associations obviously increases when one looks not just at the nearest neighbor(s) but, say, at three, four or five genes on each side from the analyzed one. However, the need to keep track of the identified neighbors and all their orthologs makes the task increasingly complex.

Practical Aspects. There are several different tools for analyzing conserved gene neighborhoods. A popular tool included in the SEED database[9], [50] tags the selected gene and displays conserved genes found in the vicinity of its orthologs ('pinned CDSs'), scoring them by the E-value of the BLAST hit. The user is given the option of choosing the size of the analyzed region (in kilobases), the number of genomes to display, and E-values for selecting the genes to show and to color the same way. This tool is most convenient for analyzing gene neighborhoods among closely related genomes; expanding it to the members of different phyla may be complicated. Another tool is available in the KEGG database, part of the KEGG Orthology[10] system [23]. Instead of BLAST E-values, as in SEED, this tool relies on the precomputed lists of orthologs and displays the members of KEGG orthologous groups located in the genome in the vicinity of the given gene. The most popular tool for studying gene neighborhoods is probably the one at the STRING[11] database [43]. It also relies on precomputed lists of orthologs and displays them over the entire phylogenetic tree. Thus, each tool has its own advantages, and by combining two or more of them, it becomes possible to analyze the gene neighborhoods in much detail and over large phylogenetic distances. Future progress in developing such tools would require creating more comprehensive ortholog databases and improvement of the phylogenetic profiling methods that would allow investigating genome neighborhoods in selected parts of the tree of life.

4.3   Gene Coexpression

General Approach. Strictly speaking, gene colocalization does not always imply coexpression. In fact, adjacent but divergently oriented genes could be part of an 'either one or another' regulatory system. The availability of genome sequences gave rise to genomic microarrays, which allowed simultaneous identification of all genes that are coexpressed in response to a specific environmental signal or in such conditions as nutritional or osmotic stress. Such data have been very useful for the specific

---

[9] http://theseed.org/

[10] http://www.kegg.jp/kegg/ko.html

[11] http://string.embl.de

conditions that they studied but microarray experiments were generally costly and narrowly targeted. Obviously, it would be very attractive to deduce gene coexpression straight from the DNA sequence, by identifying conserved transcriptional regulatory sites in front of the genes that might not even be located in the same genome neighborhood. There have been numerous attempts to predict transcription regulatory sites ab initio on the genome scale. Unfortunately, this task is quite complex and the signal-to-noise ratio is usually pretty low. A much more successful approach has been based on utilizing information about known - experimentally determined - transcriptional regulatory sites and scanning the genomes for additional instances of the same or similar sites. In the past, the sequences of regulatory sites had to be determined experimentally by DNA fingerprinting. More recently, such information has started pouring in from deep sequencing data. As a result, transcriptional profiling with probabilistic models of the likely regulatory sites has become a very promising approach to look for coexpressed genes.

Algorithmic Aspects. The typical approach includes the following steps: compiling a list of known coexpressed genes, creating a multiple alignment of the upstream regulatory sites, converting this alignment into either a frequency profile or a hidden Markov model, and using this profile or HMM to look for (additional) highly-scoring sites, preferably in the intergenic regions. In a large series of papers from Gelfand and colleagues, this approach has been used in combination with the information derived from protein sequences, such as the presence of orthologs in several different genomes [51-57], see [58,59] for review.

Practical Aspects. At this time, there are several tools for gene coexpression profiling, including Gibbs Motif Sampler [60,61] and RegPredict [62]. The first one, Gibbs Motif Sampler, is being run at the servers at the Wadsworth Center in Albany, New York[12], and at Brown University in Providence, Rhode Island[13] [63,64]. In addition, several versions of this software are available for downloading[14]. RegPredict[15] is a web service of the Lawrence Berkeley National Laboratory in Berkeley, California. It is closely associated with RegPrecise[16] and RegTransBase[17], two manually curated databases of transcriptional regulation in prokaryotes [65,66].

4.4   Protein Domain Fusions

General Approach. In some cases, adjacent genes are not just coexpressed, they may lose the stop codon that terminates the first polypeptide chain. Such cases (as well as certain gene recombination events) lead to the formation of fused genes, where a sin-

---

[12] http://bayesweb.wadsworth.org/cgi-bin/gibbs.8.pl?data_type=DNA
[13] http://ccmbweb.ccv.brown.edu/gibbs/gibbs.html
[14] http://mcmc-jags.sourceforge.net/
[15] http://regpredict.lbl.gov/
[16] http://regprecise.lbl.gov
[17] http://regtransbase.lbl.gov

gle protein consists of two or more different domains. While each domain has its own function, the fusion would be viable - and maintained in the course of evolution - only when its components are functionally linked, e.g. by participating in the same pathway or a common regulatory mechanism. Therefore, identification of fused genes offers a convenient way to deduce functional association, which is why it has been referred to as the "Rosetta stone" approach [32,67]. Obviously, protein domain fusions are only helpful when they combine a previously uncharacterized domain with a domain of known function [68]. Fusions of already characterized domains are being studied by numerous researchers for a variety of purposes but not for functional assignments, whereas fusions of uncharacterized domains are interesting but hardly ever contribute to functional analysis.

Algorithmic Aspects. Detection of gene fusions is usually performed at the protein level, through the analysis of multidomain proteins that combine on a single polypeptide chain two or protein domains that are usually found separately (widespread domain fusions, e.g. of pyrimidine biosynthesis enzymes in eukaryotes, are trivial and rarely yield new insights). The search algorithm would largely depend on whether the analyzed gene product contains an already known protein domain. If so, the analysis could be performed using the established databases of protein domains, such as Pfam[18] at the Wellcome Trust Sanger Institute or InterPro[19] at the European Bioinformatics Institute, both in Hinxon, UK, or the NCBI's Conserved Domain Database[20] databases [29,69,70]. Each of these databases allows listing all domain architectures that involve the given domain.

If, however, the analyzed gene product does not contain any protein domains that are listed in public domain databases, the only applicable way seems to be using BLAST (or PSI-BLAST, or HMMer) to find all instances of the new domain, sort the search output by length looking for the longest database hits, and then analyze those hits one-by-one to see if they contain any - known or new - conserved domains.

Analysis of meaningful protein fusions is relatively robust and is subject to few caveats. The most important of those is the existence of so-called "promiscuous" domains that associate with a wide variety of distinct proteins and do not allow any functional inferences. Another potential issue is limiting the depth of the similarity search. Tell-tale fusions of the given protein are often found only after several iterations of PSI-BLAST or JackHMMer, and the degree of sequence conservation might be fairly low. Then there is no guarantee that such domains retain the same or even marginally similar functions, particularly when fused to different partners. Thus, finding protein fusions among distant homologs makes it difficult to draw any unequivocal conclusions.

---

[18] http://pfam.sanger.ac.uk

[19] http://www.ebi.ac.uk/interpro/

[20] http://www.ncbi.nlm.nih.gov/cdd

Practical Aspects. The information on protein domain fusions is available in several databases, including FusionDB[21] at the Institut de Microbiologie de la Méditerranée in Marseille, France [71]. Still, it appears that in bacteria, a significant fraction of fused genes are fusions with the signal-transducing phosphoacceptor REC domain, DNA-binding helix-turn-helix domain, and other promiscuous domains. While it is interesting to see the variety of known protein domains that are fused with REC and therefore fall under the control of the two-component signal transduction [72] or can be found in transcriptional regulators (helix-turn-helix domain fusions), such cases do not advance the cause of functional annotation. Likewise, in eukaryotes, many domain fusions involve SH2, SH3, and other regulatory domains [73], giving no clue as to what specific activity is being regulated. On the other hand, domain fusion maps are already available for numerous domains of unknown function, DUFs in Pfam [29]. Thus, even a minor advance in understanding the function of a previously uncharacterized domain - or, say, availability of its 3D structure - can be quickly propagated to all proteins that contain this domain.

4.5   Protein-Protein Interactions

General Approach. Obviously, protein domain fusions capture only a relatively small fraction of protein-protein interactions. Some additional information on such interactions can be extracted from protein crystal structures that sometimes contain distinct protein domains and show their mutual orientation and the mode(s) of domain interactions. Such data are stored in a variety of public databases, including iPfam[22], 3did[23], DIMA[24], DOMINE[25] [74-77], and many others. However, most information on protein-protein interactions comes from experimental data. These data are being collected - and often ranked by reliability - in several aggregator databases, such as Bio-GRID[26], BindingMOAD[27], DIP[28], HitPredict[29], IntAct[30], MINT[31] [78-84], and many others. A selected list of such databases can be found in the *Nucleic Acids Research* online Molecular Biology Database Collection web site[32] [85]. Unfortunately, all experimental methods for detecting protein-protein interactions are known to bring a substantial number of false-positives. The situation has become so bad that there is even a database of known non-interacting proteins, Negatome[33] [86], designed to

---

[21] http://igs-server.cnrs-mrs.fr/FusionDB/
[22] http://ipfam.sanger.ac.uk/
[23] http://3did.irbbarcelona.org
[24] http://webclu.bio.wzw.tum.de/dima
[25] http://domine.utdallas.edu/
[26] http://www.thebiogrid.org/
[27] http://www.BindingMOAD.org
[28] http://dip.doe-mbi.ucla.edu/
[29] http://hintdb.hgc.jp/htp/
[30] http://www.ebi.ac.uk/intact/
[31] http://mint.bio.uniroma2.it/mint/
[32] http://www.oxfordjournals.org/nar/database/subcat/6/26
[33] http://mips.helmholtz-muenchen.de/proj/ppi/negatome

serve as a tool for estimating false-positive rates in protein-protein interactions experiments and tools. Accordingly, scanning the available databases for the information on protein-protein interactions is a good way to get potential clues on the function(s) of the given protein but the reliability of such clues is typically pretty low.

Practical Aspects. It generally makes sense to query the available databases not just for protein-protein interactions of the given protein but also its orthologs from other, related genomes. Some protein-protein interactions databases rank the results by reliability; incorporating these scores is generally a good idea. However, it should be noted that all those databases feed on a relatively limited number of original studies. Therefore, merely finding certain interaction in several different databases should not be used as evidence of a high-confidence interaction.

5   Combining Disparate Data into a Single Annotation

With the exception of a relatively small number of well-known and straightforward cases, functional annotations of new genes (proteins) are inherently fuzzy. One of the reasons for that is that these gene annotations are expected to be as specific and as reliable as possible. These two demands are somewhat contradictory: a very general but mostly useless annotation (e.g. a "metal-binding protein") could be made with a high degree of confidence, whereas a more specific - and more useful - annotation might not be that well-grounded and totally reliable.

The International Nucleotide Sequence Database Collaboration[34], which includes NCBI's GenBank[35], the EBI's European Nucleotide Archive[36], and the DNA Data Bank of Japan[37], uses a simple schema with two evidence qualifiers, /experiment and /inference[38], which replaced the previously used qualifiers, 'experimental' and 'non-experimental'. These two qualifiers come with controlled vocabularies[39] that specify, respectively, experimental or non-experimental evidence that supports the feature assignment[38]. These evidence codes are increasingly being used to justify functional assignments of the open reading frames in the newly sequenced genomes. As a result, it becomes much easier for the outside user to trace to the origin of the specific annotation and decide whether it is trustworthy.

It is important to note, however, that while the INSDC guidelines require the annotator to specify the evidence in the "/inference="similar to DNA sequence: INSD:AY411252.1" format[38], they impose no limits on the degree of similarity that is acceptable in that annotation. As a result, certain technically acceptable annotations may be based on extremely low similarity levels or even on previous annotations that themselves were non-experimental and highly unreliable. There have been several

---

[34] http://www.insdc.org/

[35] http://www.ncbi.nlm.nih.gov/genbank/

[36] http://www.ebi.ac.uk/ena

[37] http://www.ddbj.nig.ac.jp/

[38] http://www.ncbi.nlm.nih.gov/genbank/evidence

[39] http://www.insdc.org/documents

attempts to develop a common set of standard operating procedures for genome annotation [87], one such list is available online[40], although most links there are no longer functional. The NCBI maintains its own Prokaryotic Genome Annotation Pipeline[41] and Eukaryotic Genome Annotation Pipeline[42] projects that include certain annotation standards[43,44] designed to improve the annotation quality.

Still, there is a clear need for new computationally sound pipelines that would comb through all sorts of disparate clues discussed in the previous sections in order to a) provide the best possible annotations and b) not just list the annotation sources but also evaluate the reliability of these annotations.

For protein annotation, the UniProt web site[45] contains a variety of useful documents, including a constantly updated list of protein naming guidelines[46]. The key question is, of course, "Annotation propagation: when to cut, copy and paste?" as formulated in [88]. Several years ago we have come up with an annotation schema that included the following seven categories [89]:

1. Exact biochemical function, based on high similarity to experimentally characterized closely related homolog
2. Well defined biochemical function, unknown specificity
3. General biochemical function, based on family/superfamily assignment and/or a conserved sequence motif
4. General biological function derived from the domain organization, genome context (e.g., operons), experimental (e.g., protein-protein interactions), and/or structural genomics data (e.g., similarities to proteins with known 3D structures)
5. Certain functional insights derived from the above data
6. Widely conserved protein, expressed under certain growth condition(s)
7. Organism- or genus-specific protein, expressed under certain growth condition(s).

For the first two of the above categories, the best guidance can be found on the web site of the HAMAP project[47], which includes a set of manually created annotation rules[48] that specify the proper annotations for specific family members [90]. For the third, and particularly for the remaining categories, the decision should probably be made by a human annotator. Therefore, it is extremely important to provide that human annotator with the proper tools that simplify his/her work. In practical terms, that would mean bringing together the results of all the analyses that have been discussed above and ranking the results by their relevance and predictive value. The resulting report would probably be pretty long and confusing. As an example, the

---

[40] http://www.ncbi.nlm.nih.gov/pmc/articles/PMC3196215/table/T1/
[41] http://www.ncbi.nlm.nih.gov/genome/annotation_prok
[42] http://www.ncbi.nlm.nih.gov/books/NBK169439/
[43] http://www.ncbi.nlm.nih.gov/genome/annotation_prok/standards
[44] http://www.ncbi.nlm.nih.gov/genome/annotation_euk/process/
[45] http://www.uniprot.org/docs/
[46] http://www.uniprot.org/docs/proknameprot
[47] http://hamap.expasy.org/
[48] http://hamap.expasy.org/rules.html

report for the *Vibrio cholerae* protein VC2772 (RefSeq entry NP_232398, UniProt accession number Q9KNG7) would probably look like the following:

1. TIGRFAM04285, Nucleoid occlusion protein. Query coverage: 199/293 aa; target coverage 198/255 aa; bit score: 223.5; E-value: 1.3e-71. Family description: Nucleoid occlusion protein, a close homolog to ParB chromosome partitioning proteins including Spo0J in *Bacillus subtilis*. Confidence: High

2. SwissProt BLAST hit P26497|SP0J_BACSU, Stage 0 sporulation protein J; Query coverage: 288/293 aa; target coverage 275/282 aa; identities: 106/292; positives: 168/292; gaps: 21/292; bit score: 171; E-value: 3e-55; Confidence: High

3. PDB BLAST hit 1VZ0, Chromosome Segregation Protein Spo0j From *Thermus Thermophilus*. Query coverage: 231/293 aa; target coverage 211/230 aa; identities: 98/232; positives: 149/232; gaps: 22/232; bit score: 170; E-value: 1e-55; Confidence: High

4. TIGRFAM00180, ParB/RepB/Spo0J family partition protein. Query coverage: 179/293 aa; target coverage 186/187 aa; bit score: 177.5; E-value: 1.2e-54; Family description: Chromosomal and plasmid partition proteins related to ParB, including Spo0J, RepB, and SopB. Confidence: High

5. COG1475, Spo0J. Query coverage: 230/293 aa; target coverage 229/240 aa; bit score: 156.2; E-value: 1.1e-45; Family description: Stage 0 sporulation protein J (antagonist of Soj) containing ParB-like nuclease domain. Confidence: High

6. SUPERFAMILY SSF109709, KorB DNA-binding domain-like. Query coverage: 109/293 aa; Region: 122-230; E-value: 1.3e-33. Confidence: High

7. Pfam PF02195, ParBc. Query coverage: 89/293 aa; target coverage 88/90 aa; bit score: 109; E-value: 1.3e-29; Family description: ParB-like nuclease domain. Confidence: High

8. SUPERFAMILY SSF110849, ParB/Sulfiredoxin. Query coverage: 92/293 aa; Region: 41-132; E-value: 3.4e-28. Confidence: High

9. SwissProt BLAST hit P77174|YBDM_ECOLI, Uncharacterized protein YbdM. Query coverage: 136/293 aa; target coverage 140/209 aa; bit score: 39.3; E-value: 2e-8; Confidence: Medium

10. SwissProt BLAST hit P76068|YNAK_ECOLI, Uncharacterized protein YnaK. Query coverage: 63/293 aa; target coverage 69/87 aa; bit score: 30.4; E-value: 2e-6; Confidence: Medium

11. PDB: 1VZ0, chromosome segregation protein Spo0J from *Thermus thermophilus*.

12. PubMed: 15228524, Leonard,T.A., Butler,P.J. and Lowe,J. Structural analysis of the chromosome segregation protein Spo0J from *Thermus thermophilus*. Mol. Microbiol. 53 (2), 419-432 (2004)

13. STRING Genome neighbors: *VC_2773*, ParA family protein (257 aa), score: 0.995; *VC_2061*, ParA family protein (258 aa), score: 0.932; *gidA*, tRNA uridine 5-carboxymethylaminomethyl modification enzyme GidA (631 aa), score: 0.877; *gidB*, 16S rRNA methyltransferase GidB; specifically methylates the N7 position of guanosine (210 aa), score: 0.862; *ftsK*, putative cell division protein FtsK; DNA motor protein (960 aa), score: 0.823; *VC_A1115*, ParA family protein (407 aa), score: 0.764.

14. STRING Domain fusions: None
15. STRING Coexpression data: *atpB*, F_0F_1 ATP synthase subunit A, key component of the proton channel
16. Protein-protein interactions: ParA, a Walker-type ATPase with non-specific DNA-binding activity.

Looking at all these data, the annotator would realize that VC2772 is a DNA-binding protein that also interacts with ParA protein and participates in chromosome partitioning during cell division. Based on that, the tentative annotation would probably be as follows: Chromosome segregation protein Spo0J, contains ParB-like nuclease domain. Please note that automatic transfer of the annotation of the best database hit, Stage 0 sporulation protein Spo0J, would be an unforgivable mistake because, unlike *B. subtilis*, *Vibrio cholerae* does not form spores. This example shows some of the caveats in annotating new proteins, even those with reasonably well characterized homologs. However, there is always a hope that in the future it would be possible to create a comprehensive set of rules (expanding those already available in HAMAP[48]) that would allow a largely automated assignment of functions to a great majority of proteins encoded in any bacterial or eukaryotic genome.

6   Conclusions

In conclusion, improved functional annotation is the only feasible way to extracting information from genomic sequences and gaining a better understanding of the processes in the live cell. For numerous uncultured organisms, as well as for metagenomes, computational analysis is the only way to go. In most part, improved functional assignments would depend on the experimental characterization of the remaining unknown genes. Several recent discoveries, including the CRISPR-Cas system and the c-di-GMP, c-di-AMP-and c-di-GAMP-mediated cellular signaling in bacteria and eukaryotes, show that there could still be major gaps in our understanding of the key processes even in the relatively well-studied cells.

That said, improved algorithms for functional annotation would play a major role in generating viable hypotheses and guiding the experimental research. For many widespread uncharacterized proteins with sufficiently wide phylogenetic representation, simultaneous application of all the tools described above can be expected to generate a number of leads that would either point out the likely function or at least suggest specific experiments that would eventually allow doing so. That would indeed be an invaluable contribution of comparative genomics to genome biology and biology as a whole. Exactly this approach lies at the heart of the COMputational BRidge to EXperiments (COMBREX[49]) project, which aims at obtaining the best possible computational predictions and subjecting them to experimental verification [91,92]. This and other similar projects have a bright future, as only through combined efforts of computational, structural, and experimental biologists would it be possible to achieve a better understanding of gene function on the genome scale.

---

[49] http://combrex.bu.edu/

Acknowledgements. This study was supported by the Intramural Research Program of the National Library of Medicine at the U.S. National Institutes of Health.

References

1. Fleischmann, R.D., Adams, M.D., White, O., Clayton, R.A., Kirkness, E.F., Kerlavage, A.R., Bult, C.J., Tomb, J.-F., Dougherty, B.A., Merrick, J.M., McKenney, K., Sutton, G.G., FitzHugh, W., Fields, C., Gocayne, J.D., Scott, J., Shirley, R., Liu, L.-I., Glodek, A., Kelley, J.M., Weidman, J.F., Phillips, C.A., Spriggs, T., Hedblom, E., Cotton, M.D., Utterback, T.R., Hanna, M.C., Nguyen, D., Saudek, D.M., Brandon, R.C., Fine, L.D., Frichtman, J.L., Fuhrmann, J.L., Geoghagen, N.S.M., Gnehm, C.L., McDonald, L.A., Small, K.V., Fraser, C.M., Smith, H.O., Venter, J.C.: Whole-genome random sequencing and assembly of Haemophilus influenzae Rd. Science 269, 496–512 (1995)
2. Lander, E.S., Linton, L.M., Birren, B., Nusbaum, C., Zody, M.C., Baldwin, J., Devon, K., Dewar, K., Doyle, M., FitzHugh, W., Funke, R., Gage, D., Harris, K., Heaford, A., Howland, J., Kann, L., Lehoczky, J., LeVine, R., McEwan, P., McKernan, K., Meldrim, J., Mesirov, J.P., Miranda, C., Morris, W., Naylor, J., Raymond, C., Rosetti, M., Santos, R., Sheridan, A., Sougnez, C., Stange-Thomann, N., Stojanovic, N., Subramanian, A., Wyman, D., Rogers, J., Sulston, J., Ainscough, R., Beck, S., Bentley, D., Burton, J., Clee, C., Carter, N., Coulson, A., Deadman, R., Deloukas, P., Dunham, A., Dunham, I., Durbin, R., French, L., Grafham, D., Gregory, S., Hubbard, T., Humphray, S., Hunt, A., Jones, M., Lloyd, C., McMurray, A., Matthews, L., Mercer, S., Milne, S., Mullikin, J.C., Mungall, A., Plumb, R., Ross, M., Shownkeen, R., Sims, S., Waterston, R.H., Wilson, R.K., Hillier, L.W., McPherson, J.D., Marra, M.A., Mardis, E.R., Fulton, L.A., Chinwalla, A.T., Pepin, K.H., Gish, W.R., Chissoe, S.L., Wendl, M.C., Delehaunty, K.D., Miner, T.L., Delehaunty, A., Kramer, J.B., Cook, L.L., Fulton, R.S., Johnson, D.L., Minx, P.J., Clifton, S.W., Hawkins, T., Branscomb, E., Predki, P., Richardson, P., Wenning, S., Slezak, T., Doggett, N., Cheng, J.F., Olsen, A., Lucas, S., Elkin, C., Uberbacher, E., Frazier, M., Gibbs, R.A., Muzny, D.M., Scherer, S.E., Bouck, J.B., Sodergren, E.J., Worley, K.C., Rives, C.M., Gorrell, J.H., Metzker, M.L., Naylor, S.L., Kucherlapati, R.S., Nelson, D.L., Weinstock, G.M., Sakaki, Y., Fujiyama, A., Hattori, M., Yada, T., Toyoda, A., Itoh, T., Kawagoe, C., Watanabe, H., Totoki, Y., Taylor, T., Weissenbach, J., Heilig, R., Saurin, W., Artiguenave, F., Brottier, P., Bruls, T., Pelletier, E., Robert, C., Wincker, P., Smith, D.R., Doucette-Stamm, L., Rubenfield, M., Weinstock, K., Lee, H.M., Dubois, J., Rosenthal, A., Platzer, M., Nyakatura, G., Taudien, S., Rump, A., Yang, H., Yu, J., Wang, J., Huang, G., Gu, J., Hood, L., Rowen, L., Madan, A., Qin, S., Davis, R.W., Federspiel, N.A., Abola, A.P., Proctor, M.J., Myers, R.M., Schmutz, J., Dickson, M., Grimwood, J., Cox, D.R., Olson, M.V., Kaul, R., Shimizu, N., Kawasaki, K., Minoshima, S., Evans, G.A., Athanasiou, M., Schultz, R., Roe, B.A., Chen, F., Pan, H., Ramser, J., Lehrach, H., Reinhardt, R., McCombie, W.R., de la Bastide, M., Dedhia, N., Blocker, H., Hornischer, K., Nordsiek, G., Agarwala, R., Aravind, L., Bailey, J.A., Bateman, A., Batzoglou, S., Birney, E., Bork, P., Brown, D.G., Burge, C.B., Cerutti, L., Chen, H.C., Church, D., Clamp, M., Copley, R.R., Doerks, T., Eddy, S.R., Eichler, E.E., Furey, T.S., Galagan, J., Gilbert, J.G., Harmon, C., Hayashizaki, Y., Haussler, D., Hermjakob, H., Hokamp, K., Jang, W., Johnson, L.S., Jones, T.A., Kasif, S., Kaspryzk, A., Kennedy, S., Kent, W.J., Kitts, P., Koonin, E.V., Korf, I., Kulp, D., Lancet, D., Lowe, T.M., McLysaght, A., Mikkelsen, T., Moran, J.V., Mulder, N., Pollara, V.J., Ponting, C.P., Schuler, G., Schultz, J., Slater, G., Smit, A.F., Stupka, E., Szustakowski, J., Thierry-Mieg, D., Thierry-Mieg, J., Wagner, L.,

Wallis, J., Wheeler, R., Williams, A., Wolf, Y.I., Wolfe, K.H., Yang, S.P., Yeh, R.F., Collins, F., Guyer, M.S., Peterson, J., Felsenfeld, A., Wetterstrand, K.A., Patrinos, A., Morgan, M.J., de Jong, P., Catanese, J.J., Osoegawa, K., Shizuya, H., Choi, S., Chen, Y.J.: Initial sequencing and analysis of the human genome. Nature 409, 860–921 (2001)

3. Zhou, J., Rudd, K.E.: EcoGene 3.0. Nucleic Acids Res. 41, D613–D624 (2013)

4. Rigden, D.J., Galperin, M.Y.: Sequence analysis of GerM and SpoVS, uncharacterized bacterial 'sporulation' proteins with widespread phylogenetic distribution. Bioinformatics 24, 1793–1797 (2008)

5. Galperin, M.Y., Mekhedov, S.L., Puigbo, P., Smirnov, S., Wolf, Y.I., Rigden, D.J.: Genomic determinants of sporulation in Bacilli and Clostridia: Towards the minimal set of sporulation-specific genes. Environ. Microbiol. 14, 2870–2890 (2012)

6. Kuznetsova, E., Proudfoot, M., Sanders, S.A., Reinking, J., Savchenko, A., Arrowsmith, C.H., Edwards, A.M., Yakunin, A.F.: Enzyme genomics: Application of general enzymatic screens to discover new enzymes. FEMS Microbiol. Rev. 29, 263–279 (2005)

7. Kuznetsova, E., Proudfoot, M., Gonzalez, C.F., Brown, G., Omelchenko, M.V., Borozan, I., Carmel, L., Wolf, Y.I., Mori, H., Savchenko, A.V., Arrowsmith, C.H., Koonin, E.V., Edwards, A.M., Yakunin, A.F.: Genome-wide analysis of substrate specificities of the Escherichia coli haloacid dehalogenase-like phosphatase family. J. Biol. Chem. 281, 36149–36161 (2006)

8. Koonin, E.V., Galperin, M.Y.: Sequence - Evolution - Function. Computational Approaches in Comparative Genomics. Kluwer, Boston (2003)

9. Galperin, M.Y., Koonin, E.V.: From complete genome sequence to 'complete' understanding? Trends Biotechnol. 28, 398–406 (2010)

10. Abhiman, S., Sonnhammer, E.L.: FunShift: A database of function shift analysis on protein subfamilies. Nucleic Acids Res. 33, D197–D200 (2005)

11. Mi, H., Muruganujan, A., Thomas, P.D.: PANTHER in 2013: Modeling the evolution of gene function, and other gene attributes, in the context of phylogenetic trees. Nucleic Acids Res. 41, D377–D386 (2013)

12. Akiva, E., Brown, S., Almonacid, D.E., Barber, A.E., Custer, A.F., Hicks, M.A., Huang, C.C., Lauck, F., Mashiyama, S.T., Meng, E.C., Mischel, D., Morris, J.H., Ojha, S., Schnoes, A.M., Stryke, D., Yunes, J.M., Ferrin, T.E., Holliday, G.L., Babbitt, P.C.: The Structure-Function Linkage Database. Nucleic Acids Res. 42, D521–D530 (2014)

13. Copley, S.D.: Moonlighting is mainstream: Paradigm adjustment required. Bioessays 34, 578–588 (2012)

14. Hernandez, S., Ferragut, G., Amela, I., Perez-Pons, J., Pinol, J., Mozo-Villarias, A., Cedano, J., Querol, E.: MultitaskProtDB: A database of multitasking proteins. Nucleic Acids Res. 42, D517–D520 (2014)

15. Tatusov, R.L., Koonin, E.V., Lipman, D.J.: A genomic perspective on protein families. Science 278, 631–637 (1997)

16. Tatusov, R.L., Galperin, M.Y., Natale, D.A., Koonin, E.V.: The COG database: A tool for genome-scale analysis of protein functions and evolution. Nucleic Acids Res. 28, 33–36 (2000)

17. Altenhoff, A.M., Schneider, A., Gonnet, G.H., Dessimoz, C.: OMA 2011: Orthology inference among 1000 complete genomes. Nucleic Acids Res. 39, D289–D294 (2011)

18. Fischer, S., Brunk, B.P., Chen, F., Gao, X., Harb, O.S., Iodice, J.B., Shanmugam, D., Roos, D.S., Stoeckert, C.J.: Using OrthoMCL to assign proteins to OrthoMCL-DB groups or to cluster proteomes into new ortholog groups. Curr. Protoc. Bioinformatics ch. 6, unit 6 12 , 11–19 (2011)

19. Waterhouse, R.M., Tegenfeldt, F., Li, J., Zdobnov, E.M., Kriventseva, E.V.: OrthoDB: A hierarchical catalog of animal, fungal and bacterial orthologs. Nucleic Acids Res. 41, D358–D365 (2013)

20. Powell, S., Forslund, K., Szklarczyk, D., Trachana, K., Roth, A., Huerta-Cepas, J., Gabaldon, T., Rattei, T., Creevey, C., Kuhn, M., Jensen, L.J., von Mering, C., Bork, P.: eggnog v4.0: Nested orthology inference across 3686 organisms. Nucleic Acids Res. 42, 231–239 (2014)

21. Datta, R.S., Meacham, C., Samad, B., Neyer, C., Sjolander, K.: Berkeley PHOG: Phylo-Facts orthology group prediction web server. Nucleic Acids Res. 37, W84–W89 (2009)

22. Ostlund, G., Schmitt, T., Forslund, K., Kostler, T., Messina, D.N., Roopra, S., Frings, O., Sonnhammer, E.L.: InParanoid 7: New algorithms and tools for eukaryotic orthology analysis. Nucleic Acids Res 38, D196–D203 (2010)

23. Kanehisa, M., Goto, S., Sato, Y., Kawashima, M., Furumichi, M., Tanabe, M.: Data, information, knowledge and principle: Back to metabolism in KEGG. Nucleic Acids Res. 42, D199–D205 (2014)

24. Galperin, M.Y., Koonin, E.V.: A diverse superfamily of enzymes with ATP-dependent carboxylate-amine/thiol ligase activity. Protein Sci. 6, 2639–2643 (1997)

25. Galperin, M.Y., Bairoch, A., Koonin, E.V.: A superfamily of metalloenzymes unifies phosphopentomutase and cofactor- independent phosphoglycerate mutase with alkaline phosphatases and sulfatases. Protein Sci. 7, 1829–1835 (1998)

26. Moroz, O.V., Murzin, A.G., Makarova, K.S., Koonin, E.V., Wilson, K.S., Galperin, M.Y.: Dimeric dUTPases, HisE, and MazG belong to a new superfamily of all-alpha NTP pyrophosphohydrolases with potential "house-cleaning" functions. J. Mol. Biol. 347, 243–255 (2005)

27. Galperin, M.Y., Koonin, E.V.: Divergence and convergence in enzyme evolution. J. Biol. Chem. 287, 21–28 (2012)

28. The UniProt Consortium: Activities at the Universal Protein Resource (UniProt). Nucleic Acids Res. 42, D191–D198 (2014)

29. Finn, R.D., Bateman, A., Clements, J., Coggill, P., Eberhardt, R.Y., Eddy, S.R., Heger, A., Hetherington, K., Holm, L., Mistry, J., Sonnhammer, E.L., Tate, J., Punta, M.: Pfam: The protein families database. Nucleic Acids Res. 42, D222–D230 (2014)

30. Huynen, M.A., Snel, B.: Gene and context: Integrative approaches to genome analysis. Adv. Protein Chem. 54, 345–379 (2000)

31. Galperin, M.Y., Koonin, E.V.: Who's your neighbor? New computational approaches for functional genomics. Nat. Biotechnol. 18, 609–613 (2000)

32. Marcotte, E.M., Pellegrini, M., Ng, H.L., Rice, D.W., Yeates, T.O., Eisenberg, D.: Detecting protein function and protein-protein interactions from genome sequences. Science 285, 751–753 (1999)

33. Marcotte, E.M., Pellegrini, M., Thompson, M.J., Yeates, T.O., Eisenberg, D.: A combined algorithm for genome-wide prediction of protein function. Nature 402, 83–86 (1999)

34. Pellegrini, M., Marcotte, E.M., Thompson, M.J., Eisenberg, D., Yeates, T.O.: Assigning protein functions by comparative genome analysis: Protein phylogenetic profiles. Proc. Natl. Acad. Sci. USA 96, 4285–4288 (1999)

35. Overbeek, R., Begley, T., Butler, R.M., Choudhuri, J.V., Chuang, H.Y., Cohoon, M., de Crecy-Lagard, V., Diaz, N., Disz, T., Edwards, R., Fonstein, M., Frank, E.D., Gerdes, S., Glass, E.M., Goesmann, A., Hanson, A., Iwata-Reuyl, D., Jensen, R., Jamshidi, N., Krause, L., Kubal, M., Larsen, N., Linke, B., McHardy, A.C., Meyer, F., Neuweger, H., Olsen, G., Olson, R., Osterman, A., Portnoy, V., Pusch, G.D., Rodionov, D.A., Ruckert, C., Steiner, J., Stevens, R., Thiele, I., Vassieva, O., Ye, Y., Zagnitko, O., Vonstein, V.: The subsystems approach to genome annotation and its use in the project to annotate 1000 genomes. Nucleic Acids Res. 33, 5691–5702 (2005)

36. Overbeek, R., Fonstein, M., D'Souza, M., Pusch, G.D., Maltsev, N.: The use of contiguity on the chromosome to predict functional coupling. Silico Biol. 1 (1998)

37. Overbeek, R., Fonstein, M., D'Souza, M., Pusch, G.D., Maltsev, N.: The use of gene clusters to infer functional coupling. Proc. Natl. Acad. Sci. USA 96, 2896–2901 (1999)

38. Gaasterland, T., Ragan, M.A.: Microbial genescapes: Phyletic and functional patterns of ORF distribution among prokaryotes. Microb. Comp. Genomics 3, 199–217 (1998)

39. Rogozin, I.B., Makarova, K.S., Murvai, J., Czabarka, E., Wolf, Y.I., Tatusov, R.L., Szekely, L.A., Koonin, E.V.: Connected gene neighborhoods in prokaryotic genomes. Nucleic Acids Res. 30, 2212–2223 (2002)

40. Rogozin, I.B., Makarova, K.S., Wolf, Y.I., Koonin, E.V.: Computational approaches for the analysis of gene neighbourhoods in prokaryotic genomes. Brief Bioinform. 5, 131–149 (2004)

41. Wolf, Y.I., Rogozin, I.B., Kondrashov, A.S., Koonin, E.V.: Genome alignment, evolution of prokaryotic genome organization, and prediction of gene function using genomic context. Genome Res. 11, 356–372 (2001)

42. Yanai, I., Mellor, J.C., DeLisi, C.: Identifying functional links between genes using conserved chromosomal proximity. Trends Genet. 18, 176–179 (2002)

43. Franceschini, A., Szklarczyk, D., Frankild, S., Kuhn, M., Simonovic, M., Roth, A., Lin, J., Minguez, P., Bork, P., von Mering, C., Jensen, L.J.: STRING v9.1: Protein-protein interaction networks, with increased coverage and integration. Nucleic Acids Res. 41, 808–815 (2013)

44. Koonin, E.V., Wolf, Y.I.: Genomics of bacteria and archaea: The emerging dynamic view of the prokaryotic world. Nucleic Acids Res. 36, 6688–6719 (2008)

45. Natale, D.A., Galperin, M.Y., Tatusov, R.L., Koonin, E.V.: Using the COG database to improve gene recognition in complete genomes. Genetica 108, 9–17 (2000)

46. Koonin, E.V., Mushegian, A.R., Bork, P.: Non-orthologous gene displacement. Trends Genet. 12, 334–336 (1996)

47. Schmitt, T., Ogris, C., Sonnhammer, E.L.: FunCoup 3.0: Database of genome-wide functional coupling networks. Nucleic Acids Res. 42, 380–388 (2014)

48. Koonin, E.V., Galperin, M.Y.: Prokaryotic genomes: The emerging paradigm of genome-based microbiology. Curr. Opin. Genet. Dev. 7, 757–763 (1997)

49. Osterman, A., Overbeek, R.: Missing genes in metabolic pathways: A comparative genomics approach. Curr. Opin. Chem. Biol. 7, 238–251 (2003)

50. Overbeek, R., Olson, R., Pusch, G.D., Olsen, G.J., Davis, J.J., Disz, T., Edwards, R.A., Gerdes, S., Parrello, B., Shukla, M., Vonstein, V., Wattam, A.R., Xia, F., Stevens, R.: The SEED and the Rapid Annotation of microbial genomes using Subsystems Technology (RAST). Nucleic Acids Res. 42, D206–D214 (2014)

51. Rodionov, D.A., Mironov, A.A., Gelfand, M.S.: Transcriptional regulation of pentose utilisation systems in the Bacillus/Clostridium group of bacteria. FEMS Microbiol. Lett. 205, 305–314 (2001)

52. Rodionov, D.A., Vitreschak, A.G., Mironov, A.A., Gelfand, M.S.: Comparative genomics of thiamin biosynthesis in procaryotes. New genes and regulatory mechanisms. J. Biol. Chem. 277, 48949–48959 (2002)

53. Mironov, A.A., Koonin, E.V., Roytberg, M.A., Gelfand, M.S.: Computer analysis of transcription regulatory patterns in completely sequenced bacterial genomes. Nucleic Acids Res. 27, 2981–2989 (1999)

54. Gelfand, M.S., Koonin, E.V., Mironov, A.A.: Prediction of transcription regulatory sites in Archaea by a comparative genomic approach. Nucleic Acids Res. 28, 695–705 (2000)

55. Gelfand, M.S.: Recognition of regulatory sites by genomic comparison. Res. Microbiol. 150, 755–771 (1999)
56. Rodionov, D.A., Novichkov, P.S., Stavrovskaya, E.D., Rodionova, I.A., Li, X., Kazanov, M.D., Ravcheev, D.A., Gerasimova, A.V., Kazakov, A.E., Kovaleva, G.Y., Permina, E.A., Laikova, O.N., Overbeek, R., Romine, M.F., Fredrickson, J.K., Arkin, A.P., Dubchak, I., Osterman, A.L., Gelfand, M.S.: Comparative genomic reconstruction of transcriptional networks controlling central metabolism in the Shewanella genus. BMC Genomics 12(suppl. 1), S3 (2011)
57. Rodionov, D.A., Dubchak, I.L., Arkin, A.P., Alm, E.J., Gelfand, M.S.: Dissimilatory metabolism of nitrogen oxides in bacteria: Comparative reconstruction of transcriptional networks. PLoS Comput. Biol. 1, e55 (2005)
58. Tsoy, O.V., Pyatnitskiy, M.A., Kazanov, M.D., Gelfand, M.S.: Evolution of transcriptional regulation in closely related bacteria. BMC Evol. Biol. 12, 200 (2012)
59. Gelfand, M.S.: Evolution of transcriptional regulatory networks in microbial genomes. Curr. Opin. Struct. Biol. 16, 420–429 (2006)
60. Thompson, W., Rouchka, E.C., Lawrence, C.E.: Gibbs Recursive Sampler: Finding transcription factor binding sites. Nucleic Acids Res. 31, 3580–3585 (2003)
61. Thompson, W., McCue, L.A., Lawrence, C.E.: Using the Gibbs motif sampler to find conserved domains in DNA and protein sequences. Curr. Protoc. Bioinformatics ch. 2, unit 2 8 (2005)
62. Novichkov, P.S., Rodionov, D.A., Stavrovskaya, E.D., Novichkova, E.S., Kazakov, A.E., Gelfand, M.S., Arkin, A.P., Mironov, A.A., Dubchak, I.: RegPredict: An integrated system for regulon inference in prokaryotes by comparative genomics approach. Nucleic Acids Res. 38, W299–W307 (2010)
63. Thompson, W.A., Newberg, L.A., Conlan, S., McCue, L.A., Lawrence, C.E.: The Gibbs Centroid Sampler. Nucleic Acids Res. 35, W232–W237 (2007)
64. Newberg, L.A., Thompson, W.A., Conlan, S., Smith, T.M., McCue, L.A., Lawrence, C.E.: A phylogenetic Gibbs sampler that yields centroid solutions for cis-regulatory site prediction. Bioinformatics 23, 1718–1727 (2007)
65. Novichkov, P.S., Kazakov, A.E., Ravcheev, D.A., Leyn, S.A., Kovaleva, G.Y., Sutormin, R.A., Kazanov, M.D., Riehl, W., Arkin, A.P., Dubchak, I., Rodionov, D.A.: RegPrecise 3.0–a resource for genome-scale exploration of transcriptional regulation in bacteria. BMC Genomics 14, 745 (2013)
66. Cipriano, M.J., Novichkov, P.N., Kazakov, A.E., Rodionov, D.A., Arkin, A.P., Gelfand, M.S., Dubchak, I.: RegTransBase–a database of regulatory sequences and interactions based on literature: A resource for investigating transcriptional regulation in prokaryotes. BMC Genomics 14, 213 (2013)
67. Enright, A.J., Illopoulos, I., Kyrpides, N.C., Ouzounis, C.A.: Protein interaction maps for complete genomes based on gene fusion events. Nature 402, 86–90 (1999)
68. Doolittle, R.F.: Do you dig my groove? Nat. Genet. 23, 6–8 (1999)
69. Hunter, S., Jones, P., Mitchell, A., Apweiler, R., Attwood, T.K., Bateman, A., Bernard, T., Binns, D., Bork, P., Burge, S., de Castro, E., Coggill, P., Corbett, M., Das, U., Daugherty, L., Duquenne, L., Finn, R.D., Fraser, M., Gough, J., Haft, D., Hulo, N., Kahn, D., Kelly, E., Letunic, I., Lonsdale, D., Lopez, R., Madera, M., Maslen, J., McAnulla, C., McDowall, J., McMenamin, C., Mi, H., Mutowo-Muellenet, P., Mulder, N., Natale, D., Orengo, C., Pesseat, S., Punta, M., Quinn, A.F., Rivoire, C., Sangrador-Vegas, A., Selengut, J.D., Sigrist, C.J., Scheremetjew, M., Tate, J., Thimmajanarthanan, M., Thomas, P.D., Wu, C.H., Yeats, C., Yong, S.Y.: InterPro in 2011: New developments in the family and domain prediction database. Nucleic Acids Res. 40, D306–D312 (2012)

70. Marchler-Bauer, A., Zheng, C., Chitsaz, F., Derbyshire, M.K., Geer, L.Y., Geer, R.C., Gonzales, N.R., Gwadz, M., Hurwitz, D.I., Lanczycki, C.J., Lu, F., Lu, S., Marchler, G.H., Song, J.S., Thanki, N., Yamashita, R.A., Zhang, D., Bryant, S.H.: CDD: Conserved domains and protein three-dimensional structure. Nucleic Acids Res. 41, D348–D352 (2013)
71. Suhre, K., Claverie, J.M.: FusionDB: A database for in-depth analysis of prokaryotic gene fusion events. Nucleic Acids Res. 32, D273–D276 (2004)
72. Galperin, M.Y.: Diversity of structure and function of response regulator output domains. Curr. Opin. Microbiol. 13, 150–159 (2010)
73. Basu, M.K., Carmel, L., Rogozin, I.B., Koonin, E.V.: Evolution of protein domain promiscuity in eukaryotes. Genome Res. 18, 449–461 (2008)
74. Mosca, R., Ceol, A., Stein, A., Olivella, R., Aloy, P.: 3did: A catalog of domain-based interactions of known three-dimensional structure. Nucleic Acids Res. 42, D374–D379 (2014)
75. Finn, R.D., Miller, B.L., Clements, J., Bateman, A.: iPfam: A database of protein family and domain interactions found in the Protein Data Bank. Nucleic Acids Res. 42, D364–D373 (2014)
76. Raghavachari, B., Tasneem, A., Przytycka, T.M., Jothi, R.: DOMINE: A database of protein domain interactions. Nucleic Acids Res. 36, D656–D661 (2008)
77. Luo, Q., Pagel, P., Vilne, B., Frishman, D.: DIMA 3.0: Domain Interaction Map. Nucleic Acids Res. 39, D724–D729 (2011)
78. Licata, L., Briganti, L., Peluso, D., Perfetto, L., Iannuccelli, M., Galeota, E., Sacco, F., Palma, A., Nardozza, A.P., Santonico, E., Castagnoli, L., Cesareni, G.: MINT, the molecular interaction database: 2012 update. Nucleic Acids Res. 40, D857–D861 (2012)
79. Kerrien, S., Aranda, B., Breuza, L., Bridge, A., Broackes-Carter, F., Chen, C., Duesbury, M., Dumousseau, M., Feuermann, M., Hinz, U., Jandrasits, C., Jimenez, R.C., Khadake, J., Mahadevan, U., Masson, P., Pedruzzi, I., Pfeiffenberger, E., Porras, P., Raghunath, A., Roechert, B., Orchard, S., Hermjakob, H.: The IntAct molecular interaction database in 2012. Nucleic Acids Res. 40, D841–D846 (2012)
80. Orchard, S., Ammari, M., Aranda, B., Breuza, L., Briganti, L., Broackes-Carter, F., Campbell, N.H., Chavali, G., Chen, C., Del-Torn, N., Duesbury, M., Dumousseau, M., Galeota, E., Hinz, U., Iannuccelli, M., Jagannathan, S., Jimenez, R., Khadake, J., Lagreid, A., Licata, L., Lovering, R.C., Meldal, B., Melidoni, A.N., Milagros, M., Peluso, D., Perfetto, L., Porras, P., Raghunath, A., Ricard-Blum, S., Roechert, B., Stutz, A., Tognolli, M., van Roey, K., Cesareni, G., Hermjakob, H.: The MIntAct project–IntAct as a common curation platform for 11 molecular interaction databases. Nucleic Acids Res. 42, D358–D363 (2014)
81. Patil, A., Nakai, K., Nakamura, H.: HitPredict: A database of quality assessed protein-protein interactions in nine species. Nucleic Acids Res. 39, D744–D749 (2011)
82. Salwinski, L., Miller, C.S., Smith, A.J., Pettit, F.K., Bowie, J.U., Eisenberg, D.: The Database of Interacting Proteins: 2004 update. Nucleic Acids Res. 32, D449–D451 (2004)
83. Benson, M.L., Smith, R.D., Khazanov, N.A., Dimcheff, B., Beaver, J., Dresslar, P., Nerothin, J., Carlson, H.A.: Binding MOAD, a high-quality protein-ligand database. Nucleic Acids Res. 36, D674–D678 (2008)
84. Chatr-Aryamontri, A., Breitkreutz, B.J., Heinicke, S., Boucher, L., Winter, A., Stark, C., Nixon, J., Ramage, L., Kolas, N., O'Donnell, L., Reguly, T., Breitkreutz, A., Sellam, A., Chen, D., Chang, C., Rust, J., Livstone, M., Oughtred, R., Dolinski, K., Tyers, M.: The BioGRID interaction database: 2013 update. Nucleic Acids Res. 41, D816–D823 (2013)

85. Fernandez-Suarez, X.M., Rigden, D.J., Galperin, M.Y.: The 2014 Nucleic Acids Research Database Issue and an updated NAR online Molecular Biology Database Collection. Nucleic Acids Res. 42, D1–D6 (2014)
86. Blohm, P., Frishman, G., Smialowski, P., Goebels, F., Wachinger, B., Ruepp, A., Frishman, D.: Negatome 2.0: A database of non-interacting proteins derived by literature mining, manual annotation and protein structure analysis. Nucleic Acids Res. 42, D396–D400 (2014)
87. Angiuoli, S.V., Gussman, A., Klimke, W., Cochrane, G., Field, D., Garrity, G., Kodira, C.D., Kyrpides, N., Madupu, R., Markowitz, V., Tatusova, T., Thomson, N., White, O.: Toward an online repository of Standard Operating Procedures (SOPs) for (meta)genomic annotation. OMICS 12, 137–141 (2008)
88. Glasner, J.D., Plunkett, G., Anderson, B.D., Baumler, D.J., Biehl, B.S., Burland, V., Cabot, E.L., Darling, A.E., Mau, B., Neeno-Eckwall, E.C., Pot, D., Qiu, Y., Rissman, A.I., Worzella, S., Zaremba, S., Fedorko, J., Hampton, T., Liss, P., Rusch, M., Shaker, M., Shaull, L., Shetty, P., Thotakura, S., Whitmore, J., Blattner, F.R., Greene, J.M., Perna, N.T.: Enteropathogen Resource Integration Center (ERIC): bioinformatics support for research on biodefense-relevant enterobacteria. Nucleic Acids Res. 36, D519–D523 (2008)
89. Kolker, E., Picone, A.F., Galperin, M.Y., Romine, M.F., Higdon, R., Makarova, K.S., Kolker, N., Anderson, G.A., Qiu, X., Auberry, K.J., Babnigg, G., Beliaev, A.S., Edlefsen, P., Elias, D.A., Gorby, Y.A., Holzman, T., Klappenbach, J.A., Konstantinidis, K.T., Land, M.L., Lipton, M.S., McCue, L.A., Monroe, M., Pasa-Tolic, L., Pinchuk, G., Purvine, S., Serres, M.H., Tsapin, S., Zakrajsek, B.A., Zhu, W., Zhou, J., Larimer, F.W., Lawrence, C.E., Riley, M., Collart, F.R., Yates, J.R., Smith, R.D., Giometti, C.S., Nealson, K.H., Fredrickson, J.K., Tiedje, J.M.: Global profiling of Shewanella oneidensis MR-1: Expression of hypothetical genes and improved functional annotations. Proc. Natl. Acad. Sci. USA 102, 2099–2104 (2005)
90. Pedruzzi, I., Rivoire, C., Auchincloss, A.H., Coudert, E., Keller, G., de Castro, E., Baratin, D., Cuche, B.A., Bougueleret, L., Poux, S., Redaschi, N., Xenarios, I., Bridge, A.: HAMAP in 2013, new developments in the protein family classification and annotation system. Nucleic Acids Res 41, D584–D589 (2013)
91. Roberts, R.J., Chang, Y.C., Hu, Z., Rachlin, J.N., Anton, B.P., Pokrzywa, R.M., Choi, H.P., Faller, L.L., Guleria, J., Housman, G., Klitgord, N., Mazumdar, V., McGettrick, M.G., Osmani, L., Swaminathan, R., Tao, K.R., Letovsky, S., Vitkup, D., Segre, D., Salzberg, S.L., Delisi, C., Steffen, M., Kasif, S.: COMBREX: A project to accelerate the functional annotation of prokaryotic genomes. Nucleic Acids Res. 39, D11–D14 (2011)
92. Anton, B.P., Chang, Y.C., Brown, P., Choi, H.P., Faller, L.L., Guleria, J., Hu, Z., Klitgord, N., Levy-Moonshine, A., Maksad, A., Mazumdar, V., McGettrick, M., Osmani, L., Pokrzywa, R., Rachlin, J., Swaminathan, R., Allen, B., Housman, G., Monahan, C., Rochussen, K., Tao, K., Bhagwat, A.S., Brenner, S.E., Columbus, L., de Crecy-Lagard, V., Ferguson, D., Fomenkov, A., Gadda, G., Morgan, R.D., Osterman, A.L., Rodionov, D.A., Rodionova, I.A., Rudd, K.E., Soll, D., Spain, J., Xu, S.Y., Bateman, A., Blumenthal, R.M., Bollinger, J.M., Chang, W.S., Ferrer, M., Friedberg, I., Galperin, M.Y., Gobeill, J., Haft, D., Hunt, J., Karp, P., Klimke, W., Krebs, C., Macelis, D., Madupu, R., Martin, M.J., Miller, J.H., O'Donovan, C., Palsson, B., Ruch, P., Setterdahl, A., Sutton, G., Tate, J., Yakunin, A., Tchigvintsev, D., Plata, G., Hu, J., Greiner, R., Horn, D., Sjolander, K., Salzberg, S.L., Vitkup, D., Letovsky, S., Segre, D., DeLisi, C., Roberts, R.J., Steffen, M., Kasif, S.: The COMBREX project: Design, methodology, and initial results. PLoS Biol. 11, e1001638 (2013)

A Greedy Algorithm for Hierarchical Complete Linkage Clustering

Ernst Althaus[1], Andreas Hildebrandt[1], and Anna Katharina Hildebrandt[2]

[1] Institut für Informatik, Johannes Gutenberg-Universität, Mainz, Germany
{ernst.althaus,andreas.hildebrandt}@uni-mainz.de
[2] Max-Planck Institute for Informatics, Saarbrücken, Germany
anhild@mpi-klsb.mpg.de

Abstract. We are interested in the greedy method to compute an hierarchical complete linkage clustering. There are two known methods for this problem, one having a running time of $\mathcal{O}(n^3)$ with a space requirement of $\mathcal{O}(n)$ and one having a running time of $\mathcal{O}(n^2 \log n)$ with a space requirement of $\Theta(n^2)$, where n is the number of points to be clustered. Both methods are not capable to handle large point sets. In this paper, we give an algorithm with a space requirement of $\mathcal{O}(n)$ which is able to cluster one million points in a day on current commodity hardware.

Keywords: bioinformatics, algorithm-engineering, clustering, unsupervised machine learning.

1 Introduction

A recurring task in Bioinformatics is the separation of data points into individual clusters, which are supposed to represent distinct 'species' or 'classes' of highly similar input instances. These clusters can then recursively be interpreted as the input to a further clustering round, where clusters are merged into clusters of clusters. Repeating this process until only a single cluster remains leads to a hierarchy of clusters which can be effectively represented in a tree. The applications of these so-called hierarchical clustering schemes in practice are far too numerous to mention. In the Life Sciences alone, clustering is routinely used in, e.g., protein docking [8,11], MD simulations [18,12], analysis of gene expression data[7,17,19], genomics [1,3,13], protein folding[15,2], and many other fields.

It is possible to handle this enormous breadth of application scenarios, since the details of the application domain can be easily abstracted away: most clustering algorithms do not need to consume the original input data, but rather only require the computation of a distance or a similarity measure on the individual input points, which we call d, and an induced measure of distance or similarity of clusters, which we call D. In this sense, clustering tries to combine points of small distance (high similarity) and to separate those of large distance (small similarity). A popular strategy for hierarchical clustering thus looks as follows: start by setting the set of leaves of the cluster tree to the set of initial input points. In each step, choose the least dissimilar (the most similar) pair of nodes,

A.-H. Dediu, C. Martín-Vide, and B. Truthe (Eds.): AlCoB 2014, LNBI 8542, pp. 25–34, 2014.
© Springer International Publishing Switzerland 2014

mark them as completed, and link them to a common parent. Iterating until only a single cluster remains leads to the complete clustering tree. Please note that in some applications, computing the whole tree is not required. Instead, the user specifies an application-specific threshold on the inter-cluster distances or similarities. The computation is then terminated if no pair of clusters can be found with a similarity higher or a dissimilarity lower than the given threshold.

In this iterative hierarchical clustering scenario, two cluster distance measures are particularly popular: in single-linkage clustering, we choose to merge that pair of clusters in each step which has the closest pair of points, i.e., we want to minimize

$$D(c_1, c_2) := min_{x \in c_1, y \in c_2} d(x, y),$$

where c_1 and c_2 are two clusters, given as sets of points, and d is the given distance measure on individual points.

Single-linkage clustering allows for highly efficient algorithms, such as the classic SLINK-algorithm [16], but the structure of the resulting clusters is unsuitable for many applications: Single linkage clustering often leads to long chains of clusters which have large diameters.

In complete linkage clustering, on the other hand, we define the distance between two clusters through the distance of the pair of points with the greatest separation, i.e.,

$$D(c_1, c_2) := max_{x \in c_1, y \in c_2} d(x, y)$$

Complete linkage avoids the building-up of long cluster chains that are often problematic in single linkage clustering. More importantly, it offers a very desirable guarantee: if we terminate the clustering at a given distance threshold ϵ (or extract the clusters for this threshold from the cluster tree), we find that

$$\forall c \in \text{Clusters}(\epsilon) : d(x, y) \leq \epsilon \quad \forall x, y \in c$$

Hence, complete linkage clustering allows to separate the input set into a set of clusters such that all pairwise distances within all clusters are smaller than a given threshold.

In the following, we use n to denote the number of points to be clustered.

While this property is obviously of great relevance in many applications, a runtime of $\mathcal{O}(n^2 \log n)$ can only be achieved if $\Theta(n^2)$ memory is available[5], which cannot be provided in big data scenarios. In this case, the only known method is a naive implementation of complete linkage, which has a runtime of $\mathcal{O}(n^3)$ with a space complexity of $\mathcal{O}(n)$.

Notice that these algorithms have the following two guarantees. Given the clustering for a threshold ϵ

1. No two points within a cluster have distance more than ϵ
2. No two clusters have (complete linkage) distance of less than ϵ.

For the use in big data applications, typical approaches use one of two possible simplifications. First, the clustering methodology can be exchanged into some procedure that leads to a more efficient implementation, such as average linking[14] or Ward clustering[4]. These methods, however, do no longer guarantee that each pair of distances is smaller than the given threshold. Alternatively, it is possible to keep the distance measure of complete linkage clustering, but exchange the clustering algorithm to a more efficient scheme. Classically, this has been realized in the well-known CLINK-approach[6], which has a runtime of $\mathcal{O}(n^2)$, a memory requirement of $\mathcal{O}(n)$. While this algorithm still guarantees that two data points in a cluster have distance at most the threshold, it can produce clusters whose distance is below the threshold. Furthermore, the result of this algorithm strongly depends on the order in which the input points are processed. In practice, CLINK typically generates many more clusters than the iterative complete linkage algorithm described above. As one of the main use cases for clustering is data reduction, this deficiency renders CLINK unsuitable for many applications in practice.

In this paper, we describe an alternative implementation of the complete linkage clustering algorithm that is equivalent to the naive algorithm, but which is more efficient in practice, so that we can cluster point sets of one million points within a day.

2 Algorithm and Its Implementation

2.1 The Basic Idea

As stated above, the basic algorithm is described easily: as long as there are at least two clusters, join two clusters with the smallest distance. As stated in the introduction, we are interested in complete linkage, i.e. the distance of two clusters is the largest distance of any two points in the clusters. We consider two variants: either we compute the complete hierarchy or we are compute the clusters only as long as their diameter is below a given threshold.

The naive implementation requires $\mathcal{O}(n^3)$ time and $\mathcal{O}(n)$ space.

This can easily be improved to use only $\mathcal{O}(n^2 \log n)$ time using a priority-queue to store the distances from a cluster to all others (see e.g. [5]). The update works as follows. Assume we join clusters i and j. For a cluster k different from i and j, the distance to the new cluster is the maximum of the distances to i and j. Hence for all clusters different from i and j, we have two deletions and one insert to the priority-queue. For the new cluster, we compute the distance to a point k as the maximum of the distances to i and j and build a new priority-queue of size at most n. All operations can be performed in $\mathcal{O}(n \log n)$ in total.

The main problem for our application is the space requirement of $\Theta(n^2)$, which we can not afford as we consider millions of input points. To overcome this problem, we modify the priority-queue to store only the s nearest neighbors of a cluster (see Algorithm 1). Hence, our algorithm only requires linear space.

For a cluster k different from i and j, the update is similar to the case of priority queues of size n: if distances to i and j are stored, we keep the larger one, otherwise the distances are deleted. This takes time $\mathcal{O}(s)$.

If a priority-queue becomes empty, we refill it with s elements by computing the distances to all other clusters. We archive a running time of $\mathcal{O}(cn + s \log s)$, by first computing the maximal distance of a point of the cluster to all other points and then using a variant of quick-sort to find and sort the s smallest elements (we only recurse for the pivots of rank at most s in quick-sort).

For the new cluster, we iterate over the priority-queues of i and j and keep all clusters that are stored in both queues (with the larger distance value). We use an initialized array of size $2n$ (the maximal index of a cluster) to store all clusters with their respective distances in queue i. Then we iterate over all clusters stored in queue j using the array to decide in $\mathcal{O}(1)$, whether a cluster was stored in queue i. Finally, we again iterate over the clusters in i to restore the original state of the array. All the steps have a running time of $\mathcal{O}(s)$.

Again, if the queue is empty, we refill it with s elements.

Although the worst-case running time is $\mathcal{O}(n^3)$, as for the naive algorithm, we observed a running time of roughly $\mathcal{O}(n^2)$ (see Section 4). This is as typically only a few of the priority-queues have stored one of the two clusters joined and hence lose a member.

Data: n points
Result: a hierarchical clustering tree
make each point a leaf of the clustering tree numbered $1 \ldots n$;
for each point create and fill a bounded priority queue;
for $i = 1 \ldots n$ **do**
    iterate over the queues to find the smallest distance, say (j, k);
    create a new cluster (named $n + i$) and make it the parent of j and k;
    merge queues of j and k and use their space for queue for $n + i$;
    forall the *other queues* ℓ **do**
        remove the smaller of j and k from ℓ (if at least one exists);
        if *queue* ℓ *gets empty* **then**
            | fill queue ℓ;
        end
    end
end

Algorithm 1. The Basic Algorithm. Notice that the initial filling of the queues can be done in parallel, as well as the updates of the queues ($n + i$ and all others) in the main iteration

2.2   Some Details

As our priority-queues only store a small number of clusters, we use a simple sorted array to realize them. Furthermore, we use the array of i (or j) to store the priority queue of the new cluster to neither allocate nor free memory. We experimented with using only one of the two arrays and with using both, giving us the possibility to store more (i.e. s times the number of points in the cluster) nearest neighbors. Notice that in order to use a priority-queue with better asymptotic running time, we would need a further data-structure to store which clusters are in the queue and provide references to them.

Assuming further that the input points are members of a low-dimensional Euclidean space, we can store the points in a grid, to easily obtain candidates for the nearest neighbors. In our application, distances can be lower bounded by a three dimensional Euclidean space, which allows us to use a grid in our setting, too. In Section 4, we show the effect for different thresholds for the diameter.

If a distance stored in a priority queue is larger than the threshold, we do not have to recompute distances if the queue gets empty. Hence we mark the queue as closed. If two queues that are closed are merged, the resulting queue is marked closed too. Notice that if one of the two queues is closed, we know that only the distances stored in the closed queue are relevant when merging the queue to the other. We can use this information to compute only the relevant distances in the merged queue, but we have not implemented this idea, as we observed that this does not happen too often.

Parallelization is done trivially. In the first phase, all priority-queues can be filled in parallel and the workload is nontrivial and approximately the same for each queue. In the second phase, all priority-queues are to be updated, which has negligible workload if the queue remains non-empty and which has high workload otherwise ($\mathcal{O}(cn)$ for a cluster containing c points). Hence, partitioning the work over the individual cores is nontrivial. Nevertheless, our current implementation only uses the default partitioner of the Intel® Threading Building Blocks[1].

3   Sketch of Our Main Application

Many protein-protein docking algorithms work by first generating a large number of potential docking poses that keep the internal molecular conformation fixed or almost fixed. These candidate docking poses are then further refined by allowing some flexibility on the conformation. As this second step is computationally very expensive, it can typically not be performed for all potential candidates. Furthermore, many of these poses will be very similar, as there is an attraction to regions of low energy value. Hence, these docking poses are often first clustered to reduce their number and make the second step computationally feasible.

Often the root mean square deviation of two candidate poses is used as the measure of distance, which is defined as the sum of the squared Euclidean distances of the two positions of the atoms in the two poses. Hence, for proteins with k atoms, we are considering a $3k$ dimensional Euclidean space. In [11], we showed how the root mean square deviation between two candidate poses due to rigid transformations can be computed in constant time after some preprocessing and describe our application in more detail. The instances considered in

---

[1] https://www.threadingbuildingblocks.org/

the next section were generated from a global docking run of two monomeric subunits of hexameric hemocyanin (PDB Id 1HCY) using RosettaDock [9].

4   Experiments

All experiments were executed on a server with two six-core processors (Intel® Core™ i7–970 at 3,2 GHz) with 12 GB RAM.

Our implementation is written in C++ using Intel® Threading Building Blocks for the parallelization.

We are not aware of any alternative algorithm that can handle instances whose distance matrix does not fit into main memory and which builds a clustering tree with the guarantee that the clustering for each threshold has the two properties stated in the beginning, namely, that the distance within the clusters are at most the threshold and there are no two clusters with distance less than the threshold. To compute a clustering for a given threshold, algorithms are orders of magnitudes faster and we therefore do not compare these algorithms to ours.

As we compute exactly the same clustering tree as the algorithms described in the introduction, we do not consider the quality of the clustering tree.

To evaluate the different parameters of the algorithm, we experimented with a rather small inputs of 100.000 points either distributed uniformly at random in the unit-cube in three-dimensional Euclidean space or from our application. To compute averages and standard deviations, we made ten independent runs in each setting. As the standard deviations for the random instances are very small, we do not give the numbers in the tables. Only when measuring the running-time dependency of the input size, did we increase the size of the point set.

In Figure 1, we show the time that the algorithm needs in order to make a certain number of unions. In order to join the first two points into a cluster, all pairs of distances have to be computed. Afterwards, the necessary distances are stored in the small priority queues and the joins were performed quite fast. When the small priority queues get empty, we have to refill them which increases the average time for a join of two clusters. Although, the number of refills of the priority queues does not increase more in the last iterations, the re-computation of distances becomes expensive, as we have to compute the distance of each node in a cluster to all other nodes. Therefore the time per iteration increases. Furthermore, the parallelization is less effective at this time (partially due to our native implementation). Especially the very last iterations are quite expensive. For large instances and the parallel version of our implementation, doing the last 1.000 joins takes about twice as long as for the first 1.000 joins (although all pairs of distances are computed). Notice that once all distances fit into main memory, we can use the known algorithm which would take basically the time needed to compute all pairs of distances of the remaining clusters, which can be performed by computing all pairs of distances of the input points once.

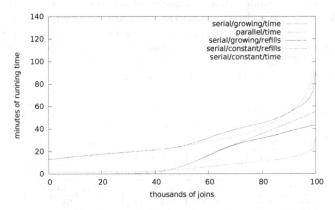

Fig. 1. We show the running time (in minutes) required to make n joins (in thousands) and the number of refills of the priority queues (in thousands) for different versions of our program: using growing chunks or not and using parallelization or not. The instance has 100.000 points and is from our application

Notice that the naive algorithms recomputes all pairs of distances in each iteration and hence would basically need n times the time of the first iteration of our algorithm. For the instance with 100.000 points, this is roughly 200 hours for the serial version and 20 hours for the parallel version. For the instance with 1.000.000 points, this is roughly 10 years for the serial version and one year for the parallel version.

Furthermore, we show the effect of our naive parallelization in Figure 1. Parallelization pays off, especially in the first phase when all pairs of distances are computed. The effect of the parallelization is hampered by the trivial parallelization and the therefore resulting communication overhead. With our machine, we save a factor of 2-3 with our naive parallelization.

We evaluated the influence of the size of the priority queue on the running time and the space requirement (see Table 1). If the queue is very small, we have to recompute distances very often, if it is large, its maintenance becomes expensive. A good trade-off is rather small (around 16) and hence the total space requirement is very low (approximately 32.7 Mb for clustering one million random points). Notice that the total running time for the growing priority is somewhat smaller. A more careful analysis shows that the gain is only in the very last iterations (see Figure 1). Hence it remains unclear which version is to be preferred, if the last iterations are done with by storing all pairs of distances.

Notice that the running time for the instances from our application is about four times as large as for a random instance. This is mainly due to the more expensive distance computation. Furthermore, we observed that the number of distance computations is slightly larger for the benchmarks from the application.

Table 1. We show the running time (min.seconds) and the number of distance evaluations (in billion) together with the respective standard deviation for different sizes of the priority queue, either having constant size queues or having size growing linearly with the number of points in the cluster. Furthermore, we show the space requirement (in megabyte) of our algorithm. The first table are random instances with 100.000 points whereas the instances for the second table came from our application.

size of the queue	1	2	4	8	16	32	64
constant size priority queues							
time (min)	92.8	67.1	51.3	39.4	33.5	32.0	35.1
distance eval. (billion)	393	276	201	146	111	86	69
growing priority queues							
time (min)	63.4	36.9	27.0	26.9	24.8	43.6	50.5
distance eval. (billion)	183	127	144	72	50	46	44
memory usage	21.8	22.7	24.6	27.3	32.7	45.2	71.0

size of the queue	1	4	8	16	32	64
constant size priority queues						
time (min)	283 ± 93	163 ± 47	134 ± 30	119 ± 21	105 ± 21	106 ± 12
distance eval (billion)	424 ± 6.5	214 ± 3.2	159 ± 2.9	125 ± 2.8	101 ± 4.1	82 ± 2.1
growing priority queues						
time (min)	126 ± 34	101 ± 17	92 ± 15	89 ± 6	81 ± 8	88 ± 8
distance eval (billion)	121 ± 2.2	88 ± 1.8	77 ± 3.1	67 ± 1.6	54 ± 4.1	50 ± 2.2
memory usage	164	167	169	175	187	213

Table 2. We show the running time (min.seconds) for different distance thresholds once using the grid and once without the grid. The first table shows the numbers for random instances with 100.000 points whereas the second shows the numbers for instances of our application. In the second table the threshold is scaled such that the centers of the molecules are in the unit cube.

threshold	0.5	0.25	0.1	0.05
random with grid	63.15	30.37	12.26	9.23
random without grid	21.53	21.28	19.32	14.33
biological with grid	221 ± 23	59.5 ± 5.1	13.6 ± 2.3	5.6 ± 1.6
biological without grid	65.5 ± 8.2	40.7 ± 7.1	20.3 ± 5.4	11.4 ± 1.8

Storing the points in a grid only pays off, if a large fraction of the distance-computations can be saved, as shown in Table 2 (for an instance with 100.000 points, only if about 1% of the distances computations remain). The reason is the much worse cache-efficiency. Considering larger instances, the cells of the grid will contain more points and hence the cache-efficiency becomes better. The size of the grid plays only a minor role.

In Figure 2 we show the increase of running time with the number of points. The observed running time for both versions is around n^2.

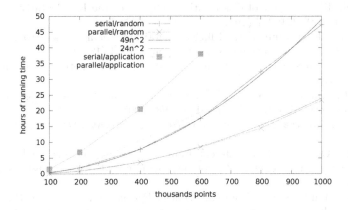

Fig. 2. We show the running time (in hours) required to cluster n points (in thousands) for the serial and parallel version of our program using growing queues. Furthermore, we show $49n^2$ and $24n^2$ as a reference

5   Conclusion

Complete-linkage clustering is an important tool with many applications in diverse fields of data mining. Classically, implementers can choose between two different strategies which trade off running time against memory. In big data scenarios, one of these is typically far too slow, while the other requires far too much space. In this work, we have presented a simple albeit effective scheme to reduce the running time of a space efficient implementation of complete-linkage clustering. The algorithm can easily be parallelized on shared-memory machines. Our experiments demonstrate that, in real-world scenarios, our algorithm indeed leads to greatly improved running-times while only mildly increasing memory usage. The algorithm will be integrated into the next major release of the open-source BALL library [10], where it will be used in docking- and structure prediction scenarios. In future work, we want to improve the algorithm, e.g. by productively use more memory, by a better parallelization or by a better engineering of the algorithm as indicated in the paper.

References

1. Bao, E., Jiang, T., Kaloshian, I., Girke, T.: Seed: Efficient clustering of next-generation sequences. Bioinformatics 27(18), 2502–2509 (2011), http://bioinformatics.oxfordjournals.org/content/27/18/2502.abstract
2. Bu, D., Li, S.C., Li, M.: Clustering 100,000 protein structure decoys in minutes. IEEE/ACM Transactions on Computational Biology and Bioinformatics 9(3), 765–773 (2012)
3. Chong, Z., Ruan, J., Wu, C.I.: Rainbow: An integrated tool for efficient clustering and assembling rad-seq reads. Bioinformatics 28(21), 2732–2737 (2012), http://bioinformatics.oxfordjournals.org/content/28/21/2732.abstract

4. Cormack, R.: A review of classification. Journal of the Royal Statistical Society, Series A 134(3), 321–367 (1971)
5. Day, W.H., Edelsbrunner, H.: Efficient algorithms for agglomerative hierarchical clustering methods. Journal of Classification 1, 1–24 (1984)
6. Defays, D.: An efficient algorithm for a complete link method. Computer Journal 20, 364–366 (1977)
7. Ernst, J., Nau, G.J., Bar-Joseph, Z.: Clustering short time series gene expression data. Bioinformatics 21(suppl. 1), i159–i168 (2005),
   http://bioinformatics.oxfordjournals.org/content/21/suppl_1/i159.abstract
8. Feliu, E., Oliva, B.: How different from random are docking predictions when ranked by scoring functions? Proteins: Structure, Function, and Bioinformatics 78(16), 3376–3385 (2010)
9. Gray, J., Moughan, S., Wang, C., Schueler-Furman, O., Kuhlman, B., Rohl, C., Baker, D.: Protein-protein docking with simultaneous optimization of rigid-body displacement and side-chain conformations. J. Mol. Biol. 331(1), 281–299 (2003)
10. Hildebrandt, A., Dehof, A.K., Rurainski, A., Bertsch, A., Schumann, M., Toussaint, N., Moll, A., Stockel, D., Nickels, S., Mueller, S., Lenhof, H.P., Kohlbacher, O.: BALL - Biochemical Algorithms Library 1.3. BMC Bioinformatics 11(1), 531 (2010)
11. Hildebrandt, A.K., Diezen, M., Lengauer, T., Lenhof, H.P., Althaus, E., Hildebrandt, A.: Efficient computation of root mean square deviations under rigid transformations (submitted)
12. Jamroz, M., Kolinski, A.: Clusco: Clustering and comparison of protein models. BMC Bioinformatics 14(1), 62 (2013)
13. Miele, V., Penel, S., Duret, L.: Ultra-fast sequence clustering from similarity networks with silix. BMC Bioinformatics 12(1), 116 (2011),
    http://www.biomedcentral.com/1471-2105/12/116
14. Murtagh, F.: Complexities of hierarchic clustering algorithms: The state of the art. Computational Statistics Quarterly 1, 101–113 (1984)
15. Shortle, D., Simons, K.T., Baker, D.: Clustering of low-energy conformations near the native structures of small proteins. Proceedings of the National Academy of Sciences 95(19), 11158–11162 (1998), http://www.pnas.org/content/95/19/11158.abstract
16. Sibson, R.: SLINK: An optimally efficient algorithm for the single-link cluster method. The Computer Journal 16(1), 30–34 (1973)
17. Sivriver, J., Habib, N., Friedman, N.: An integrative clustering and modeling algorithm for dynamical gene expression data. Bioinformatics 27(13), i392–i400 (2011),
    http://bioinformatics.oxfordjournals.org/content/27/13/i392.abstract
18. Torda, A.E., van Gunsteren, W.F.: Algorithms for clustering molecular dynamics configurations. J. Comput. Chem. 15(12), 1331–1340 (1994),
    http://dx.doi.org/10.1002/jcc.540151203
19. Wang, Y., Xu, M., Wang, Z., Tao, M., Zhu, J., Wang, L., Li, R., Berceli, S.A., Wu, R.: How to cluster gene expression dynamics in response to environmental signals. Briefings in Bioinformatics 13(2), 162–174 (2012),
    http://bib.oxfordjournals.org/content/13/2/162.abstract

Vester's Sensitivity Model for Genetic Networks with Time-Discrete Dynamics

Liana Amaya Moreno[1], Ozlem Defterli[2], Armin Fügenschuh[1], and Gerhard-Wilhelm Weber[3]

[1] Department of Mechanical Engineering
University of the Federal Armed Forces Hamburg
Holstenhofweg 85, 22043 Hamburg, Germany
[2] Department of Mathematics and Computer Science
Faculty of Art and Sciences, Çankaya University
06810 Ankara, Turkey
[3] Institute of Applied Mathematics, Middle East Technical University
06531 Ankara, Turkey

Abstract. We propose a new method to explore the characteristics of genetic networks whose dynamics are described by a linear discrete dynamical model $x_{t+1} = Ax_t$. The gene expression data x_t is given for various time points and the matrix A of interactions among the genes is unknown. First we formulate and solve a parameter estimation problem by linear programming in order to obtain the entries of the matrix A. We then use ideas from Vester's *Sensitivity Model*, more precisely, the *Impact Matrix*, and the determination of the *Systemic Roles*, to understand the interactions among the genes and their role in the system. The method identifies prominent outliers, that is, the most active, reactive, buffering and critical genes in the network. Numerical examples for different datasets containing mRNA transcript levels during the cell cycle of budding yeast are presented.

Keywords: Linear Programming, Parameter Estimation, Discrete Dynamical System, Sensitivity Analysis, Genetic Networks, Operational Research, Systems Biology.

1 Introduction

The recent availability of big amounts of gene expression data has enhanced the study of genetic networks, which is a challenging and promising topic. The main goal of these studies is to understand and estimate the dynamical interrelations among the genes in the network. Several approaches were developed in the recent years to model the regulatory interactions. They differ by the mathematical techniques that came to application: modeling by graphs, Bayesian networks, Boolean networks, discrete and continuous dynamical systems (or systems of ordinary differential equations). For more information we refer to Jong [14], Bansal et al. [17], or Ay et al. [1], for instance, and the references therein. Furthermore, suitable

A.-H. Dediu, C. Martín-Vide, and B. Truthe (Eds.): AlCoB 2014, LNBI 8542, pp. 35–46, 2014.
© Springer International Publishing Switzerland 2014

mathematical tools for the understanding of the networks are constantly broadening the spectrum of possibilities to study such systems, see Weber [11], Defterli et al. [19] and [20], Defterli [18], Weber [12], Yee et al. [16], Jong et al. [15].

In this present work, we use a discrete linear model to describe the dynamics of the network. We assume that the expression level of any gene at a certain point in time is the result of the weighted sum of the expression level of all the other genes at the previous point in time only [7], with respect to a given time discretization. Denote by $G := \{1, \ldots, n\}$ the set of genes and by T the finite set of time steps. The set T' contains all but the last time step of T. Then

$$x_{t+1,i} = \sum_{j \in G} a_{i,j} x_{t,j}, \qquad \forall\, i \in G, t \in T', \tag{1}$$

where $x_t = (x_{t,1}, x_{t,2}, \ldots, x_{t,n})^\top \in \mathbb{R}^n$ is a vector with the expression level of the genes in G at time step t. Here, $A = (a_{i,j}) \in \mathbb{R}^{n \times n}$ is a matrix, where the influence coefficient $a_{i,j}$ represents the ability of gene j to regulate gene i. To solve such models, linear regression has frequently been used, see for example Zhang et al. [22], Someren et al. [6], Someren et al. [7], or Bansal et al. [17]. These methods are facing the problem that there is usually much more genes than time steps, which leads to non-unique (or multiple) possible solutions. However, these methods do not allow for a control on the obtained solution. In our method, we compute the matrix A by means of linear programming. We minimize the L_1-norm of the matrix A, in order to get the simplest matrix that can explain the system's dynamical behavior (cf. Occam's razor).

Our core question is, is there more information that can be retrieved from analyzing the matrix A? We try to gain insights into the behavior of the individual genes and their role in the system as a total. In the economical literature, we found a method called *Sensitivity Model* (or *Paper Computer*), proposed by Vester and von Hesler in 1980 [9] (usually referred to as Vester's model). Vester's model has its background in the area of network thinking, combining elements from system dynamics, fuzzy logic and bio-cybernetics. It is important to clarify that in spite of the name, the model has nothing to do with the concept of sensitivity analysis associated with optimization problems. The main purpose of this model is to provide a system dynamic modeling tool capable of handling and analyzing complex systems by setting out the inner structures, hence enabling further interventions (control/regulation) on the system. Cole [3], Wolf et al. [5], and Neumann et al. [10] present some of the most common areas of use of Vester's model. Nevertheless, the model has a context-independent structure, so that it can be applied to new areas, as we demonstrate in this work.

In the context of genetic networks we want to understand the regulatory interactions from a systemic point of view. Hereto, we adapt Vester's ideas to characterize the genes by analyzing them based on their relations and their effect on the system as whole. We aim to obtain information about the structure and the functioning of the genetic network.

2 The Method

Our method consists of two steps. In the first step, we compute the matrix A of the discrete dynamical system, using linear programming. In the second step, we analyze A to derive dynamical properties of the genes involved in the network, using Vester's Sensitivity Model.

2.1 Linear Programming

We consider a linear model that describes the interactions among the genes according to equation (1). Our objective then is to estimate the entries of A. For this, we use a dataset containing measured gene expression data x_t, within a finite time horizon of discrete time steps $t \in T$. The parameter $x_{t,i}$ contains gene expression data for gene $i \in G$ at time step $t \in T$. The entries of the matrix A are decision variables $a_{i,j} \in \mathbb{R}$. The auxiliary variables $a_{i,j}^+$ and $a_{i,j}^-$ are used to linearize the nonlinear matrix norm $\| \cdot \|_1$.

We intend to find a matrix A that describes the dynamics of the network, therefore explaining the interaction among the genes. Applying the principle of parsimony, of all the possible matrices that can be used for this, we want to find one with the "simplest structure", that is, a matrix whose entires are the smallest possible in absolute value. This could result in the estimation of a sparse matrix, which according to [22] should be the case for regulatory systems. With this in mind, we want to minimize the following objective function:

$$\|A\|_1 = \sum_{i,j \in G} |a_{i,j}| = \sum_{i,j \in G} \left(a_{i,j}^+ + a_{i,j}^- \right). \tag{2}$$

We linearize the expression $|a_{i,j}|$ by

$$a_{i,j} = a_{i,j}^+ - a_{i,j}^-, \qquad \forall\, i,j \in G. \tag{3}$$

Equation (1) gives already the first constraint of the linear model, providing the relation between gene expression data subsequent in time,

$$x_{t+1,i} = \sum_{j \in G} a_{i,j} \cdot x_{t,j}, \qquad \forall\, i \in G,\ t \in T'. \tag{4}$$

Moreover, we impose a certain condition on the structure of the matrix A, i.e., the diagonal must be zero. This is necessary to apply Vester's method, where this setting is required for the *Impact Matrix* (see below for a more detailed explanation). For our application, it translates to the fact that a gene cannot influence itself. Hence

$$a_{i,i} = 0, \qquad \forall\, i \in G. \tag{5}$$

Summing it up, the linear programming model then is as follows:

$$\min (2), \text{ subject to } (3), (4), (5). \tag{6}$$

2.2   Vester's Sensitivity Model

Vester's Sensitivity Model for an analysis of dynamical systems is described by a recursive structure of 9 steps altogether, see Vester [8]. In the following, we briefly describe the principal steps proposed by Vester in his model that are also relevant to our method.

Initially one has to describe the system, that means, express it in terms of the key elements (or variables) that are the most relevant. (In our application that means to select a set of genes G that should be studied as a system.) In general, this results in a reduction of the complexity providing a much more compact (practical) representation of the system. Next, the relations between the selected representative variables are studied in order to determine the magnitude of the influences among them. These influences are of great importance since they determine the behavior of the system. (In our application, this step is practically carried out by solving the linear program (6).)

Our motivation to use *Vester's Sensitivity Model* in the context of genetic networks arose from the formulation of the linear programming model. Not knowing the matrix A at the beginning means that we have no information about the interaction among the genes, consequently the inner structure of the network is unknown for us. Furthermore, even if we estimate such a matrix, we need to achieve certain level of knowledge regarding the structure. Only then we would be able to understand how the system works, this means, understand its behavior and understand what causes the system to behave as it does. Vester's Sensitivity Model states that these questions can be answered by analyzing the *Systemic Role* that the variables, in our case the genes, have. In turn, the *Systemic Role* is determined by the *Indices of Influence*, they summarize the information about the magnitude and the character of the interactions among the genes, and are calculated from the *Impact Matrix*. In this way, we can interpret the results form the linear programming model and achieve our initial objective, that is, to gain insights into the biological processes in genetic network.

The *Impact Matrix* is the matrix that contains the information of all the interactions among the variables in the system. The entries of this matrix reflect the influence of the variable i on the variable j. They are calculated by measuring the (pairwise) effect on the variables when the others change, therefore the name *Impact Matrix*. It is important to remark that variables do not influence themselves, therefore entires along the diagonal are not meaningful. A scale form 0 to 3 was initially proposed by [8] for measuring the effect of one variable in one another answering to the question: If the variable x changes, how does the variables y change?

This scaling does not distinguish between positive and negative influences, that is, only the magnitude of the influence irrespective of the sign is measured. With these values the *Impact Matrix* is built and the following indicators (**AS** and **PS**) and the *Indices of Influence* (**Q** and **P**) are calculated for each variable:

– Active Sum (**AS**): the sum along rows, indicates how large is the effect of the variable on the others.

- Passive Sum (**PS**): the sum along columns, indicates how sensitive is the variable, how does it react to changes in the system.
- Quotient (**Q**): the ratio **AS/PS**.
- Product (**P**): the product **AS*PS**.

The quotient **Q**, determines how dominant or influenceable a variable is. The larger/smaller the quotient is, the more active/reactive character the variable has, respectively. On the other hand, the product **P** determines how participative a variable is. The larger/smaller the product, the more critical/buffering it is. Hence, the character of each variable is determined by the pair (**Q, P**). Highly active variables will be located in the upper left corner of the *System Role plot*, a plot whose axes are **PS** and **AS**. Highly reactive variables will be located in the lower right corner of the same plot. Analogously, highly critical and buffering variables will be found in the upper right and lower left corner, respectively. The rest of the variables (genes) will be located in the area in between these four locations. In other words, the variables are characterized by their dominance/influenceability, and by how participative they actually are, revealing the potential of each of them.

To conclude, let us point out what we use from Vester's Sensitivity Model in our proposed method, and how it is used. In first place, we regard the genetic network as the complex system to study, where the variables are the genes in the datasets and the interactions among them are the influence coefficients $a_{i,j}$ to estimate. That means, we use the estimated matrix A as *Impact Matrix*. This leads to a different scaling with respect to the one proposed in [8], some other scalings can also be found in [13,3]. Moreover, since the influence coefficients come from the solution of a linear programming problem without any constraint on the sign of the decision variables, it is possible that some entries in the matrix are negative. Vester's *Impact Matrix* has only positive entries and therefore we work with the absolute value of the estimated matrix A, without affecting the forthcoming analysis, given that the *Impact Matrix* has information only about the magnitude of the interactions, and not about their sign. We then calculate the indicators **AS, PS** and the *Indices of Influence* **Q, P**. Afterwards, we identify the most active, reactive, critical and buffering genes in the network, and thus determine their *Systemic Role*. Finally, the corresponding interpretation of their role is briefly discussed.

3   Data

Datasets containing mRNA transcript levels during the cell cycle of budding yeast are considered here, Cho et al. [21], Zhang [22]. The data was collected at 17 time points taken by 10 minute (min.) intervals covering nearly two full cell cycles and containing 5 phases (Early G1, Late G1, S, G2 and M). The complete yeast cell cycle dataset has 6220 genes [21] and shows fluctuation of their expression levels during the 17 time points. Cho et al. [21] identified from this dataset 416 genes based on their peak times and grouped them into the five cell cycle phases. Finally a subset of 384 genes was classified into only one phase (some genes peak at more than one phase during the cell cycle, see Yeung et al. [16]). Yet a smaller subset with 23 genes was chosen in [18] and studied to analyze and anticipate the time discrete dynamics of the corresponding subnetwork. The genes of this subnetwork cover all the cell cycle phases.

We use the initial network with 384 genes form [21] and also the subnetwork with 23 genes as in [18]. The data is normalized across each cell cycle.

3.1   Data Analysis

Correlation analysis has been widely used in the study of gene environment networks with the aim of exploring gene expression data and thus providing insights into regulatory mechanism. Moreover, clustering techniques make use of correlation coefficients to define similarity measures and therefore discover patters in groups of genes with similar expression level data [16], Bendor et al. [2], Someren et al. [7]. It is not in the scope of this work to carry out a correlation analysis of the genes in the mentioned datasets, nevertheless we use a correlation analysis to explore their features to have a clear picture of what can be expected. Figure 1 depicts the correlation coefficients between the genes in the two different datasets. The genes were sorted by phase. As expected, for the majority of the genes in the same cycle phase, high correlation coefficients are observed. The black dashed lines show the grouping by phase. Genes in the same group not only peak in the same cycle phase, but their dynamics along the whole time horizon is also similar (see Figure 2), which was observed for both datasets. Given that the dynamics in each phase is similar, we took the mean expression level as a representative for all the genes in the same phase and plotted it for the different representatives along the time horizon, see Figure 3. On the other hand, if we calculate the correlation coefficients of the 17 time points where the data was collected, see Figure 4, a periodic behavior is observed. This behavior is not surprising since the time horizon covers almost 2 cell cycles [21], meaning that the expression level of each gene has two observable peaks, each one happening in the phase it was classified into, see Figure 3. It is clear that

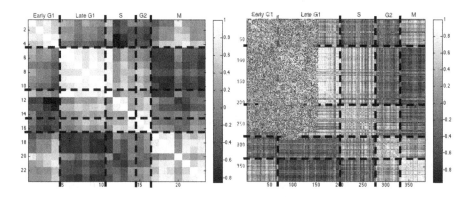

Fig. 1. Gene correlation matrix 23 (left) and 384 (right)

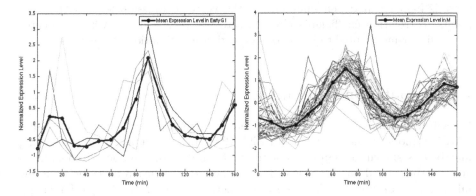

Fig. 2. Genes peaking in *Early G1* for the dataset with 23 genes and in *M* for the dataset with 384 genes

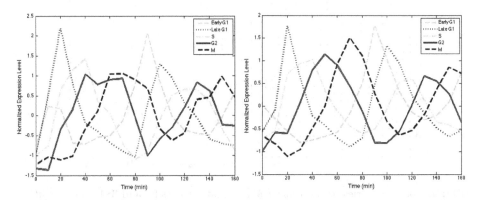

Fig. 3. Mean value for the expression level of genes in each phase; 23 genes (left) and 384 genes (right)

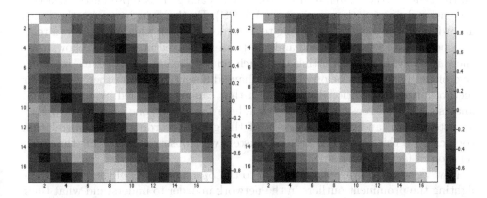

Fig. 4. Time-point correlation matrix; 23 genes (left) and 384 genes (right)

there is high linear dependency among the vector expression level at the time points (correlation coefficients close to zero are very seldom), which indicates that a linear model might give a good enough approximation of the dynamics of the system.

4   Computational Results

We solved the linear programming problem (6) for the selected datasets using AMPL as modeling language and CPLEX 12.5 Optimizer as solver. The resulting matrix A is shown in Figure 5. For the first dataset, the number of nonzeros is 368, for the second dataset it is 6144. For the latter, solution times below a few minutes are currently needed, hence our method could be applicable for bigger datasets with many more genes and time steps.

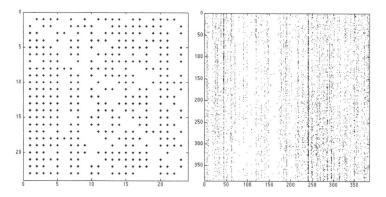

Fig. 5. Nonzero entries in the *Impact Matrix*; 23 genes (left) and 384 genes (right)

For each one of the genes we compute the values **AS**, **PS**, **Q** and **P**. We plot all the genes in the *Systemic Role plot*, where each gene is represented by a two dimensional point with coordinates (**AS, PS**). Furthermore, we color the genes according to the phase they belong to, in order to check whether there is a relation between the role of a gene and the phase it peaked in. Figure 6 shows that there is no evidence supporting that such relation exists, that means, irrespective of the phase they belong to, they have different roles. In other words, our method provides additional information that cannot be gained by a correlation analysis. Figure 7 reassures this position, here we selected one gene, marked with an asterisk (*), and colored the rest of the genes according to the correlation coefficient to this gene. Also here no clusters were observed indicating that such relation in fact exists.

Now we want to draw the attention to the *Systemic Roles* of the genes, highlighting the prominent outliers in the network in order to understand what their character reveals. Let us start stating that for the 23 dataset a clear configuration for the roles can be observed. The relatively most active gene belongs to

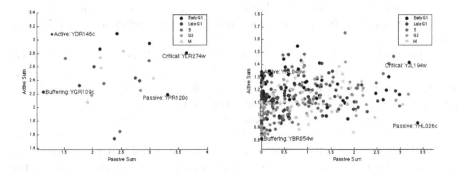

Fig. 6. *Systemic Role plot* colored by phase; 23 genes (left) and 384 genes (right)

the *M* phase and is YDR146c. It regulates many other genes and is regulated just by few. Being a transcription factor [21], it influences the other genes by contributing to specific biochemical processes. As for the interpretation of its role, we can say it is a dominant gene. It could be used as lever; if changes were to happen to its gene expression level, significant changes could be produced on the other genes and therefore the cell cycle could present alterations. This is the prototype of gene that can be used to trigger some desired effect.

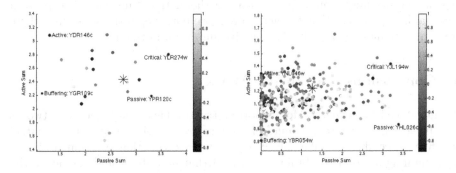

Fig. 7. *Systemic Role plot* colored by correlation coefficient to the gene marked with asterisk (*); 23 genes (left) and 384 genes (right)

Gene YPR120c belongs to the *Late G1* phase. It emerges as the most reactive gene, which means that its expression level is highly sensitive to alterations in the expression level of other genes, but the cell cycle would not change much with changes in its expression level. Such kind of genes are consider to have a damping function, since the effects will not propagate in the system, in spite of being highly sensitive. Its function is as cell cycle regulator [21], this might be the reason behind its highly reactive character. This gene needs a lot of information from the system, i.e., from other genes and therefore is influenced by relatively many.

As the most critical we observe gene YLR274w belonging to the *Early G1* phase. It has highly regulator capabilities but at the same time is highly regulated, which makes it a "risky" gene and difficult to control. This goes in accordance with its function in the cell cycle as DNA replicator [21].

Finally the most buffering gene turned out to be YGR109c, it also belongs to the *Late G1* phase. This gene is the one with the lowest "activity" in the system (inert element), it does not regulate nor is regulated by others. Such genes, could be regarded as "stable", since their state is the one changing the least through the time horizon. If we compare it with gene YPR120c acting both as cycle regulators [21], we can say that the regulatory function of the first one is relatively smaller. In turn it does not use much information from the other genes and thus it is more difficult to change its expression level.

In general we can say that the network as a whole has a buffering character, most of the genes tend to be located in the lower left corner which correspond to the buffering part of the *Systemic Role plot*. Moreover, if we describe the system with just one point in this plot, taking for example the mean value, this point would be closest to the buffering corner than to any other corner in the plane.

A similar analysis can be made for the dataset with 384 genes as for the most active, reactive, buffering and critical genes. Both, the pattern in the *Impact Matrix* and the fact that the matrix is more sparse than the one in the 23 case, explain the accumulation of genes near to the Active Sum axis. The pattern (vertical lines) suggest that the genes are more reactive, in the sense that there are relatively more genes with low passive sum than in the first dataset.

5   Conclusions and Future Work

We introduced a new method for exploring structural features of genetic networks with linear time-discrete dynamics of the form $x_{t+1} = Ax_t$, where A is a matrix of influence coefficients. We solved a parameter estimation problem by linear programming in order to obtain a matrix A that describes the dynamics of the network and has the minimum L_1 norm possible for two different datasets (with gene expression data given at various time points). Features of the datasets were investigated using correlation analysis. Afterwards we used some of the central ideas of the *Sensitivity Model* proposed by Vester [8] to understand the interactions among the genes in the network and characterize each one of them according to their role in the system. More precisely, we let the estimated matrix A be the *Impact Matrix* that explains the influences among the variables (genes) in the system (genetic network). We then proceeded to calculate the *Indices of Influence* **Q** and **P**, which reflect the active/reactive and critical/buffering character of the genes. No evidence of a relation between the *Systemic Role* of the genes and their phase group nor the correlation coefficient corresponding to a specific gene was observed. Finally, and with the aim understanding what the

Systemic role concept brings into play, we identified the most active, reactive, critical and buffering genes in the network and analyzed the interpretation given by the Vester's Sensitivity Model to such elements, i.e., as levers (active), risk factors (critical), measuring sensors (reactive) and inert elements (buffering).

In our future studies, we will extend this method in different ways. On one hand we would like to make use of genetic network features to make a more realistic estimation of the matrix A. Furthermore, will use different models to describe the dynamics of the network such as piecewise linear differential equations, also known as hybrid systems, and later on, stochastic hybrid systems with jumps (cf. Temoçin et al. [4]). It would also be interesting to consider different types of norms during estimation process. With respect to the ideas used form Vester, one could think of developing a more accurate definition of roles considering more extensively the possible combinations of the indexes Q and P, in such a way that more specific features of genetic networks are taken into account.

Acknowledgments. We are grateful to Prof. Dr.-Ing. Wolfram Funk for partially funding our conference participation.

References

1. Ay, A., Arnosti, D.N., Steegenga, W.T., Sijbers, A.M., Dechering, K.J., Reinders, M.J.T.: Mathematical Modeling of Gene Expression: A Guide for the Perplexed Biologist. Crit. Rev. Biochem. Mol. Biol. 46(2), 137–151 (2011)
2. Ben-Dor, A., Shamir, R., Yakhini, Z.: Clustering Gene Expression Patterns. Computational Biology 6(3-4) (1999)
3. Cole, A.: The Influence Matrix Methodology: A Technical Report. LC0506/175, Foundation for Research, Science and Technology, FRST (2006)
4. Temoccin, B.Z., Weber, G.W.: Optimal Control of Stochastic Hybrid System with Jumps: A Numerical Approximation. Computational and Applied Mathematics 259, 443–451 (2014)
5. Wolf, C., Person, F., Jelse, K.: A Logistic Analysis with the Sensitivity Model Prof. Vester. Tech. Rep. B2048, IVL Swedish Environmental Research Institute Ltd (2012)
6. van Someren, E.P., Vaes, B.L.T., Steegenga, W.T., Sijbers, A.M., Dechering, K.J., Reinders, M.J.T.: Least Absolute Regression Network Analysis of the Murine Osteoblast Differentiation Network. Bioinformatics 22(4), 477–484 (2006)
7. van Someren, E.P., Wessels, L.F.A., Reinders, M.J.T.: Linear Modeling of Genetic Networks from Experimental Data. In: Proc. Int. Conf. Intell. Syst. Mol. Biol., vol. 8, pp. 355–366 (2000)
8. Vester, F.: Die Kunst vernetzt zu denken. Ideen und Werkzeuge fuer einen neuen Umgang mit Komplexitaet. Deutsche Verlags-Anstalt GmbH Stuttgart (1999)
9. Vester, F., von Hesler, A.: Sensitivitaetsmodell. Bundesminister d. Innern, Bonn (1980)
10. Neumann, G., Düring, D.: Methodology to Understand the Role of Knowledge Management in Logistic Companies. LogForum 4(5) (2008)

11. Weber, G.W., Kropat, E., Akteke-Öztürk, B., Görgülü, Z.K., Guo, D.: A Survey on OR and Mathematical Methods Applied on Gene-Environment Networks. CEJOR 17, 315–341 (2009)
12. Weber, G.W., Defterli, O., Kropat, E.: Qualitative Simulation of Genetic Regulatory Networks Using Piecewise-Linear Models. European Journal of Operational Research 211(1), 1–14 (2011)
13. Bossel, H.: System, Dynamik, Simulation. Modellbildung, Analyse und Simulation komplexer Systeme. Books on Demand GmbH Norderstedt/Germany (2004)
14. de Jong, H.: Modeling and Simulation of Genetic Regulatory Systems: A Literature Review. Computational Biology 9(1), 67–107 (2002)
15. de Jong, H., Gouzé, L.J., Page, C., Sari, T., Geiselmann, J.: Qualitative Simulation of Genetic Regulatory Networks Using Piecewise-Linear Models. Bulletin of Mathematical Biology 66, 301–340 (2004)
16. Yeung, K.Y., Ruzzo, W.L.: An Empirical Study on Principal Component Analysis for Clustering Gene Expression Data. Tech. Rep. UWCSE20001103, Department of Computer Science & Engineering University of Washington (2000)
17. Bansal, M., Belcastro, V., Ambesi-Impiombato, A., di Bernardo, D.: How to Infer Gene Networks from Expression Profiles. Molecular Systems Biology 3(78) (2007)
18. Defterli, O.: Modern Mathematical Methods in Modeling and Dynamics of Regulatory Systems of Gene-Environment Networks. Ph.D. thesis, Graduate School of Natural Sciences, Department of Mathematics, Middle East Technical University (August 2011)
19. Defterli, O., Fügenschuh, A., Weber, G.W.: New Discretization and Optimization Techniques with Results in the Dynamics of Gene-Environment Networks. In: 3rd Global Conference on Power Control & Optimization (February 2010)
20. Defterli, O., Fügenschuh, A., Weber, G.W.: Modern Tools for the Time-Discrete Dynamics and Optimization of Gene-Environment Networks. Commun. Nonlinear Sci. Numer. Simulat. 16, 4768–4779 (2011)
21. Cho, R.J., Campell, M.J., Winzeler, E.A., Steinmetz, L., Conway, A., Wolfsberg, T.G., Gabrielian, L.A.E., Landsman, D., Lockhart, D.J., Davis, R.W.: A Genome-Wide Transcriptional Analysis of the Mitotic Cell Cycles. Molecular Cell 2, 65–73 (1998)
22. Zhang, S.Q., Ching, W.K., Tsing, N.K., Leung, H.Y., Guo, D.: A New Multiple Regression Approach for the Construction of Genetic Regulatory Networks. Artificial Intelligence in Medicine 48(2-3), 153–160 (2010)

Complexity and Polynomial-Time Approximation Algorithms around the Scaffolding Problem

Annie Chateau[1,2] and Rodolphe Giroudeau[1]

[1] LIRMM - CNRS UMR 5506 - 161 rue Ada 34090 Montpellier, France
[2] Institut de Biologie Computationnelle
95 rue de la Galéra 34000 Montpellier, France
{annie.chateau,rodolphe.giroudeau}@lirmm.fr

Abstract. We explore in this paper some complexity issues inspired by the contig scaffolding problem in bioinformatics. We focus on the following problem: given an undirected graph with no loop, and a perfect matching on this graph, find a set of cycles and paths covering every vertex of the graph, with edges alternatively in the matching and outside the matching, and satisfying a given constraint on the numbers of cycles and paths. We show that this problem is $\mathcal{NP}$-complete, even in bipartite graphs. We also exhibit non-approximability and polynomial-time approximation results, in the optimization versions of the problem.

Keywords: Complexity, Polynomial-Time Approximation, Scaffolding.

1 Introduction

We investigate the complexity of a problem inspired by a problem from bioinformatics, namely the contig scaffolding problem. When a new genome is sequenced, it is not possible, due to technological issues, to read the whole sequence directly from the DNA molecule. Instead, the sequence is built through different steps, each of them presenting algorithmic challenges. We focus here on the contig scaffolding step, which consists, given a set of sequences of various lengths called *contigs*, to infer the order and the orientation of the contigs along the genome, using a set of possibly inconsistent pairing information. First described in [10], it was presented as a problem of path merging in a particular kind of graphs, and was stated as $\mathcal{NP}$-complete. Further studies describe different types of heuristics and computational approaches ([4,5,9]), or compare the accuracy of the exact approach to the heuristics ([7]), but none of them further investigate the complexity aspects of the problem, especially in terms of approximation. We focus here on this latter aspect.

A more general framework was recently proposed in [2], where the problem is presented as the resolution of consecutive ones property problem on matrices encoding hypergraphs. The authors include multiplicity ranges on the contigs, meaning that a contig may be repeated in the scaffolds, and propose interesting approximation results in the case where there are no constraint on the number of paths and cycles. This approach can use phylogenetic information, instead of the

A.-H. Dediu, C. Martín-Vide, and B. Truthe (Eds.): AlCoB 2014, LNBI 8542, pp. 47–58, 2014.

classical use of paired fragments. Also recently, several types of information have been mixed to infer scaffolds from the assembly data, for instance in [1], where the chromatin structure of the chromosomes has been added to the classical mate-pair information to complete the human, mouse and drosophila genomes. This general trend highlights the need of flexible tools to efficiently solve 'scaffolding like' problems.

Here we propose an alternative formalization of the problem, which is inspired by a generalization of the very well known Traveling Salesman Problem. This model is more general than the one proposed in [10], and allows to integrate the desired structure of the genome (number of circular or linear chromosomes). This is a preliminary work intended to be completed and extended to several directions, by the introduction of the multiplicities of the contigs, and the use of several kinds of information sources (multi-criteria). We expose the first, simple, problem and its optimization versions. One of them is directly relative to the scaffolding problem, but we also try to enlarge this context.

A very huge literature has been provided concerning the Traveling Salesman Problem and its variants. We refer the reader to [13] for an overview on the domain. Concerning the more general problem of finding a cover with a fixed number of vertex disjoint cycles and paths, the papers especially focus on feasibility criteria, like sufficient conditions, typically on the degrees of the vertices, for the graph to admit such a cover (see, for instance, [3]). The cases where the numbers of paths and cycles are not fixed define a wide range of possibilities. Indeed, finding an optimal cover by disjoint cycles is a polynomial problem, when the number of cycles is not fixed ([15]). On the contrary, finding an optimal cover by disjoint paths with at least two edges is $\mathcal{NP}$-complete ([14]). Also, the problem to infer the number of cycles of a cycle cover, known as cycle packing, has been already studied: finding the minimal number of cycles which are necessary to cover a graph is $\mathcal{NP}$-complete ([11]). Anyway, concerning the problem of finding, and optimizing, a spanning subgraph with a fixed number of cycles and paths, this is to our knowledge the first study of this kind of problem in terms of complexity and approximability.

In this article, we study a variation of the scaffolding problem in the framework of classical complexity. Our contribution is summarized in Table 1.

This article is organized as follows: the next section is devoted to formal description of the SCAFFOLD problems. In Section 3 and 4 we pay attention to computational complexity and non-approximability results whereas in Section 5 we design a polynomial-time approximation algorithm for the maximization problem.

2   Formal Description of the Problems

In what follows, we consider $G = (V, E)$ a undirected graph with an even number n of vertices and without self loops. We suppose that there exists a perfect matching in G, denoted by M^*. Let $w : E \to \mathbb{N}$ be a weight function on the edges. In the bioinformatic context, edges in M^* represent the contigs, and the

Table 1. Synthetic complexity/approximation results for SCAFFOLD problems. Notice that all complexity results can be extended to the bipartite case.

Problems	Complexity		Approximability			
	Decision	Ref.	Min	Ref.	Max	Ref.
(σ_p, σ_c)−SP	$\mathcal{NP}$-C	Th. 2	Inapprox.	Th. 4	Ratio 3	Th. 9
(σ_p, σ_c)−SP with $l_p \geq 2$ and $l_c = 6$	$\mathcal{NP}$-C	Th. 3	Inapprox.	Th. 4	Ratio 3	Th. 9
$(0, 1)$−SP	$\mathcal{NP}$-C	Th. 2	Inapprox.	Th. 4	Ratio 2	Th. 10

other edges figure the contiguity information given by the mate-pairs, or any other kind of information.

In order to model the genomic structure by fixed numbers of linear chromosomes (paths) and circular ones (cycles), the class of considered problems are parameterized by two integers, respectively denoted by σ_p and σ_c.

Definition 1. *In the following, an* alternating-cycle *(resp.* alternating-path*) in G, relatively to a perfect matching[1] M^* of G, is a cycle (resp. a path) such that its edges alternatively belong to M^* or not.*

Notice that an alternating-cycle has necessarily an even number of vertices. The class of the (σ_p, σ_c)−SCAFFOLD and MIN/MAX−(σ_p, σ_c)−SCAFFOLD PROBLEMS are defined as follows:

(σ_p, σ_c)−SCAFFOLD PROBLEM (SP):
Instance: Let $G = (V, E)$ be a graph with $2n$ vertices. Let M^* be a perfect matching of G, and $(\sigma_p, \sigma_c) \in \mathbb{N} \times \mathbb{N} \backslash \{(0, 0)\}$.
Question: Does it exist a vertex disjoint collection of exactly σ_p alternating-paths and σ_c alternating-cycles, covering the vertices of G?

MIN/MAX-(σ_p, σ_c)−SCAFFOLD PROBLEM:
Instance: Let $G = (V, E, w)$ be a graph with $2n$ vertices. Let M^* be a perfect matching of G, and $(\sigma_p, \sigma_c) \in \mathbb{N} \times \mathbb{N} \backslash \{(0, 0)\}$.
Question: Find a vertex disjoint collection of exactly σ_p alternating-paths and σ_c alternating-cycles, covering the vertices of G, and of minimal (resp. maximal) total weight.

In the case where the weight on the edges represents, for instance, the number of mate-pairs that support the contiguity of two contigs, in their relative orientation, the MAX−(σ_p, σ_c)−SCAFFOLD PROBLEM corresponds to the problem of finding an order and an orientation of the contigs with maximal support, and forming exactly σ_p linear chromosomes and σ_c circular chromosomes.

3   Computational Complexity

We explore in this section the complexity of (σ_p, σ_c)−SP (decision problem). We show that the problem remains $\mathcal{NP}$−complete, even in bipartite graphs. In this paragraph, we use a reduction from the HAMILTONIAN CIRCUIT PROBLEM ([8]).

---

[1] Represented for instance by a list of n edges $\{2i, 2i + 1\}$.

HAMILTONIAN CIRCUIT PROBLEM (HC):
Instance: $G = (V, A)$
Question: Does G contain a Hamiltonian circuit?

The reduction relies on the following polynomial transformation, illustrated by Figure 1:

Transformation 1. *Let $G = (V, A)$ an instance of the HC problem. We consider the following graph $G' = (\{u_i, i \in \{1, 2, 3, 4\}, u \in V\}, E)$:*

- *$\forall u \in V$, we consider the path of length four $\mathcal{P}_{4,u} = u_1 - u_2 - u_3 - u_4$, (the set of all paths of length four is denoted by $\mathcal{P}_4 = \cup_{u \in V} \mathcal{P}_{4,u}$),*
- *$\forall (u, v) \in A$, we add an edge $\{u_4, v_1\}$ in E.*

We consider the perfect matching M^* on G', consisting in the edges of the kind $\{u_1, u_2\}$ and $\{u_3, u_4\}$, $\forall u \in V$.

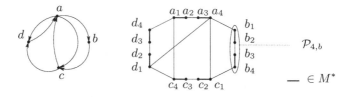

Fig. 1. Transformation of an instance of HC (left) to an instance of (σ_p, σ_c)–SP by Transformation 1 (right). Stronger edges belong to the perfect matching.

Notice that the graph G' is bipartite: the first set of the partition is constituted by vertices $u_i, i \in \{2, 4\}, \forall u \in V$ and the second by the other vertices.

Theorem 2. *The problem (σ_p, σ_c)–SP is $\mathcal{NP}$–complete, even if the graph is bipartite.*

Proof. The problem is clearly in $\mathcal{NP}$. Let $G = (V, A)$ an instance of the HC problem, and G' the graph produced by Transformation 1. Notice that G' is a bipartite graph by construction. We consider $(\sigma_p + \sigma_c)$ copies of G', denoted by $G'_1, \ldots G'_{\sigma_p + \sigma_c}$. If $\sigma_p = 0$, then we connect sequentially the copies of G' with one edge, joining two inner vertices of one path $\mathcal{P}_4$ respectively in G'_i and G'_{i+1}. If $\sigma_p \geq 1$, we connect all the copies to a single vertex v, and add a vertex v', and an edge $\{v, v'\} \in M^*$ (See Figure 2). If $\sigma_c = 0$, we use reduction from the DIRECTED HAMILTONIAN PATH (DHP) instead of the HC in the following discussion. In any case, the instance of (σ_p, σ_c)–SP produced from G is a bipartite graph.

- We suppose that there exists a solution of (σ_p, σ_c)–SP. It is easy to see that in σ_c copies of G', we have a alternating Hamiltonian cycle. We consider one of them, giving an Hamiltonian cycle in G', denoted by $\mathcal{C}'$. By construction, $\mathcal{C}'$ contains all edges of $\mathcal{P}_4$ since this is the only way to cover the vertices u_2 and u_3. Notice that the incoming and outcoming edges of a path $p \in \mathcal{P}_4$, and $\{u_2, u_3\}$, do not belong to M^*. On the contrary, all edges of type $\{u_1, u_2\}$ and $\{u_3, u_4\}$ belong to M^*. Therefore, by compressing all paths of type $\mathcal{P}_4, \forall u \in V$ in $\mathcal{C}'$ and considering an arc $\{u, v\}$ if $\{u_4, v_1\} \in \mathcal{C}'$ (resp. $\{v, u\}$ if $\{u_1, v_4\} \in \mathcal{C}'$), we obtain a positive solution for the HC.
- Conversely, let $\mathcal{C}$ be an Hamiltonian Circuit in G, it is easy to construct a feasible solution for the (σ_p, σ_c)–SP in the transformed graph, by considering $\mathcal{C}' = M^* \cup \{\{u_2, u_3\}, \forall u \in V\} \cup \{\{v_4, u_1\}$ with an arc $(v, u) \in \mathcal{C}\}$ in σ_c of the copies, taking the same without one edge not belonging to M^* in $(\sigma_p - 1)$ other copies, and in the remaining copy, connecting v to the inner circuit to form a path (see Figure 2).

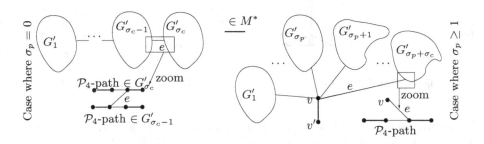

Fig. 2. Construction of an instance (σ_p, σ_c)–SP based on Transformation 1

In order to explore the cases where the length of the cycles and paths are also fixed, we considered the following particular case: the length of the σ_p paths, denoted by l_p is at least 2, and the length of the σ_p cycles is 6. Thus we now consider a reduction from the PARTITION INTO TRIANGLES [8].

PARTITION INTO TRIANGLES:
Instance: $G = (V, E)$, with $|V| = 3q, q \in \mathbb{N}$
Question: Can the vertices of G be partitioned into q disjoint sets $V_1, V_2, \ldots, V_q$, each containing exactly three vertices, such that for each $V_i = \{u_i, v_i, w_i\}, 1 \leq i \leq q$, all three of the edges $\{u_i, v_i\}$, $\{u_i, w_i\}$, $\{w_i, v_i\}$ belong to E?

Again, we define polynomial-time transformations from an instance of PARTITION INTO TRIANGLES (PT) to an instance of (σ_p, σ_c)–SP with $l_p \geq 2$ and $l_c = 6$ (see Figure 3), and (σ_p, σ_c)–SP with $l_p \geq 2$ and $l_c = 12$, for bipartite graphs respectively (see Figure 4).

Transformation 2. *Let $G = (V, E)$ be an instance of* PARTITION INTO TRIAN-GLES. *We consider the graph $G' = (V' = V_0 \cup V_1 \cup V_2, E' = E_0 \cup E_1 \cup E_2 \cup E_3 \cup E_4)$:*

- *We consider two copies of G denoted by $G_0 = (V_0, E_0)$ and $G_1 = (V_1, E_1)$ with vertices respectively denoted by x^0 and x^1 for $x \in V$.*
- *$\forall x \in V$, $\{x^0, x^1\} \in E_2$.*
- *$\forall \{x, y\} \in E$, $\{x^0, y^1\} \in E_3$ and $\{x^1, y^0\} \in E_3$.*
- *Lastly, we add σ_p paths of length l_p from σ_p vertices of G_1 denoted by $x_j^1 - x_j^2 - \ldots - x_j^{l_p}$. We denote by V_2 and E_4 respectively the set of vertices and the set of edges of these paths.*

The perfect matching M^* consists in the edges of E_2 and the edges of the form $\{x_j^{2i}, x_j^{2i+1}\}$ in the additional paths.

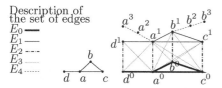
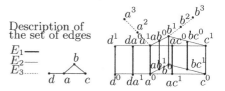

Fig. 3. Transformation 2   **Fig. 4.** Transformation 3

Transformation 3. *Let $G = (V, E)$ be an instance of* PARTITION INTO TRI-ANGLES. *We consider the graph $G' = (V' = V_0 \cup V_1 \cup V_2, E' = E_1 \cup E_2 \cup E_3)$:*

- *For each vertex $x \in V$, we add two vertices, denoted by x^0 and x^1, in V'. This set of vertices is denoted by V_0.*
- *For each edge $\{x, y\} \in E$, we add two vertices xy^0 and xy^1. This set of vertices is denoted by V_1.*
- *$\forall x \in V$, $\{x_i^0, x_i^1\} \in E_1$, and for all $\{x, y\} \in E$, $\{xy^0, xy^1\} \in E_1$.*
- *$\forall \{x, y\} \in E$, $\{x^0, xy^1\}$, $\{xy^1, y^0\}$, $\{x^1, xy^0\}$ and $\{xy^0, y^1\}$ belong to E_2.*
- *Lastly, we add σ_p paths of length l_p from σ_p vertices of G_1 denoted by $x_j^1 - x_j^2 - \ldots - x_j^{l_p}$. We denote by V_2 and E_3 respectively the set of vertices and the set of edges of these paths.*

The perfect matching consists in the edges of E_1 and the edges of the form $\{x_j^{2i}, x_j^{2i+1}\}$ in the additional paths.

Theorem 3. *The (σ_p, σ_c)–SP with $l_p \geq 2$ and $l_c = 6$ (resp. (σ_p, σ_c)–SP with $l_p \geq 2$ and $l_c = 12$ in a bipartite graph) is $\mathcal{NP}$–complete.*

Proof. The problem is clearly in $\mathcal{NP}$. Let $G = (V, E)$ be an instance of PARTI-TION INTO TRIANGLES. We consider the graph G' obtained from G by Trans-formation 2.

- We suppose that there exists a positive solution for the (σ_p, σ_c)−SP with $l_p \geq 2$ and $l_c = 6$ in the graph G' i.e. all vertices are covered by either a alternating-cycle of length six or by a alternating-path of length l_p. The paths are obviously the paths V_2. Thus the alternating-cycles are covering G' restricted to $G_0 \cup G_1 \cup E_2 \cup E_3$. Any cycle is of size six, and contains three edges of E_2 alternating with three edges of either E_0, E_1 or E_3. The edges of E_2 correspond to vertices in the original graph G, and two such edges can be connected by an edge of E_0, E_1 or E_3 if and only if there is an edge in the original graph between the corresponding vertices. Thus, a alternating-cycle of size six corresponds to a triangle in G. Then, the solution of the (σ_p, σ_c)−SP with $l_p \geq 2$ and $l_c = 6$ in the graph G' defines a partition into triangles in G.
- Conversely, we suppose that it exists in G a partition into triangles, let us construct a positive solution for the (σ_p, σ_c)−SP with $l_p \geq 2$ and $l_c = 6$ in the graph G'. The alternating-paths are given by the additional paths described above, and for a triangle $\{x, y, z\}$, we consider the alternating-cycle of length six $\{x_0, y_0, y_1, z_0, z_1, x_1, x_0\}$. It is clear that all alternating-cycles and the alternating-paths cover the vertices of G'.

Concerning the bipartite case, using the same argument as previously on the graph obtained by Transformation 3, it is clear that the alternating-cycles of size 12 in G' are necessarily of the form $(a^0, ab^1, ab^0, b^1, b^0, bc^1, bc^0, c^1, c^0, ca^1, ca^0, a^1, a^0)$.

Clearly, all previous problems remain $\mathcal{NP}$−complete in the optimization version, even in complete graphs.

4   Inapproximability Results for MIN−(σ_p, σ_c)−SP

Theorem 4. *The problem* MIN−$(0, 1)$−SP *(resp.* MIN−$(1, 0)$−SP*) is non-approximable, unless* $\mathcal{P} = \mathcal{NP}$*, even if the graph is bipartite.*

Proof. We use a similar proof to the proof of non-approximability for the TSP in [12]. We suppose that there exists a polynomial-time approximation algorithm A for this problem, with approximation ratio $\rho > 0$. Let $G = (V, E)$ be an instance of HC with order n. We build an instance of MIN−$(0, 1)$−SP in the following way:

- We use the adaptation of Transformation 1 to transform the graph G into a graph G', like in the Proof of Theorem 2. We add all missing edges of the kind $\{u_1, v_4\}$ for any $u, v \in V$. Let $G'' = (V'', E'')$ be the graph resulting of the transformation. Notice that G'' is bipartite, since the vertices u_1 and v_4 belong to different sets of the bipartition described in Transformation 1.
- The distance $d(v_i, v_j)$ on an edge $(v_i, v_j) \in E''$ is defined as follows:

$$d(v_i, v_j) = \begin{cases} 1 \text{ if } \{v_i, v_j\} \in E \\ 0 \text{ if } \{v_i, v_j\} \in \mathcal{P}_4 \\ \rho \times n \text{ for every other edge } \in E'' \end{cases}$$

Clearly, considering the weight of the solution S given by Algorithm A gives a polynomial-time decision algorithm for the HC: G has an Hamiltonian circuit if and only if $w(S) \leq \rho \times n$.

Corollary 5. *The problem* $\mathrm{MIN}-(\sigma_p, \sigma_c)-\mathrm{SP}$ *is non-approximable, unless* $\mathcal{P} = \mathcal{NP}$, *even if the graph is bipartite.*

Due to the lack of place, the proof is omitted, but it relies on the ideas of proofs of Theorem 2 and Theorem 4.

5   A Polynomial-Time Approximation Algorithm for $\mathrm{MAX}-(\sigma_p, \sigma_c)-\mathrm{SP}$

In this section, we focus on the maximization version of the problem. In what follows, we consider only complete graphs or order $2n$, and that the sufficient and necessary condition $n \geq \sigma_p + 2\sigma_c$ to the existence of a solution in a complete graph holds. Moreover, we suppose that the weights are non-negative. Notice that this latter condition is reached if the weight of an edge represents a number of mate-pairs. We describe in this case a polynomial-time algorithm and prove its approximation ratio.

Let M^* be the initial perfect matching in the graph. We compute a maximum weight matching of maximum cardinality in $G \backslash M^*$, for which there exists an algorithm running in $O(n^3)$ [6]. We obtain a cycle cover of maximal weight, with alternating-cycles (see Figure 6). We note $S_{M^{**}} = \{C_1 \ldots C_h\}$ the set of alternating-cycles in this cycle cover. We define a polynomial-time algorithm which transforms this cover into a solution of the $\mathrm{MAX}-(\sigma_p, \sigma_c)-\mathrm{SP}$. First, we determine a set of edges that we can remove in order to build the required cycles and paths.

Definition 6. *For each cycle* $C_i, \forall i \in \{1, \ldots, h\}$, *we note* $t_i = |C_i \backslash M^*|$. *We define* removable sets of edges *for* C_i *the following way:*

- *If* t_i *is even, then we take alternatively one of every two consecutive edges from* $C_i \backslash M^*$, *such as to alternate edges belonging to this set and edges that do not, on the cycle. There are two such sets.*
- *If* t_i *is odd, then we take two consecutive edges from* $C_i \backslash M^*$, *called* twin edges, *and do the same as previous, starting from these edges. Thus, there are exactly* t_i *such sets.*

For each cycle C_i, *we define a* minimal removable set of edges, *noted* R_i, *as a removable set of edges of* C_i *with minimal total weight. In the case where* t_i *is odd, we denote by* e_i^* *the twin edge of* R_i *of minimal weight (see Figure 5).*

Lemma 7. *For each cycle* $C_i, i \in \{1, \ldots, h\}$, *with* $|C_i \backslash M^*| = t_i$, *and a minimal removable set of edges* R_i *of* C_i, *we have:* $w(R_i) \leq w(C_i \backslash (M^* \cup R_i))$ *(resp.* $w(R_i) \leq w(C_i \backslash (M^* \cup R_i)) + w(e_i^*)$ *) if* t_i *is even (resp. odd). In any case, we have* $w(R_i) \leq 2.w(C_i \backslash (M^* \cup R_i))$.

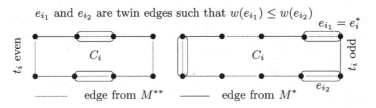

e_{i_1} and e_{i_2} are twin edges such that $w(e_{i_1}) \leq w(e_{i_2})$

$e_{i_1} = e_i^*$

t_i even

C_i

C_i

t_i odd

e_{i_2}

$\cdots\cdots$ edge from M^{**} —— edge from M^*

Fig. 5. Removable set of edges in the cycles of $S_{M^{**}}$. The removable sets are figured by circled edges.

Proof. If t_i is even, then by Definition 6 it is clear that $w(R_i) \leq w(C_i \backslash (M^* \cup R_i))$.

We suppose now that t_i is odd. The set of edges $R_i' = \{e_i^*\} \cup (C_i \backslash (M^* \cup R_i))$ also defines a removable set of edges. By minimality of R_i, we get the second inequality.

The third inequality is trivial in the case where t_i is even. Now, we suppose that t_i is odd. It is sufficient to show that $w(e_i^*) \leq w(C_i \backslash (M^* \cup R_i))$. Suppose that $w(e_i^*) > w(C_i \backslash (M^* \cup R_i))$. Let e_{i_1} and e_{i_2} be the twin edges of R_i with $e_{i_1} = e_i^*$. Clearly, $w(e_{i_2}) \geq w(e_i^*) > w(C_i \backslash (M^* \cup R_i))$. By hypothesis, $w(R_i) \geq w(e_{i_2}) + w(e_i^*) > w(C_i \backslash (M^* \cup R_i)) + w(e_i^*) \geq w(R_i)$, which is not possible.

Algorithm 1. Maximal cycle cover algorithm for $\text{MAX}-(\sigma_p, \sigma_c)-\text{SP}$

$M^{**} \leftarrow$ A maximal weight perfect matching for $G \backslash M^*$;
$S_{M^{**}} \leftarrow M^{**} \cup M^*$;
$h \leftarrow$ the number of cycles in $S_{M^{**}}$;
for *cycle* $C_i \in S_{M^{**}}$ **do**
| $\quad R_i \leftarrow$ A minimal removable edges set of C_i;
end
$R \leftarrow \bigcup_{i=1}^{h} R_i$;
Remove the edges of R from the cycles;
Build σ_c cycles by cycling σ_c paths of length four and, if needed, by using two paths of length two;
Build the σ_p paths by merging, if needed, the remaining paths;

Lemma 8. *Algorithm 1 gives a feasible solution.*

Proof. We consider the number of cycles and paths that we can build from C_i by removing edges in R_i and merge back the pieces. It can be any pair $(x, y) \in \mathbb{N} \times \mathbb{N} \backslash \{(0, 0)\}$ where x is the number of cycles, and y the number of paths, satisfying $y \leq -2x + t_i$.

This property is additive, meaning that if we consider two cycles of sizes t_i and t_j, we can realize any pair $(x, y) \in \mathbb{N} \times \mathbb{N} \backslash \{(0, 0)\}$ where x is the number of cycles, and y the number of paths, satisfying $y \leq -2x + t_i + t_j$.

Thus, if we consider the whole set of cycles, we can realize any pair $(x, y) \in \mathbb{N} \times \mathbb{N}\setminus\{(0,0)\}$ where x is the number of cycles, and y the number of paths, satisfying $y \leq -2x + \sum_{i=1}^{h} t_i$. Since $\sum_{i=1}^{h} t_i = \sum_{i=1}^{h} |C_i\setminus M^*| = |M^*|$, and $|M^*| = n \geq \sigma_p + 2\sigma_c$, we have: $\sigma_p \leq -2\sigma_c + \sum_{i=1}^{h} t_i$. Consequently, it is possible to realize a solution by removing only edges belonging to R and recombining the remaining pieces.

Theorem 9. *The Algorithm 1 provides a solution for the* $\text{MAX}-(\sigma_p, \sigma_c)-\text{SP}$ *in complete graphs with non-negative weights, with an approximation ratio of three and a time complexity $\mathcal{O}(n^3)$. The bound is tight.*

Proof. Algorithm 1 finishes with σ_p paths and σ_c cycles, by construction of the sets R_i. It runs in time complexity $\mathcal{O}(n^3)$, which is the initial cycle cover computation complexity. The other operations can be executed in $\mathcal{O}(n^2)$.

We denote by S_{opt} an optimal solution for the $\text{MAX}-(\sigma_p, \sigma_c)-\text{SP}$, and S_H the solution given by the heuristic described above.

We have $w(S_{M^{**}}) = \sum_{i=1}^{h} w(C_i) = \sum_{i=1}^{h}(w(C_i\setminus R_i) + w(R_i))$. Using Lemma 7, we know that $w(R_i) \leq 2.w(C_i\setminus R_i)$ and so $w(S_{M^{**}}) \leq 3. \sum_{i=1}^{h} w(C_i\setminus R_i)$. Since S_{opt} also gives a matching of $G\setminus M^*$, we have $w(S_{opt}) \leq w(S_{M^{**}}) \leq 3. \sum_{i=1}^{h} w(C_i\setminus R_i)$. Finally, since $\forall i \in \{1\ldots h\}, C_i\setminus R_i \subset S_H$, we have $w(S_{opt}) \leq 3.w(S_H)$. Moreover, the bound is tight, consider the graph on Figure 6.

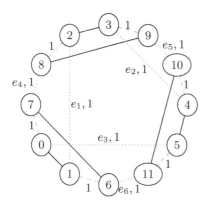

Not appearing edges are edges of weight 0. We suppose that $\sigma_p = 0$ and $\sigma_c = 3$. Suppose that the maximum weighted matching of maximum cardinality algorithm produces the two cycles $C_1 = (0, 1, 2, 3, 4, 5)$ and $C_2 = (6, 7, 8, 9, 10, 11)$ (dotted edges), of total weight six. By taking $R_1 = \{e_1, e_2\}$ and $R_2 = \{e_5, e_6\}$, the re-merging operation would give a solution of weight two: $\{(0, 1, 5, 4), (7, 6, 11, 10), (2, 3, 8, 9)\}$. But an optimal solution to the problem, also of weight six, would be obtained with the dashed edges.

Fig. 6. The bound of three is tight for Algorithm 1

The approximation ratio could be improved in the TSP-like problem.

Theorem 10. *There exists a polynomial-time approximation algorithm for the* $\text{MAX}-(0, 1)-\text{SP}$ *in complete graphs with non-negative weights, with an approximation ratio of two and a time complexity $\mathcal{O}(n^3)$. The bound is tight.*

Proof. Let $S_{M^{**}} = \{C_1, \ldots, C_h\}$ be the alternating-cycles given by the maximum weight matching of maximal cardinality algorithm. For each $i \in \{1 \ldots h\}$, let e_i be an edge of minimal weight in $C_i \backslash M$. We remove all the edges e_i, and connect the cycles in order to obtain only one cycle, by arbitrary edges. As the same way as previously S_{opt} (resp. S_H) designs an optimal solution (resp. the heuristic described above) for the MAX$-(0,1)-$SP.

We have $w(S_{M^{**}}) = \sum_{i=1}^{h} w(C_i) = \sum_{i=1}^{h} (w(C_i \backslash \{e_i\}) + w(e_i))$.

Since e_i is an edge of minimal weight in C_i, which counts at least two edges outside M^*, we have $w(e_i) \leq w(C_i \backslash \{e_i\})$. Then: $w(S_{M^{**}}) \leq 2. \sum_{i=1}^{h} w(C_i \backslash \{e_i\})$.

Furthermore, since S_{opt} also gives a matching of $G \backslash M^*$, we have $w(S_{opt}) \leq w(S_{M^{**}}) \leq 2. \sum_{i=1}^{h} w(C_i \backslash \{e_i\})$. Finally, since $\forall i \in \{1 \ldots h\}, C_i \backslash \{e_i\} \subset S_H$, we obtain $w(S_{opt}) \leq 2.w(S_H)$. Moreover, the bound is tight, consider the graph on Figure 7.

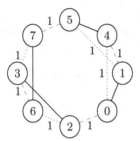

Not appearing edges are edges of weight 0. Suppose that the weighted matching of maximum cardinality algorithm produces the two cycles with dotted edges. One optimal solution of the MAX$-(0,1)-$SP in this graph is given by the dashed edges. Both solutions have weight four. But the solution produced by the heuristic has weight only two.

Fig. 7. The bound of two is tight for the MAX$-(0,1)-$SP

6 Conclusion

In this article, we investigate problems inspired by the contig scaffolding problem in bioinformatics, from the point of view of complexity and approximation algorithms with guaranteed performance ratio. In such a context, we propose some negative results in term of complexity ($\mathcal{NP}$-completeness/non-approximability) and we design efficient polynomial-time approximation algorithms.

From a combinatorial point of view, further interesting questions would be to compare the computation time of several exact methods, on randomly generated and real instances, and to continue the exploration of the approximation algorithms for the maximization problem.

From a bioinformatic point of view, several perspectives have to be explored. First, we have to improve the relevance of the problem by considering a relaxed version: indeed, our complete graphs allow weight equal to 0, which have no meaning for the scaffolding problem. Then, we have to consider now the following problem, trying to find a cover of maximal weight, by at least σ_p paths and at most σ_c cycles not containing weights equal to 0, such that the structure of the given solution is as close as possible to the desired structure. Furthermore, we have now to extend the model to take into account the multiplicities of the repeated contigs, and possibly mix multiple sources of information and thus, multi-criteria. Finally, we intend to verify on simulated data and real data, if a good approximation ratio improves the biological quality of the results.

Acknowledgments. This work was supported by the Institut de Biologie Computationnelle (IBC) and Défi MASTODONS SePhHaDe from CNRS.

References

1. Burton, J.N., Adey, A., Patwardhan, R.P., Qiu, R., Kitzman, J.O., Shendure, J.: Chromosome-scale scaffolding of de novo genome assemblies based on chromatin interactions. Nature Biotechnology, 1119–1125 (November 2013)
2. Chauve, C., Patterson, M., Rajaraman, A.: Hypergraph covering problems motivated by genome assembly questions. In: Lecroq, T., Mouchard, L. (eds.) IWOCA 2013. LNCS, vol. 8288, pp. 428–432. Springer, Heidelberg (2013)
3. Chiba, S., Fujita, S.: Covering vertices by a specified number of disjoint cycles, edges and isolated vertices. Discrete Mathematics 313(3), 269–277 (2013)
4. Dayarian, A., Michael, T.P., Sengupta, A.M.: SOPRA: scaffolding algorithm for paired reads via statistical optimization. BMC Bioinformatics 11, 345 (2010)
5. Donmez, N., Brudno, M.: SCARPA: scaffolding reads with practical algorithms. Bioinformatics 29(4), 428–434 (2013)
6. Gabow, H.N.: An efficient implementation of Edmonds' algorithm for maximum matching on graphs. Journal of the ACM 23(2), 221–234 (1976)
7. Gao, S., Sung, W., Nagarajan, N.: Opera: reconstructing optimal genomic scaffolds with high-throughput paired-end sequences. Journal of Computational Biology 18(11), 1681–1691 (2011)
8. Garey, M.R., Johnson, D.S.: Computers and Intractability: A Guide to the Theory of NP-Completeness. W. H. Freeman & Co., New York (1979)
9. Gritsenko, A.A., Nijkamp, J.F., Reinders, M.J., Ridder, D.D.: GRASS: a generic algorithm for scaffolding next-generation sequencing assemblies. Bioinformatics, 1429–1437 (2012)
10. Huson, D.H., Reinert, K., Myers, E.W.: The greedy path-merging algorithm for contig scaffolding. Journal of ACM 49(5), 603–615 (2002)
11. Krivelevich, M., Nutov, Z., Salavatipour, M.R., Yuster, J.V., Yuster, R.: Approximation algorithms and hardness results for cycle packing problems. ACM Transaction on Algorithms 3(4) (November 2007)
12. Sahni, S., Gonzalez, T.: P-complete approximation problems. Journal of ACM 23(3), 555–565 (1976), http://doi.acm.org/10.1145/321958.321975
13. Shmoys, D.B., Lenstra, J.K., Kan, A.H.G.R., Lawler, E.L.: The Traveling Salesman Problem: a guided tour of combinatorial optimization. John Wiley & Sons (1985)
14. Steiner, G.: On the k-path partition of graphs. Theoretical Computer Science 290, 2147–2155 (2003)
15. Tutte, W.T.: A short proof of the factor theorem for finite graphs. Canadian Journal of Mathematics 6, 347–352 (1954)

Heuristics for the Sorting by Length-Weighted Inversions Problem on Signed Permutations

Thiago da Silva Arruda, Ulisses Dias, and Zanoni Dias

University of Campinas, Institute of Computing
Av Albert Einstein, 1251, Campinas/SP - Brazil
{thiago.arruda@students.,udias@,zanoni@}ic.unicamp.br

Abstract. Genome Rearrangement is a field that addresses the problem of finding the minimum number of global operations that transform a given genome into another. In this work, we deal with inversion events, which occur when a segment of DNA sequence in the genome is reversed. In our model, each inversion costs the number of elements in the reversed segment. We present a new heuristic for this problem and we show that our method outperforms a previous approach. Our method uses the metaheuristic called Greedy Randomized Adaptive Search Procedure (GRASP) that has been routinely used to find solutions for combinatorial optimization problems. In essence, we implemented an iterative process in which each iteration receives a feasible solution whose neighborhood is investigated for a better solution. We use as initial solution a sequence of inversions of minimum length when each inversion costs one unit, which is a problem that already has several polynomial time algorithms. In almost every case, we were able to improve that initial solution by providing a less-costly sequence of inversions.

Keywords: Genome rearrangement, length-weighted inversion, GRASP.

1 Introduction

Genome comparison among closely related species has revealed that organisms share large blocks of DNA sequence and that the ordering of these blocks may change across species. The main explanation for the variability in chromosome structure is that genomes undergo large scale mutational events, also called rearrangements, during the evolutionary process. A genome rearrangement event called inversion occurs when a segment of DNA sequence in the genome is reversed. In bacteria, inversions have long been recognized as one of the most frequently observed rearrangements.

Darling, Miklós and Ragan [9] studied eight *Yersinia* genomes and found that inversions are shorter than expected under a neutral model. The understanding that short inversions are preferred is not entirely new. Previous results suggested that the sequence of operations that most likely happened during the evolution may not involve the movement of many long sequences [7]. Sankoff [17] found evidence that the frequency of short segment inversions is high in the evolution

A.-H. Dediu, C. Martín-Vide, and B. Truthe (Eds.): AlCoB 2014, LNBI 8542, pp. 59–70, 2014.

of microbial genomes, whereas it seems less prevalent in plants and animals. Lefebvre *et al.* [15] and Sankoff *et al.* [18] found a large excess of short inversions, specially those involving a single gene, which contrasted with the null hypothesis that the two endpoints of an inversion occur by random and independently.

While a large body of literature exists on mathematical problems related to the computation of the inversion distance between two genomes, these works generally do not take into account the length of the reversed segments, *i.e.*, each inversion costs one unit. The first polynomial time algorithm was presented by Hannenhalli and Pevzner [13] and other algorithms were later devised in order to simplify it or to improve its running time [5,6].

Few algorithms consider the number of elements in each inversion. In 2002, Ajana *et al.* [2] developed a randomized heuristic that allows the user to choose a secondary criterion to be used with the Hannenhalli and Pevzner [13] algorithm. In other words, Hannenhalli and Pevzner algorithm can generate many equally optimal solutions, so the Ajana *et al.* algorithm selects, at random, one of the shortest allowable inversions at each step.

In 2004, Swidan *et al.* [19] created a length-sensitive model where the cost of reversing a subsequence of length l is given by the function $f(l) = l^\alpha$, for $\alpha \geq 0$. For $\alpha = 1$, we do not know any NP-hardness proof and there is no evidence that a polynomial algorithm can be devised. Swidan *et al.* were able to guarantee the approximation ratio $O(\lg n)$. We work on this problem and we show a method that finds better solutions in most of the cases.

In a previous work, we presented a greedy algorithm for the sorting by length-weighted inversion problem when gene orientation is not taken into account [3]. Here, we deal with a more biologically relevant problem since we are considering both orientation and length of each inversion.

Our method uses the metaheuristic introduced by Feo and Resende [10] called Greedy Randomized Adaptive Search Procedure (GRASP), which has been routinely used to find solutions for combinatorial optimization problems [1,8,11,14, 16]. The reader is referred to the review written by Festa and Resende [12] for a more detailed discussion about GRASP.

In essence, GRASP searches for good solutions as an iterative process. Each iteration uses an initial solution s in a discrete set of feasible solutions S, and its neighborhood is investigated until a local minimum is found. Our approach follows this model and also implements other basic components of GRASP as described by Feo and Resende [10].

The rest of this paper is organized as follows. Section 2 is a brief introduction to the problem. It provides the concepts and definitions used throughout the text. Section 3 presents our search procedure. The section is self-contained, and we assume no previous knowledge about the concepts of GRASP. Section 4 analyses the model using a practical approach. We base our analysis on the improvement that was achieved by our model when an optimum solution for the sorting by inversions problem is used as initial solution.

2   Definitions

We represent a pair of single-chromosome genomes by permutations of integer numbers, where each number represents a block of DNA sequence shared between the two genomes, and the signs represent gene orientation. Let n be the number of conserved blocks, we assume that one of the genomes is represented by the identity permutation $\iota = (1 \ 2 \ \dots \ n)$ and the other genome is represented by the permutation $\pi = (\pi_1 \ \pi_2 \ \dots \ \pi_n)$, for $\pi_i \in \mathbb{I}$, $0 < |\pi_i| \leq n$ and $i \neq j \leftrightarrow |\pi_i| \neq |\pi_j|$.

The following functions can be applied to identify any element i in π.

Definition 1. *Position:* $p(\pi, i) = k \leftrightarrow |\pi_k| = i$, $p(\pi, i) \in \{1, 2, \dots, n\}$.

Definition 2. *Sign:* $s(\pi, i) = \begin{cases} 1, & \text{if } \pi_i < 0. \\ 0, & \text{if } \pi_i > 0. \end{cases}$

Given a permutation π, we extend it with two elements $\pi_0 = 0$ and $\pi_{n+1} = n + 1$. The extended permutation is still denoted π. A pair of elements π_i and π_{i+1}, with $0 \leq i \leq n$, is a breakpoint if $\pi_{i+1} - \pi_i \neq 1$. The identity permutation ι is the only permutation with no breakpoints.

Sometimes it is useful to treat permutations as functions such that $\pi(i) = \pi_i$ and $\pi(-i) = -\pi(i)$. The composition of two permutations π and σ is the permutation $\pi\sigma = (\pi_{\sigma(1)} \ \pi_{\sigma(2)} \ \dots \ \pi_{\sigma(n)})$. We can see the composition as the relabeling of elements in π according to elements in σ. Note that ι is the neutral element such that $\pi\iota = \iota\pi = \pi$. We define π^{-1} as the permutation such that $\pi\pi^{-1} = \pi^{-1}\pi = \iota$. The inverse permutation is the function such that $\pi^{-1}_{\pi(i)} = i$.

An inversion $\rho(i, j)$ is a rearrangement event that reverses the order and the signs of a consecutive section of a genome: $\pi\rho(i, j) = (\pi_1 \ \dots \ \pi_{i-1} \ -\pi_j \ \dots \ -\pi_i \ \pi_{j+1} \ \dots \ \pi_n)$, such that $1 \leq i \leq j \leq n$.

The inversion distance $d(\pi)$ between an arbitrary genome π and the identity ι is the minimum number t of operations $\rho_1, \rho_2, \dots \rho_t$ such that $\pi\rho_1\rho_2 \dots \rho_t = \iota$.

The length-weighted inversion distance $d_{lw}(\pi)$ between π and ι is a minimization problem written as "$d_{lw}(\pi) = \min \sum_{k=1}^{t'} cost(\rho_k)$", where $cost$ is an objective function given by $cost(\rho(i, j)) = j - i + 1$ and $\pi\rho_1\rho_2 \dots \rho_{t'} = \iota$. For this problem we have $t' \geq d(\pi)$.

Let us consider the permutation $\pi = (+2 \ +1 \ -5 \ -4 \ -3)$. Three inversions are enough to sort π, as $\pi\rho(2, 5)\rho(1, 4)\rho(1, 5) = \iota$, which implies $d(\pi) = 3$. The cost for this sequence is $cost(\rho(2, 5)) + cost(\rho(1, 4)) + cost(\rho(1, 5)) = 4 + 4 + 5 = 13$. The length-weighted inversion problem aims to decrease the overall cost, so we allow the number of inversions to increase if that helps us achive our goal. For instance, the sequence $\pi\rho(1, 2)\rho(3, 5)\rho(1, 1)\rho(2, 2)$ has 4 inversions and costs 7.

3   The Meta-heuristic

Our heuristic is an iterative process in which each iteration starts from a feasible solution whose neighborhood is investigated for a better solution. If a better solution is found, it is kept as the current solution and used in the next iteration. Below we state the main points of our heuristic.

1. An initial solution is choosen and it impacts on the quality of the final solution.
2. Let s be a solution, we set how different from s another solution s' should be in order to be considered in the same neighborhood as s. Our definition for neighborhood is given in Section 3.1.
3. We implemented a method to build a new solution in the neighborhood of s. The main aspects of our method are described in Section 3.2 and an auxiliary function is presented in Section 3.3.

3.1   Neighborhood

Let S be a discrete set of feasible solutions for the sorting by length-weighted inversions problem, we represent each solution $s \in S$ as a sequence of permutations that starts with π and ends with ι. Each permutation in the sequence differs from the previous one by one inversion.

Definition 3. *A solution $s \in S$ is a sequence of permutations $s = < s_0, s_1, \ldots, s_m >$ such that $s_k = s_{k-1}\rho_k$, $1 \leq k \leq m$, $s_0 = \pi$ and $s_m = \iota$.*

Definition 4. *Let $s = < s_0, s_1, \ldots, s_m >$ be a solution in S, we define $sub(s, p, q)$ as the subsequence $sub(s, p, q) = < s_p, s_{p+1}, \ldots, s_q >$, for $0 \leq p \leq q \leq m$.*

The neighborhood $N(s)$ is a set of solutions. We define an element $s' \in N(s)$ as follows.

Definition 5. *A solution $s' = < s'_0, s'_1, \ldots, s'_{m'} >$ is in $N(s)$ iff $sub(s, 0, p) = sub(s', 0, p')$ and $sub(s, q, m) = sub(s', q', m')$, for $0 \leq p < q \leq m$ and $0 \leq p' < q' \leq m'$.*

Observe that $N(s)$ has a wide number of elements. In fact, by this definition, any solution s' is in $N(s)$ since we can assign $p = 0$ and $q = m$. In order to make this definition useful, we need to further constrain the values that might be assigned to p and q. Therefore, we propose a new definition for neighborhood that we represent by $N_f(s)$, where f is how far p and q are from each other.

Definition 6. *Let f be a natural number, a solution $s' = < s'_0, s'_1, \ldots, s'_{m'} >$ is in the neighborhood $N_f(s)$ iff $sub(s, 0, p) = sub(s', 0, p')$, $sub(s, q, m) = sub(s', q', m')$ and $q - p + 1 = f$, for $0 \leq p < q \leq m$ and $0 \leq p' < q' \leq m'$. We say that the subsequence $sub(s, p, q)$ is a frame of size f.*

3.2   Local Search

Let $s = < s_0, s_1, \ldots, s_m >$ be a solution in S and $sub(s, p, q)$ be a frame, the sequence of inversions that transform s_p into s_q is the same that would be used to transform $s_q^{-1} s_p$ into ι. That is because we can relabel the entire frame using composition in order to create the sequence $< s_q^{-1} s_p, s_q^{-1} s_{p+1}, \ldots, s_q^{-1} s_q >$, where $s_q^{-1} s_q = \iota$. Therefore, we conclude that creating a new sequence for the window $sub(s, p, q)$ is equivalent to sorting the instance $s_q^{-1} s_p$ by length-weighted inversion. For conciseness, in Section 3.3 we will talk of solving the sorting by length-weighted inversion problem where we mean creating a new frame to replace $sub(s, p, q)$.

Let $sub(s, p, q)$ be a frame we want to replace such that $q - p + 1 = f$, we find a solution for the permutation $s_q^{-1} s_p$ and this solution can be easily transformed back into a sub-sequence going from s_p to s_q. If this new sub-sequence costs less than $sub(s, p, q)$, we replace the frame $sub(s, p, q)$ for the new sub-sequence in order to create a new solution $s' \in N_f(s)$.

We have $m - f + 1$ different frames of size f in s, and some frames are more likely to be improved than others, so we created a method where the frame that will be improved must be selected in a two-step process.

1. We calculate the cost for each frame $sub(s, p, q) = < s_p, s_{p+1}, \ldots, s_q >$ by summing the cost of each inversion ρ_k such that $s_k = s_{k-1} \rho_k$, $p < k \le q$. Then we select a limited number of high-cost frames defined as a parameter named `frame_limit`. The frames that were not selected will be discarded.
2. The frame must be selected by a random process called roulette wheel selection mechanism, which is very common in Genetic Algorithm techniques [4]. The pseudocode is presented in Algorithm 1. The variable `Elements` receives a list of frames selected in item 1. Let $costs$ be an array with the cost of each frame in variable `Elements` and min be the index of the lowest cost, the score for each frame is given by $score[i] \leftarrow costs[i] - costs[min] + 1$. After that, a random number R is generated in the range defined by the sum of all costs. Finally, we select the first frame in the array such that when all previous scores are added it gives us at least R.

Algorithm 1. selectByRouletteWheel

Data: $Elements, Scores$
$sum\_scores \leftarrow sum(Scores)$
$R \leftarrow \mathbf{random}(1, sum\_scores)$
$k \leftarrow 0$
$curSum \leftarrow 0$
while $curSum < R$ **do**
    $curSum \leftarrow curSum + score[k]$
    $k \leftarrow k + 1$
end while
return $Elements[k]$

After selecting the frame, we use the algorithm described in Section 3.3 in hope of finding an improvement. Then, we repeat the process again a limited number of times that should be set beforehand. As reasonable, the more iterations, the better the answer provided in the end. A tradeoff between solution quality and computational time is necessary.

3.3   Building Solutions

Here, we present a greedy randomized approach to find a solution for any permutation π. Starting from a sequence of permutations that has only π as element, we construct a solution one element at a time. Each step gathers a set of candidate inversions that can be used to extend the partial solution.

Each candidate inversion is ranked based on its likelyhood of producing sequences of short length inversions. In other words, we estimate the benefit of each inversion. We assess the benefit based on two aspects of a permutation: number of oriented pairs $(nop(\pi))$ [6] and entropy $(ent(\pi))$.

Definition 7. *An oriented pair (π_i, π_j) is a pair of consecutive integer $|\pi_i| = |\pi_j| \pm 1$ with opposite signs.*

The concept of oriented pair was used by Bergeron [6] to define an algorithm to the sorting by inversions problem. The inversion induced by an oriented pair (π_i, π_j) is $\rho(i, j-1)$, if $\pi_i + \pi_j = 1$, and $\rho(i+1, j)$, if $\pi_i + \pi_j = -1$. The induced inversion guarantees that π_i and π_j will be placed side by side and that they will not be a breakpoint.

The Bergeron's algorithm is based on using induced inversions that maximize the number of oriented pairs in the next permutation, which is possible as long as π has negative elements. If π has only positive elements, then we have the so-called "hurdle" as defined by Hannenhalli and Pevzner [13], which is a configuration with no oriented pair, $nop(\pi) = 0$. In these cases, we use the operations "hurdle cutting" and "hurdle merging" as proposed by Bergeron [6].

The second important aspect for assessing the benefit of an inversion is the entropy. In essence, entropy is the computation of how far each element is from its final position plus a penalty for negative elements.

Definition 8. *Entropy:* $ent(\pi) = \sum\limits_{i=1}^{n} |i - p(\pi, i)| + \sum\limits_{i=1}^{n} s(\pi, i)$

We have $ent(\pi) = 0$ if, and only if, $\pi = \iota$. Therefore, we should pick inversions that decrease the entropy. We should also pick inversions that increase the number of oriented pairs following the line of reasoning introduced by Bergeron. In this case, the ratio $enop(\pi) = ent(\pi)/nop(\pi)$ joins both concepts together. Next we introduce two functions for calculating the benefit of an inversion. One function uses the entropy ent and the other uses both entropy and number of oriented pairs as defined by $enop$.

Definition 9. *Let ρ be an inversion, we define its benefit in two different ways:*
$\delta_1(\pi, \rho) = \frac{ent(\pi) - ent(\pi\rho)}{cost(\rho)}$
$\delta_2(\pi, \rho) = \frac{enop(\pi) - enop(\pi\rho)}{cost(\rho)}$

Both benefit functions have the desirable property that if we keep applying positive benefit inversions on π, we will eventually reach the identity permutation. However, we have found some permutations such that no positive benefit inversion is possible. Thus, we decided to consider only the inversions that are induced by an oriented pair when π has negative elements, so we guarantee that at least one breakpoint will be removed. When π has no negative elements, as we explained earlier, the operations "hurdle cutting" and "hurdle merging" [6] are used.

When we have inversions induced by oriented pairs available, the inversion that will be used in order to extend the partial solution must be approved in a two-step process.

1. We calculate the benefit of each inversion ρ induced by an oriented pair, then we select a limited number of inversions having the highest benefits. The exact number of inversions moving to the second phase is defined as a parameter named `inversion_limit`. The inversions that were not selected will be discarded.
2. The inversions that were not discarded in item 1 must also be selected by the roulette wheel mechanism described in Algorithm 1. Each permutation has a selection likelihood proportional to its benefit. The variable `Elements` receives a list of inversions selected in the previous step and the variable `Scores` receives the benefit of each inversion.

Since we have two functions to calculate the benefit of an inversion (δ_1 and δ_2), we run the entire process once for each function and we choose the final solution as the one that has the minimum cost.

4   Experimental Results

We have implemented our heuristic in `C++`. We use the program `GRIMM` [21] to find the optimum solution for the sorting by inversions problem, which is then used as initial solution. Below we set the parameters that were mentioned in Section 3. Using these parameters, the time spent processing each permutation is shown in Table 1.

1. The frame size f plays a very important role and the best outcome is achieved if we change f at runtime. In our experiments, we loop f through the sequence $< 14, 12, \ldots, 4 >$. We run our method 150 times for each value of f. Overall, our method will run 900 iteration steps.
2. In Section 3.2 we mentioned the parameter `frame_limit`. This parameter is set as a fraction of the available frames. In our experiments, we allowed 75% of the best ranked frames to move to the second phase. We tried several values for `frame_limit` like 25%, 50% and 75%. Our final conclusion is that 75% is the value that leads to the best results.
3. In Section 3.3 we mentioned the parameter `inversion_limit`. This parameter is fixed no matter the number of inversions available. In our experiments, we allowed only 5 of the best ranked inversions to move to the second phase. We tried several values for `inversion_limit` like 3, 5, 10, 15 and 20. Our final conclusion is that 5 is the value that leads to the best results

Table 1. Average time (in seconds) to process each permutation of a given size

Size	10	15	20	25	30	35	40	45	50	55
Time	0.6	1.5	3.1	5.1	7.3	9.9	12.2	14.5	16.6	21.7

Size	60	65	70	75	80	85	90	95	100
Time	25.1	27.7	32.1	32.1	27.5	29.3	31.2	32.8	34.4

To find the theoretical time complexity of our implementation, we compute the complexity of each part:

1. We use GRIMM to obtain an initial solution and it runs in $O(n^4)$.
2. The number of iterations is a parameter and we will use l to refer to it.
3. In order to select a frame we need $O(m \log(m))$ time, where m is the number of permutations in the initial solution. Recall that the first step to select a frame consists of choosing a limited number of high-cost frames, which we achieve by sorting the frame list. Therefore, the stated complexity follows. Since $m = O(n)$, we have that this step runs in $O(n \log(n))$.
4. We need $O(fn^2)$ to build a new solution to replace the frame. Each step we gather $O(n)$ inversions induced by oriented pairs and we need $O(n)$ time to compute the benefit for each one of them. Besides, the new frame will have a size proportional to f. Therefore, the stated complexity follows.

Our final complexity is $O(Initial\_Solution + Iterations * (Search\_Frame + Improve\_Frame))$, which gives us $O(n^4 + l(n \log(n) + fn^2))$. After some simplifications, we conclude that our implementation runs in $O(n^4 + fln^2)$. This complexity can be reduced to $O(fln^2)$ if we use the algorithm proposed by Tannier and Sagot [20] to obtain the initial solution, which runs in $O(n\sqrt{n \log n})$.

The main quality measure used in our experiments is the difference in cost between the sequence produced by our implementation and the initial solution produced by GRIMM.

For n in the set $\{10, 15, \ldots, 100\}$, we generated 1000 random permutations and ran our implementation on them. Figure 1 shows how often our approach improves the initial solution. We were able to improve it in 94.0% of the test cases. When we consider only larger permutations such that $n \geq 50$, we notice that 99.3% of the initial solutions were improved.

The colors in Figure 1 represent frame sizes and, therefore, bars show us which size was responsible for the first improvement in the initial solution. As an example, for $n = 100$ we were able to improve 99.8% of the initial solutions. In 94.4% of the cases the first improvement occured when $f = 14$ and in 4.4% of the cases the first improvement occured when $f = 12$.

Figure 2 shows the percentage of improvement on average obtained by using our method. Let $S_{initial}$ be the initial solution and S_{final} be the final solution produced by our greedy randomized approach, we define improvement as the difference in cost between $S_{initial}$ and S_{final} divided by the cost of $S_{initial}$, in

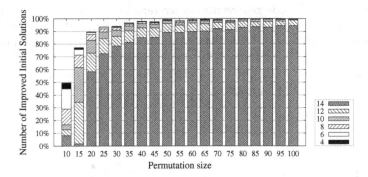

Fig. 1. This graph reports the percentage of times our heuristic improved the initial solution. Overall, the initial solution was improved in 94.0% of the cases. If we consider only larger permutations such that $n \geq 50$, we observe that 99.3% of the initial solutions were improved. In our graph, colors represent frame sizes and colored bars show us which frame size was responsible for the first improvement in the initial solution.

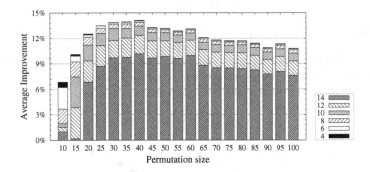

Fig. 2. This graph shows the percentage of improvement obtained by using our method. We observe that our solutions cost around 12.6% less than the initial solution. Colors represent frame sizes and colored bars show us which frame size was responsible for the improvement on average.

other words, $\%\_of\_improvement = 100 * \frac{cost(S_{initial}) - cost(S_{final})}{cost(S_{initial})}$. We observe that our solutions cost around 12.6% less than the initial solution.

We continue our analysis by comparing our results against a previous algorithm for the sorting by length-weighted inversion problem developed by Swidan et al. [19]. They created a length-sensitive model where the cost of reversing a subsequence of length l is given by the function $f(l) = l^\alpha$, for $\alpha \geq 0$. They were able to guarantee the approximation ratio $O(\lg n)$ when $\alpha = 1$, which is the same cost function we use in this paper. Our analysis shows that our method found solutions that cost less than those provided by Swidan et al. in 97.5% of the cases. When we consider only larger permutations such that $n \geq 50$, we notice that in 99.9% of the cases our solutions cost less than those provided by Swidan et al.

In Figure 3, the Y-axis represents an average of the costs for the dataset of 1000 permutations of each size in the set $\{10, 15, \ldots, 100\}$ and the X-axis represents permutation size. The label `Swidan` represents the algorithm proposed by Swidan et al. [19]. As we can see, `Swidan` is outperformed by both `GRIMM` and our approach (labelled as `GRASP`). On average, our solutions cost 23.3% less than the solutions provided by `Swidan`.

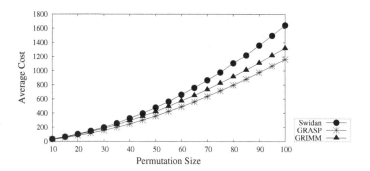

Fig. 3. This graph shows a comparative analysis among three algorithms. The Y-axis represents the average cost and the X-axis represents the permutation size. The label `Swidan` represents the algorithm proposed by Swidan et al. [19]. `GRIMM` is an optimum solution for the sorting by inversions problem when each inversion costs one unit [21]. `GRASP` represents our approach presented in this paper.

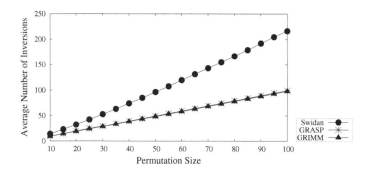

Fig. 4. This graph shows the average number of inversions of solutions provided by each of the three algorithms used in our analysis. The Y-axis represents average number of inversions and X-axis represents permutation size. The labels for the algorithms are the same used in Figure 3.

In Figure 4 we show the number of inversions used by each approach. `Swidan` returns results with a lot of inversions, which is reasonable since they are not trying to minimize how many inversions are being used. `GRIMM` represents the optimum value in the graph, since it is an algorithm developed for the sorting by inversion problem.

We observe that our approach (labelled as GRASP) increases in only 1.8% the number of inversions provided by GRIMM. This is a desirable property because we do not want our solution to be far away from the most parsimonious solutions. The same analysis for Swidan shows that it provides solutions that have on average 97.6% more inversions than GRIMM.

5 Conclusions

In this paper we presented a greedy randomized search procedure for the problem of finding a sequence of inversions that has the minimum length-weighted cost.

Our model has some parameters that can be easily set and tuned. The parameters impact the time spent processing the results and the solution quality. We have a dataset of 19,000 instances, so we decided to set the parameters in order to make the heuristic run faster. That setting was able to outperform a previous approach and significantly improve the initial solution.

We were able to improve the initial solution in most of the cases during our experiments. The larger the size of the permutation, the easier it is to improve the initial solution. For permutations of size greater or equal to 50, we improved the initial solution in more than 99.3% of the cases. In addition, our solutions cost 12.6% less than the initial solution and 23.3% less than the previous algorithm.

We analysed the number of inversions provided by our heuristic and we showed that our solutions are still close to the most parsimonious scenarios. Indeed, our solutions have on average 1.8% more inversions than the minimum number. A similar analysis showed that the previous algorithm has on average 97.6% more inversions than the minimum number.

Acknowledgments. This work was made possible by a Postdoctoral Fellowship from FAPESP to UD (number 2012/01584-3) and by project fundings from CNPq to ZD (numbers 477692/2012-5 and 483370/2013-4). The authors also thank the Center for Computational Engineering and Sciences at Unicamp for financial support through the FAPESP/CEPID Grant 2013/08293-7. FAPESP and CNPq are Brazilian research funding agencies.

References

1. Aiex, R.M., Binato, S., Resende, M.G.C.: Parallel grasp with path-relinking for job shop scheduling. Parallel Computing 29, 2003 (2002)
2. Ajana, Y., Lefebvre, J.F., Tillier, E.R.M., El-Mabrouk, N.: Exploring the set of all minimal sequences of reversals - an application to test the replication-directed reversal hypothesis. In: Guigó, R., Gusfield, D. (eds.) WABI 2002. LNCS, vol. 2452, pp. 300–315. Springer, Heidelberg (2002)
3. Arruda, T.S., Dias, U., Dias, Z.: Heuristics for the sorting by length-weighted inversion problem. In: Proceedings of the International Conference on Bioinformatics, Computational Biology and Biomedical Informatics, pp. 498–507. ACM (2013)

4. Bäck, T.: Evolutionary algorithms in theory and practice: evolution strategies, evolutionary programming, genetic algorithms. Oxford University Press, Oxford (1996)
5. Bader, D.A., Moret, B.M.E., Yan, M.: A linear-time algorithm for computing inversion distance between signed permutations with an experimental study. Journal of Computational Biology 8(5), 483–491 (2001)
6. Bergeron, A.: A very elementary presentation of the Hannenhalli-Pevzner theory. Discrete Applied Mathematics 146, 134–145 (2005)
7. Blanchette, M., Kunisawa, T., Sankoff, D.: Parametric genome rearrangement. Gene 172(1), C11–C17 (1996)
8. Cano, R.G., Kunigami, G., Souza, C.C., Rezende, P.J.: A hybrid grasp heuristic to construct effective drawings of proportional symbol maps. Computers and Operations Research 40(5), 1435–1447 (2013)
9. Darling, A.E., Miklós, I., Ragan, M.A.: Dynamics of genome rearrangement in bacterial populations. PLoS Genetics 4(7), 1000128 (2008)
10. Feo, T., Resende, M.G.C.: Greedy randomized adaptive search procedures. Journal of Global Optimization 6(2), 109–133 (1995)
11. Feo, T.A., Pardalos, M.: A greedy randomized adaptive search procedure for the 2-partition problem. Operations Research (1994)
12. Festa, P., Resende, M.: Grasp: basic components and enhancements. Telecommunication Systems 46(3), 253–271 (2011)
13. Hannenhalli, S., Pevzner, P.A.: Transforming cabbage into turnip: polynomial algorithm for sorting signed permutations by reversals. Journal of the ACM 46(1), 1–27 (1999)
14. Laguna, M., Martí, R.: Grasp and path relinking for 2-layer straight line crossing minimization. INFORMS Journal on Computing 11, 44–52 (1999)
15. Lefebvre, J.F., El-Mabrouk, N., Tillier, E., Sankoff, D.: Detection and validation of single gene inversions. Bioinformatics 19, i190–i196 (2003)
16. Ribeiro, C.C., Uchoa, E., Werneck, R.F.: A hybrid grasp with perturbations for the steiner problem in graphs. INFORMS Journal on Computing 14, 200–202 (2001)
17. Sankoff, D.: Short inversions and conserved gene cluster. Bioinformatics 18(10), 1305–1308 (2002)
18. Sankoff, D., Lefebvre, J.F., Tillier, E., Maler, A., El-Mabrouk, N.: The distribution of inversion lengths in bacteria. In: Lagergren, J. (ed.) RECOMB-WS 2004. LNCS (LNBI), vol. 3388, pp. 97–108. Springer, Heidelberg (2005)
19. Swidan, F., Bender, M., Ge, D., He, S., Hu, H., Pinter, R.: Sorting by length-weighted reversals: Dealing with signs and circularity. In: Sahinalp, S.C., Muthukrishnan, S.M., Dogrusoz, U. (eds.) CPM 2004. LNCS, vol. 3109, pp. 32–46. Springer, Heidelberg (2004)
20. Tannier, E., Sagot, M.-F.: Sorting by reversals in subquadratic time. In: Sahinalp, S.C., Muthukrishnan, S.M., Dogrusoz, U. (eds.) CPM 2004. LNCS, vol. 3109, pp. 1–13. Springer, Heidelberg (2004)
21. Tesler, G.: Grimm: genome rearrangements web server. Bioinformatics 18(3), 492–493 (2002), http://bioinformatics.oxfordjournals.org/content/18/3/492.abstract

On Low Treewidth Graphs and Supertrees

Alexander Grigoriev[1], Steven Kelk[2], and Nela Lekić[2]

[1] Department of Quantitative Economics
[2] Department of Knowledge Engineering (DKE)
Maastricht University, P.O. Box 616, 6200 MD Maastricht, The Netherlands
{a.grigoriev,steven.kelk,nela.lekic}@maastrichtuniversity.nl

Abstract. Compatibility of unrooted phylogenetic trees is a well studied problem in phylogenetics. It asks to determine whether for a set of k input trees $T_1, ..., T_k$ there exists a larger tree (called a supertree) that contains the topologies of all k input trees. When any such supertree exists we call the instance compatible and otherwise incompatible. It is known that the problem is NP-hard and FPT, although a constructive FPT algorithm is not known. It has been shown that whenever the treewidth of an auxiliary structure known as the display graph is strictly larger than the number of input trees, the instance is incompatible. Here we show that whenever the treewidth of the display graph is at most 2, the instance is compatible. Furthermore, we give a polynomial-time algorithm to construct a supertree in this case. Finally, we demonstrate both compatible and incompatible instances that have display graphs with treewidth 3, highlighting that the treewidth of the display graph is (on its own) not sufficient to determine compatibility.

Keywords: Phylogenetic tree, unrooted compatibility, supertree, display graph, treewidth.

1 Introduction

One of the central challenges within computational evolutionary biology is to infer the evolutionary history of a set of contemporary species (or more generally, *taxa*) X using only the genotype of the contemporary species. This evolutionary history is usually modeled as a *phylogenetic tree*, essentially a tree in which the leaves are bijectively labeled by the elements of X and the internal nodes of the tree represent (hypothetical) ancestors [11].

There is already an extensive literature available on the extent to which different optimization criteria on the space of phylogenetic trees (e.g. likelihood, parsimony) are able to identify the "true" evolutionary history. In any case it is well-known that most of these problems are NP-hard, and this intractability is a serious obstacle when constructing phylogenetic trees for large numbers of taxa. This has been one of the motivations behind *supertree* methods [2]. Here the goal is to first construct phylogenetic trees for small (overlapping) subsets of X and then to puzzle the partial trees together into a single tree on X that

A.-H. Dediu, C. Martín-Vide, and B. Truthe (Eds.): AlCoB 2014, LNBI 8542, pp. 71–82, 2014.

contains all the topologies of the partial trees, in which case we say the partial trees are *compatible*, or to conclude that no such tree exists.

The computational complexity landscape of the compatibility problem is uneven. In the case that all the partial trees are rooted (i.e. in which the flow of evolution is assumed to be away from a designated root, towards the taxa) the problem is polynomial-time solvable, using the algorithm of Aho [1]. However, in the case of unrooted trees the problem is NP-hard, even when all the partial trees have at most 4 taxa [12]. Nevertheless, due to the fact that many tree-building algorithms actually construct *un*rooted trees, and because of the risk of distorting the underlying phylogenetic signal through a poor choice of root location, it remains attractive to try and solve this NP-hard variant of the problem directly.

In this article we approach the unrooted compatibility problem from a graph-theoretical angle. There is a recent trend in this direction, which to a large extent can be traced back to a seminal paper of Bryant and Lagergren [4]. They observed that there is a relationship between the compatibility question and the *treewidth* of an auxiliary graph known as the *display graph*. The display graph is obtained by identifying the taxa of the input trees, and treewidth is an intensely well-studied parameter in the algorithmic graph theory literature (see e.g. [3]). Low (or bounded) treewidth often facilitates algorithmic tractability, and given that it is a measure of "distance from being a tree", it is tempting to try and exploit this tractability in questions pertaining to phylogenetic compatiblity and incongruence. Bryant and Largergren observed that for k unrooted trees to be compatible, it is necessary (but not sufficient) that the display graph has treewidth at most k. The upper bound on the treewidth that this condition generates, subsequently makes it possible to formulate and answer the compatibility question in a computationally efficient way. However, this efficiency is purely theoretical in nature, obtained via the indirect route of monadic second order logic [5], and it remains a challenge to succinctly characterize phylogenetic compatibility. Since Bryant and Largergren various other authors have picked up this thread (e.g. [9]), with particular attention for triangulation-based approaches (see e.g. [13,10,14]) although the question remains: what *exactly* is the role of treewidth in compatibility?

Here we take a step forward in understanding the link between treewidth and compatibility. We prove that if the display graph of a set of unrooted binary trees has treewidth at most 2, then the input trees are compatible, and this holds for any number of input trees. In other words, it is not necessary to look deeper into the structure of the display graph, compatibility is immediately guaranteed. The proof of this, based on graph separators and graph minors, is surprisingly involved. Moreover, we describe a simple polynomial-time algorithm to construct a supertree for the input trees, when this condition holds. We also show that in some sense this result is "best possible": we show how to construct both compatible and incompatible instances that have display graphs of treewidth 3, for any number of trees. This confirms that the treewidth of the display graph cannot, on its own, fully capture phylogenetic compatibility, and that auxiliary information is indeed necessary if we are to obtain a complete characterization.

2   Preliminaries

Let X be a finite set. An *unrooted phylogenetic X-tree* is a tree whose leaves are bijectively labeled by the elements of set X. It is called *binary* when all its inner nodes (nonleaf nodes) are of degree 3. An unrooted binary phylogenetic tree on four leaves is called a *quartet*. In the remainder of the article we focus almost exclusively on unrooted binary trees, often writing simply trees or X-trees for short.

We call elements of X *taxa* or *leaves*. For some X-tree T and some subset $X' \subseteq X$ we denote by $T(X')$ the subtree of T induced by X' and by $T|X'$ the tree obtained from $T(X')$ by suppressing vertices of degree 2. Furthermore, we say a tree S *displays* a tree T if T can be obtained from a subgraph of S by suppressing vertices of degree two.

Given a set X a *split* is defined as a bipartition of X. If we label the components of the partition by A and B, then we can denote the split by $A|B$. Note that each edge of an X-tree naturally induces a split. If $A|B$ is a split induced by an edge of a tree T, then we say that T *contains* split $A|B$. We use $ab|cd$ to denote the quartet in which taxa a and b are on one side of the internal edge and c and d are on the other. We write $ab|cd \in T$ if T displays $ab|cd$.

Given a set $\mathcal{T}$ of k trees $T_1, ..., T_k$ we wish to know if there exists a single tree S that displays T_i for all $i \in \{1, ...k\}$. A tree that displays all the input trees, if such a tree exists, is called a *supertree*. When a supertree does exist we call the instance *compatible*, otherwise *incompatible*. A supertree is not necessarily unique. To see when such a tree is unique and many more details on this topic we refer the reader to [7] or [11].

The *display graph* $D(\mathcal{T})$ of a set of trees $\mathcal{T}$ is the graph obtained from the disjoint union of trees in $\mathcal{T}$ by identifying vertices with the same taxon labels. Note that $D(\mathcal{T})$ can be disconnected if and only if the trees in $\mathcal{T}$ can be bipartitioned into two sets $\mathcal{T}_1, \mathcal{T}_2$ such that $X(\mathcal{T}_1) \cap X(\mathcal{T}_2) = \emptyset$, where $X(T)$ refers to the set of taxa of T. In such a case $\mathcal{T}$ permits a supertree if and only if both $\mathcal{T}_1$ and $\mathcal{T}_2$ do. Hence for the remainder of the article we focus on the case when $D(\mathcal{T})$ is connected.

Before we can start discussing our result we need a few graph theoretic definitions. Let $G = (V, E)$ be an undirected graph. For any two subsets of vertices $A, B \subseteq V$ and any $Z \subseteq V$ we say Z *separates* sets A and B in G if every path in G that starts at some vertex $u \in A$ and ends at some vertex $v \in B$ contains a vertex from Z. Such a set Z is called an (A, B)-*separator*, or simply a *separator*. A graph M is a *minor* of a graph G if M can be obtained from a subgraph of G by contracting edges.

The treewidth of a graph G, denoted $tw(G)$, has a somewhat technical definition. We give it here for completeness although for the main result it is sufficient to note that trees have treewidth 1, and that graphs with treewidth at most 2 are exactly those graphs that do not have a K_4-minor (where K_4 is the complete graph on 4 vertices). We will also use the well-known fact that if M is a minor of G, $tw(M) \leq tw(G)$.

Let G be a graph, T a tree and $(B_t)_{t \in T}$ a family of subsets of $V(G)$, also called *bags*, indexed by vertices of T. We say T is a *tree-decomposition* of G if the following conditions are satisfied:

(T_1) $V(G) = \cup_{t \in T} B_t$;
(T_2) for every edge $e \in G$ these exists a bag B_t in T such that both endpoints of e lie in B_t;
(T_3) $B_u \cap B_v \subseteq B_w$ whenever vertices u, v, w of T are such that w is on a path from u to v in T.

The width of a tree-decomposition is the size of its largest bag minus one. The *treewidth* of a graph G, also denoted $tw(G)$, is the minimum width over all possible tree-decompositions of G.

For remaining graph theory terminology we refer to standard texts such as [6].

3   Main Results

Lemma 1. [8, Corollary 1]. *Let T_1 and T_2 be two unrooted phylogenetic trees on the same set of taxa X. Then T_1 and T_2 are compatible if and only if there do not exist four taxa $a, b, c, d \subseteq X$ such that $ab|cd \in T_1$ and $ac|bd \in T_2$.*

Lemma 2. *Let D be the display graph of the two quartets $ab|cd$ and $ac|bd$. Then D has K_4 as a minor. Hence, $tw(D) \geq 3$.*

Proof. Let $Q_1 = ab|cd$ and $Q_2 = ac|bd$. Both Q_1 and Q_2 have exactly two inner nodes, denote them u, v and w, z respectively. Then it is immediate to see that vertices u, v, w, z of D form a K_4 minor (obtained by suppressing leaves a, b, c, d which all have degree 2 in D). □

Theorem 3. *Let T_1 and T_2 be two unrooted phylogenetic trees. Let D be the display graph of T_1 and T_2. Then T_1 and T_2 are compatible if and only if $tw(D) \leq 2$.*

Proof. Let T_1 and T_2 be two trees on taxa sets X and X' respectively. Let $X^* = X \cap X'$. Then T_1 and T_2 are compatible if and only if $T_1|X^*$ and $T_2|X^*$ are compatible [11]. Thus we only have to consider two trees T_1 and T_2 on the same set of taxa X. Let $D(T_1, T_2)$ be their display graph. Suppose for the sake of contradiction that $tw(D(T_1, T_2)) \leq 2$ while T_1 and T_2 are incompatible. From Lemma 1, T_1 and T_2 contain incompatible quartets Q_1 and Q_2 (w.l.o.g. let T_i display Q_i) and since Q_i is displayed in T_i, it is also displayed in $D(T_1, T_2)$, so $D(Q_1, Q_2)$ is a minor of $D(T_1, T_2)$. Since $D(Q_1, Q_2)$ is a minor of G, and using Lemma 2, $tw(G) \geq tw(D(Q_1, Q_2)) \geq 3$, contradicting the fact that $tw(D(T_1, T_2)) \leq 2$. This compleA wtes our proof in one direction; for the other see [4]. □

In the following main theorem we emphasize that the trees in $\mathcal{T}$ do not need to be on the same set of taxa, but that for this proof the input trees do need to be binary.

Theorem 4. *Let $\mathcal{T}$ be a set of k binary unrooted phylogenetic trees $T_1, ..., T_k$ and let D be their display graph. If $tw(D) \leq 2$, then $T_1, ..., T_k$ are compatible, in which case a supertree can be constructed in polynomial time.*

Proof. We give a constructive proof in which we will build a supertree S for $\mathcal{T}$. The idea is to find an appropriate separator of D and to reduce the problem into smaller instances of the same problem i.e. an induction proof. The induction will be on the cardinality of $X = \cup_{T_i \in \mathcal{T}} X(T_i)$. For the base case observe that an instance with $|X| \leq 3$ is trivially compatible.

Before we start the construction we apply a number of operations on D that are safe to do, in the sense that they preserve (in)compatibility of the instance and do not cause the treewidth of D to rise. We remove any taxon that has degree 1 in D and contract any inner vertex that has degree 2 in D. This clearly affects neither the compatibility nor the treewidth. Furthermore, for every tree T_i with $i \in \{1, ...k\}$ that has fewer than 4 leaves, we exclude it from the display graph. Such a tree carries no topological information and thus does not change the compatibility, while removing something from a graph cannot increase its treewidth. The *cleaning up* procedure means that we apply all these operations on D repeatedly until we cannot apply them anymore. In other words, we can assume D to have treewidth exactly 2, that all inner vertices of D have degree 3, that all taxa have degree at least 2 and that no tree has fewer than 4 taxa.

Consider a planar embedding of the display graph $D(\mathcal{T})$. This exists and can be found in polynomial time because $D(\mathcal{T})$ has treewidth at most 2. The *boundary* of a face F of $D(\mathcal{T})$, denoted $B(F)$, is the set of edges and vertices that are incident to the interior of the face. We say that two distinct faces F_1, F_2 are *minimally adjacent* if the following three conditions hold: (1) F_1 and F_2 are adjacent; (2) $B(F_1) \cap B(F_2)$ is isomorphic to a path containing at least one edge; (3) the internal vertices of the path $B(F_1) \cap B(F_2)$ all have degree 2 in $D(\mathcal{T})$, and the two endpoints of the path each have degree 3 or higher in $D(\mathcal{T})$. Due to space considerations we omit the proof, but it can be shown that if the treewidth of D is 2 we can always find two such faces, neither equal to the outer face, in polynomial time.

Let F_1 and F_2 be two minimally adjacent faces of D, neither equal to the outer face. Denote by $p(u, v)$ the path $B(F_1) \cap B(F_2)$ they share. (After locating F_1 and F_2 this path can easily be found in polynomial time). By definition u and v must have degree at least 3 in D. Also, by minimal adjacency of F_1 and F_2 and due to cleaning up, none of the interior nodes of $p(u, v)$ can be internal tree nodes. Moreover, since we removed all trees on fewer than four taxa, at most one leaf can appear as an interior node of the path. Such a leaf can only exist if both u and v are inner nodes of some trees. Now, u and v can either be both leaves, both inner nodes or one of them a leaf another an inner node. These are the three cases we have to consider.

Case(i) is when both u and v are leaves. We claim this cannot happen. In this case, path $p(u,v)$ must be an edge. But if it is an edge it is connecting two leaves and will have already been removed during cleaning up.

Case(ii) is when u is a leaf and v is an inner node. Again we have that path $p(u,v)$ must be an edge (u,v) which both faces share. Let x, respectively y, be any vertex other than u or v on the boundary of F_1, respectively F_2. See Figure 1(a). We claim that any path between x and y must contain either u or v. In particular, suppose there exists a path $p(x,y)$ such that $u,v \notin p(x,y)$. Let x' and y' be vertices on $p(x,y)$ such that the subpath $p(x',y')$ is the shortest subpath of $p(x,y)$ with the property that both of its endpoints are on the boundaries of F_1 and F_2, respectively. See Figure 1(a). Then D contains a K_4 minor formed by vertices u,v,x' and y'. This is a contradiction on D having treewidth 2. So we have that any path between x and y passes through either u or v. Thus $\{u,v\}$ is a separator of D.

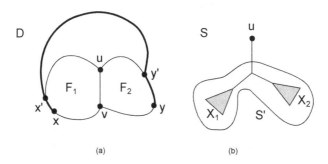

(a)   (b)

Fig. 1. (a) Two minimally adjacent faces F_1 and F_2 in D. The vertices u,v,x',y' induce a K_4 minor. (b) A supertree as constructed in case (ii).

Removing u and v from the vertex set of D disconnects it and divides the set of taxa into two sets X_1 and X_2, such that $X = X_1 \cup X_2 \cup \{u\}$. We claim that supertree S as shown in Figure 1(b), where S' is a supertree of $T_1, ..., T_k$ restricted to taxa set $X \setminus \{u\}$, displays all k input trees $T_1, ..., T_k$. To prove this we have to show two things. One, that the supertree S' exists (and that it has an edge corresponding to split $X_1|X_2$) and two, that all quartets in $T_1, ..., T_k$ are also in S. (The latter is sufficient because a set of unrooted trees is compatible if and only if the set of quartets displayed by the trees is compatible).

To prove the first claim let $X' := X \setminus \{u\}$ and notice that by induction the instance $T_1, ..., T_k|X'$ is compatible and thus has a supertree. We now claim that there exists some supertree of $T_1, ..., T_k|X'$, call it S', which contains split $X_1|X_2$. First of all notice that (a restriction of) $X_1|X_2$ must be a split in every input tree restricted to X'. To see this we show that there does not exist a quartet $ab|cd$ with $a,c \in X_1$ and $b,d \in X_2$ in any of the input trees (prior to removal of u and v). Suppose such a quartet did exist in some tree. Then there would exist edge-disjoint paths $p(a,b)$ and $p(c,d)$ in D, where the interior nodes of these paths are internal tree nodes. Since removing u and v from D disconnects it (such that X_1 and X_2 are subsequently in separate components), it must be that those paths had to use either u or v. Since u is a taxon it cannot be used for this purpose. So both paths had to use inner vertex v. However, this contradicts the edge-disjointness of the two paths. Hence quartet $ab|cd$ cannot be displayed by any tree.

We conclude from this that in each $T_i|X'$ there exists an edge e that induces a split $A|B$, such that $A \subseteq X_1$ and $B \subseteq X_2$. Furthermore both X_1 and X_2 must contain at least one taxon each. (This follows because edge (u, v) belongs to some input tree T, and walking from u to v along the boundary of F_1 whilst avoiding edge (u, v) necessitates entering and leaving T via its taxa, which in turn means that some taxon not equal to u must exist on the part of the boundary of F_1 not shared by F_2. The same argument holds for F_2.) As such, in each $T_i|X'$ it is possible to contract (the subtree induced by) X_1 and/or X_2 into a single "meta-taxon".

Let T^* (respectively, T^{**}) be the set of trees obtained by taking the trees on X' and contracting all the X_2 (respectively, X_1) taxa into a single meta-taxon W_2 (respectively, W_1). Note that contracting in this way cannot increase the treewidth of D and that $1 \leq |X_i| < |X|$ for $i \in \{1, 2\}$. Hence, by induction supertrees of T^* and T^{**} exist. Finally, construct supertree S' with split $X_1|X_2$ from two supertrees for T^* and T^{**} by adding an edge between W_1 and W_2 and afterwards suppressing W_1 and W_2. (The function of W_1 and W_2 was precisely to ensure that we would know how to glue the two separately constructed supertrees together).

To see the second claim note that since S' is a supertree of $T_1, ..., T_k$ restricted $X \setminus \{u\}$ we only have to show that quartets of $T_1, ..., T_k$ that contain taxon u are displayed by S. So w.l.o.g. let $a \in X_1, b, c \in X_2$. Then if quartet $au|bc$ is displayed by some input tree T it is also clearly displayed by the supertree S. We claim quartets $ub|ac$ or $uc|ab$ cannot exist in any of the input trees. These two quartets are the same up to relabeling so let's consider quartet $ub|ac$ induced by some tree T sitting inside D. Then $p(u, b)$ and $p(a, c)$ are edge-disjoint and contain no taxa. As argued before $p(a, c)$ must pass through v. But since (u, v) is an edge it follows that it must belong to the same tree T, and therefore v also lies on the path $p(u, b)$. But then it is not possible that T displays $ub|ac$, contradiction.

Case(iii) is when both u and v are inner nodes. We could have that $p(u, v)$ is an edge, in which case u and v are inner nodes of the same tree, or we could have that $p(u, v)$ contains a single taxon t. Note that in the latter case u and v are inner nodes of two different trees and taxon t must have degree 2 in D due to the minimal adjacency of F_1 and F_2. The argument for $\{u, v\}$ being a separator of D goes through in this case as well regardless of $p(u, v)$ being an edge or a path containing a single taxon t. We again denote by X_1 and X_2 the two sets of taxa that emerge from splitting D by removing u and v (and t if it exists on (u, v)).

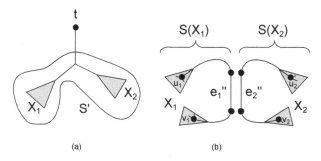

Fig. 2. (a) A supertree constructed in case (iii) when there exists a taxon t on the common boundary of the two faces. (b) Construction of a supertree in case (iii) when the common boundary of the two faces is a single edge.

Subcase 1. Consider first the subcase when some taxon $t \in p(u, v)$. As before we have to show that there exists some S', a supertree of $T_1, ..., T_k$ restricted to $X' := X \setminus \{t\}$ with split $X_1|X_2$, and that the supertree S as shown in Figure 2(a) displays all quartets induced by $T_1, ..., T_k$. The proof for this case is almost identical to that of case (ii) modulo some minor differences. Due to space limitations we omit the proof.

Subcase 2. The last thing to consider is the subcase when (u, v) is an edge while both u and v are inner nodes (necessarily of the same tree T). Let X_1 and X_2 be two disjoint sets of taxa that result from splitting D after removing u and v. We claim that $|X_1| \geq 2$ and $|X_2| \geq 2$. This follows directly from $u, v \in T$: any cycle that links them together must leave the tree T via some taxon a and re-enter it via a (necessarily different) taxon b. Since u and v belong to both faces F_1 and F_2 it follows that the boundaries of these two faces must each contain (at least) two taxa. The two taxa on the boundary of (w.l.o.g) F_1 are still in the same connected component after deletion of $\{u, v\}$, but are not in the same connected component as the taxa from the boundary of F_2, so $|X_1| \geq 2$ and $|X_2| \geq 2$.

Now we claim that the tree shown in Figure 2(b) is a supertree of $T_1, ..., T_k$. Let's first explain what that image means. Note that apart from the tree T in which the internal edge $e = (u, v)$ can be found, all other trees have taxa sets either completely contained inside X_1 or completely contained inside X_2. This is the case because otherwise there would be a path from some element in X_1 to some element in X_2, contradicting the fact that $\{u, v\}$ is a separator. The idea is to cut T into two parts, one on X_1, one on X_2, recursively build supertrees of $T_1, ..., T_k|X_1$ and $T_1, ..., T_k|X_2$ and join them as indicated in the figure.

Now, consider the display graph D. Suppose we delete the edge $e = (u, v) \in T$, and replace it with two edges $e_1 = (u_1, v_1)$ and $e_2 = (u_2, v_2)$ (where u_i and v_i are u and v duplicated). Because $\{u, v\}$ is a separator, this creates two disjoint display graphs, one on X_1 and one on X_2. These are minors of the original display graph so have treewidth at most 2, and they are smaller instances of the problem. So by induction supertrees of these smaller instances exist. Let $S(X_1)$ be a supertree on X_1 and $S(X_2)$ be a supertree on X_2. All trees except T will be displayed by the disjoint union of $S(X_1)$ and $S(X_2)$, because only T has taxa from both X_1 and X_2. What is left to explain is how to glue $S(X_1)$ and $S(X_2)$ into a supertree S such that S displays T as well.

Note that $S(X_i)$ contains an image of edge e_i. The image need not be an edge in $S(X_i)$, it could also be a path, whose endpoint we denote by u_i and v_i in Figure 2(b). Take any edge on path $p(u_i, v_i)$, call it e_i', and subdivide it twice to create two adjacent degree-2 vertices; let e_i'' be the edge between them. Now, by identifying e_1'' and e_2'' we ensure that we get a supertree that displays (all the quartets in) T, as well as all the other trees.

This completes the case analysis. Polynomial time is achieved because all relevant operations (recognizing whether a graph has treewidth at most 2, finding a planar embedding, finding two minimally adjacent faces, finding the separator $\{u, v\}$, and all the various tree manipulation operations) can easily be performed in (low-order) polynomial time.   □

We now give a summary of the algorithm implicitly described in the above proof. Given k input phylogenetic trees, construct their display graph D and clean it up. We start by verifying in polynomial time that the treewidth of D is at most 2. Let F_1 and F_2 be any two minimally adjacent faces of D (if two such faces do not exist, then the instance is trivially compatible). By definition of minimal adjacency we know that the intersection of borders of the two faces must be isomorphic to a path $p(u, v)$ containing at least one edge. Denote by X_1 and X_2 are two sets of taxa obtained from separating D by removing $\{u, v\}$.

We saw that we can w.l.o.g. assume v to be an inner node. If u is a leaf, then we construct a supertree S as shown in figure 1(b) and recursively solve two smaller instances with input trees $T_i|X_1$ and $T_i|X_2$ for $i \in \{1, ..., k\}$. (Note that in the actual algorithm we also add an extra "meta-taxon" into each of the two smaller instances which tells us where to graft the two solutions back together). Otherwise, u is an inner node. In this case, path $p(u, v)$ can either contain a taxon t or be an edge. When it contains a taxon t a supertree S is given in figure 2(a) and we recursively solve two smaller instances with input trees $T_i|X_1$ and $T_i|X_2$ for $i \in \{1, ..., k\}$ (note that in this case the two taxa sets X_1 and X_2 are obtained after removing $\{u, v, t\}$ from D). When u is an inner node and $p(u, v)$ is an edge, then we construct a supertree as in figure 2(b) and recursively solve two smaller instances $T_i|X_1$ and $T_i|X_2$ for $i \in \{1, ..., k\}$. We continue until the instance is trivially compatible.

4   Beyond Treewidth 2

Two incompatible quartets induce a display graph with treewidth 3, so treewidth 3 cannot guarantee compatibility. However, it is natural to ask whether treewidth 3 guarantees compatibility if the number of input trees becomes sufficiently large. Unfortunately, the answer to that question is no. Namely, for any number of trees there exists a compatible instance with $tw(D) = 3$ and an incompatible instance with $tw(D) = 3$, as we now demonstrate. Figure 3 shows the display graph of k trees with leaves denoted as black dots and vertices of K_4 minors with red dots (note that some leaves, for example z, can also be a vertex of a K_4 minor). Note that vertices a, b, c, z form a K_4 minor in $D(T_1, T_2, T_3)$, vertices b, c, d, q form a K_4 minor in $D(T_2, T_3, T_4)$, vertices d, e, f, s form a K_4 minor in $D(T_4, T_5, T_6)$ and so on. Now note that all those K_4 minors are attached together by a sequence of series and parallel compositions inside $D(T_1, ..., T_k)$. So we can conclude that the treewidth of the display graph of k trees as shown in figure 3 is 3. (Equivalently, we can describe a tree decomposition in which all bags have size at most 4). Compatibility of this instance can be verified without too much difficulty (details omitted).

Now we need to show the same for an incompatible instance. In Figure 4 trees T_1, T_2, T_3 are incompatible, thus the whole instance is incompatible. Furthermore, trees $T_4, ..., T_k$ are chosen to be the same as in Figure 3, so are compatible and $tw(D(T_4, ..., T_k)) = 3$. We have verified that $tw(D(T_1, T_2, T_3)) = 3$. Since $D(T_1, T_2, T_3)$ and $D(T_4, ..., T_k)$ are attached in series to form $D(T_1, ..., T_k)$ we conclude $tw(D(T_1, ..., T_k)) = 3$.

It is not difficult to generalize these constructions for any treewidth higher than 3, and any number of trees.

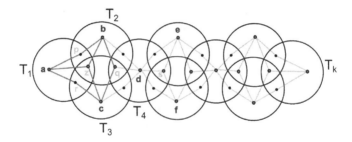

Fig. 3. Display graph of an instance with k input trees. Red (larger) vertices are inner nodes while black (smaller) vertices are leaves. The treewidth of D is 3 and the instance is compatible.

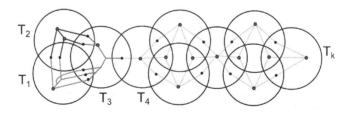

Fig. 4. Display graph of an instance with k input trees. The treewidth of D is 3 and the instance is incompatible.

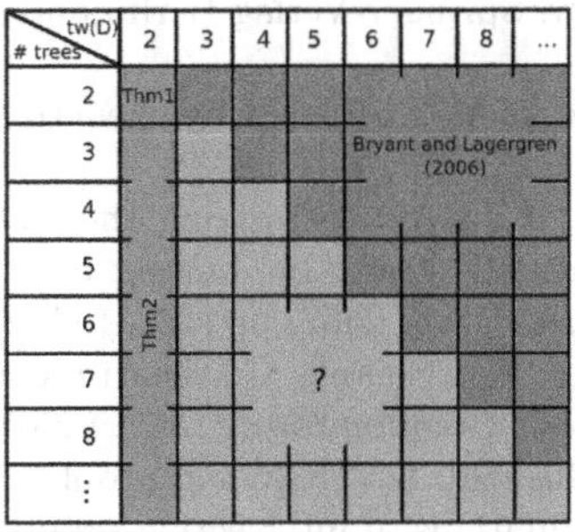

Fig. 5. The green (respectively, red) area shows which combinations of (number of input trees, treewidth of display graph) are always compatible (respectively, incompatible). The grey area indicates that both compatible and incompatible instances exist for this combination of parameters.

5 Conclusion

Figure 5 summarizes our results. The red area is due to result of Bryant and Lagergren which proves that that any instance on k trees whose display graph has treewidth strictly greater than k must be incompatible. The green area is due to our result. What we are left with is the grey area in which (as demonstrated by the constructions in the previous section) we cannot conclude anything about compatibility of the instances based only on treewidth of the display graph and the number of trees, at least not with the current results. An obvious open question is whether existing characterizations (such as legal triangulations [13]) can be specialized to yield simple and efficient combinatorial algorithms in the case of treewidth 3 or higher.

Acknowledgements. Nela Lekić was supported by an NWO Vrije Competitie grant.

References

1. Aho, A., Sagiv, Y., Szymanski, T., Ullman, J.: Inferring a tree from lowest common ancestors with an application to the optimization of relational expressions. SIAM Journal on Computing 10(3), 405–421 (1981)
2. Bininda-Emonds, O.: Phylogenetic Supertrees: Combining Information to Reveal the Tree of Life. Kluwer Academic Publishers (2004)
3. Bodlaender, H.L., Koster, A.M.C.A.: Treewidth computations I. Upper bounds. Information and Computation 208(3), 259–275 (2010)
4. Bryant, D., Lagergren, J.: Compatibility of unrooted phylogenetic trees is FPT. Theoretical Computer Science 351(3), 296–302 (2006)
5. Courcelle, B.: he monadic second-order logic of graphs. I. Recognizable sets of finite graphs. Information and Computation 85(1), 12–75 (1990)

6. Diestel, R.: Graph Theory. Springer-Verlag Berlin and Heidelberg GmbH & Company KG (2000)
7. Dress, A., Huber, K.T., Koolen, J.: Basic Phylogenetic Combinatorics. Cambridge University Press (2012)
8. Ganapathy, G., Warnow, T.J.: Approximating the complement of the maximum compatible subset of leaves of k trees. In: Jansen, K., Leonardi, S., Vazirani, V.V. (eds.) APPROX 2002. LNCS, vol. 2462, pp. 122–134. Springer, Heidelberg (2002)
9. Grünewald, S., Humphries, P.J., Semple, C.: Quartet compatibility and the quartet graph. Electronic Journal of Combinatorics 15(1) (2008)
10. Gysel, R., Stevens, K., Gusfield, D.: Reducing problems in unrooted tree compatibility to restricted triangulations of intersection graphs. In: Raphael, B., Tang, J. (eds.) WABI 2012. LNCS, vol. 7534, pp. 93–105. Springer, Heidelberg (2012)
11. Semple, C., Steel, M.: Phylogenetics. Oxford University Press (2003)
12. Steel, M.: The complexity of reconstructing trees from qualitative characters and subtrees. Journal of Classification 9(1), 91–116 (1992), doi:10.1007/BF02618470
13. Vakati, S., Fernández-Baca, D.: Graph triangulations and the compatibility of unrooted phylogenetic trees. Applied Mathematics Letters 24(5), 719–723 (2011)
14. Vakati, S., Fernández-Baca, D.: Characterizing compatibility and agreement of unrooted trees via cuts in graphs. CoRR abs/1307.7828 (2013)

On Optimal Read Trimming in Next Generation Sequencing and Its Complexity

Ivo Hedtke[1,2], Ioana Lemnian[2], Matthias Müller-Hannemann[2],
and Ivo Grosse[2,3]

[1] Department of Mathematics and Computer Science, Osnabrück University
Albrechtstrasse 28, 49076 Osnabrück, Germany
ivo.hedtke@uni-osnabrueck.de
[2] Institute of Computer Science, Martin-Luther-University Halle-Wittenberg
Von-Seckendorff-Platz 1, 06120 Halle, Germany
{lemnian,muellerh,grosse}@informatik.uni-halle.de
[3] German Centre for Integrative Biodiversity Research (iDiv) Halle-Jena-Leipzig
Deutscher Platz 5e, 04103 Leipzig, Germany

Abstract. Read trimming is a fundamental first step of the analysis of next generation sequencing (NGS) data. Traditionally, read trimming is performed heuristically, and algorithmic work in this area has been neglected. Here, we address this topic and formulate three constrained optimization problems for block-based trimming, i.e., truncating the same low-quality positions at both ends for all reads and removing low-quality truncated reads. We find that the three problems are $\mathcal{NP}$-hard. However, the non-random distribution of quality scores in NGS data sets makes it tempting to speculate that quality constraints for read positions are typically satisfied by fulfilling quality constraints for reads. Based on this speculation, we propose three relaxed problems and develop efficient polynomial-time algorithms for them. We find that (i) the omitted constraints are indeed almost always satisfied and (ii) the algorithms for the relaxed problems typically yield a higher number of untrimmed bases than traditional heuristics.

Keywords: Next Generation Sequencing, Trimming, $\mathcal{NP}$-completeness, Polynomial-Time Algorithms.

1 Introduction

Next generation sequencing (NGS) technologies are of increasing importance in the life sciences and have enabled scientists to perform a plethora of studies that would otherwise be impossible [9]. Despite continuous improvements of sequencing technologies over the last decade, the resulting reads contain and will continue to contain sequencing errors. Hence, besides preprocessing steps like adapter and duplicate removal, eliminating low-quality bases plays a crucial role for downstream NGS data analyses, and different approaches have been proposed for truncating low-quality positions and removing low-quality reads [3].

Window-based approaches treat all reads independently of each other, simply scan a window of a given size over each read, and retain those windows whose

A.-H. Dediu, C. Martín-Vide, and B. Truthe (Eds.): AlCoB 2014, LNBI 8542, pp. 83–94, 2014.

quality lies above a given threshold [13]. Block-based approaches, in contrast, do not treat all reads independently of each other, but truncate a certain number of positions at both ends of each read and subsequently remove those truncated reads whose quality falls below a given threshold [1].

Both of these approaches have their strengths and weaknesses, and both of them are widely used in practice, so we refrain from advocating one approach over the other. However, from an algorithmic perspective the optimization problem imposed by window-based approaches is trivial, whereas the optimization problem imposed by block-based approaches is not, so we focus on the latter.

In this work, we formulate three constrained optimization problems corresponding to popular variants of block-based trimming (Section 2). We study their computational complexity and find that all of them are $\mathcal{NP}$-hard (Section 3). Hence, we propose three relaxed problems by omitting one set of constraints from each of the three hard problems, and we present polynomial-time algorithms for their solution (Section 4). We find by empirical studies that these algorithms typically yield a higher number of "surviving bases" than traditional heuristics for block-based trimming (Section 5).

2   Problems and Notation

In this section we introduce the used notation and formulate block-based trimming as a constrained optimization problem.

The goal of block-based trimming is to (i) truncate low-quality positions at both ends of the reads and to (ii) remove low-quality reads. Traditionally, block-based trimming is done heuristically by performing steps (i) and (ii) sequentially using tools such as FastX [8], PRINSEQ [12], or NGS QC toolkit [11]. For step (i) we need a scalar measure for the quality of each position across all N reads of a given NGS data set. Examples for popular quality measures are the mean quality, the median quality, or any other percentile of the quality distribution. Likewise, we need a scalar measure for the quality of each read for step (ii), and it is common practice to use the same quality measure for both steps.

Based on a chosen quality measure Q and a chosen threshold T for that quality measure, we can define steps (i) and (ii) as follows. In step (i), flag all positions whose quality Q exceeds threshold T, select the longest interval of flagged positions, and truncate the leftover positions at the 5' ends and 3' ends of all reads, yielding a set of N truncated reads. In step (ii), select all truncated reads whose quality Q exceeds threshold T, and remove all remaining reads. Obviously, performing steps (i) and (ii) *sequentially* may yield suboptimal trimming results compared to a hypothetical trimming approach that performs both steps *simultaneously*.

With the goal of developing such a trimming approach that provably yields optimal trimming results, we formulate the following constrained optimization problem. Consider trimmed data sets of $\tilde{N}$ truncated reads of length $\tilde{L} = r - \ell + 1$ ranging from position ℓ to position r, consider the *column* constraints (i) that the $\tilde{L}$ *column* qualities Q exceed threshold T at all $\tilde{L}$ positions, and consider the *row* constraints (ii) that the $\tilde{N}$ *row* qualities Q exceed threshold T for all of the $\tilde{N}$ truncated reads. We now formulate the problem of optimal block-based trimming as the constrained optimization problem to find that trimmed data set – out of all trimmed data sets that satisfy the column constraints (i) and the row constraints (ii) – that maximizes $\tilde{N} \times \tilde{L}$, the number of remaining bases of the trimmed data set.

In order to address this constrained optimization problem, we translate it in several ways into problems on integer or binary grids. The quality sequences corresponding to the reads are the *rows* of the grid, while the *columns* represent the positions, so the size of the grid equals the number of sequenced bases. The integer grid corresponds to the PHRED-scores [5,6] of the reads. By applying a threshold T they can be transformed into a binary grid where each cell encodes whether the quality score of that particular base is greater than or equal to threshold T. In the positive case, the cell entry is one, otherwise zero.

We can now formulate the constrained optimization problem as follows: Given a binary grid G or a grid M of positive integers with N rows and L columns, find the biggest block B (size(B) := #rows(B) $\times$ #columns(B)) specified by a left border ℓ_B and a right border r_B of contiguous columns and a set of selected rows $R \subseteq \{1, \ldots, N\}$ of G or M where

z_r-z_c-**zeros:**   each row of B contains at most z_r zeros and the selected rows contain at most z_c zeros per column.

p_r-p_c-**percent:** each row of B contains at most p_r percent zero entries and the selected rows contain at most p_c percent zero entries per column.

m_r-m_c-**mean:**   each row of B has a mean value of m_r or higher and each column in B has a mean value of m_c or higher.

Notation. We call a block $B \subseteq G$ that fulfills the z_r-z_c-zeros property z_r-z_c-*zeros feasible* or simply *feasible*, and we use the same notion for p_r-p_c-percent and m_r-m_c-mean.

3 Complexity

In this section we investigate the computational complexity of the three problems z_r-z_c-zeros, p_r-p_c-percent, and m_r-m_c-mean. We show that the corresponding decision versions are $\mathcal{NP}$-complete. Obviously, all three problems are in $\mathcal{NP}$.

Theorem 1. *Deciding whether z_r-z_c-zeros has a solution B with* size(B) $\geq k$ *is $\mathcal{NP}$-complete.*

Proof. We give a reduction from RESTRICTED EXACT COVER BY 3-SETS (RXC3) ([7, Appendix A]): Given a set $X = \{1, \ldots, 3q\}$ and a collection $\mathcal{C}$ of 3-element subsets of X such that each element in X appears in exactly three subsets of $\mathcal{C}$. Does $\mathcal{C}$ contain an exact cover for X, i.e., a subcollection $\mathcal{C}' \subseteq \mathcal{C}$ such that every element of X occurs in exactly one member of $\mathcal{C}'$?

Given an arbitrary instance $(X, \mathcal{C})$ of RXC3 we construct in polynomial time an instance (G, z_r, z_c) of the corresponding decision problem of z_r-z_c-zeros as follows.

We identify the elements of X with the columns of a grid G'. Hence, we choose $L' := 3q$. There are $N' := 3q$ sets $C_i \in \mathcal{C}$. We define corresponding grid rows by setting

$$G'[i][j] := \begin{cases} 0 & \text{if } j \in C_i, \\ 1 & \text{otherwise.} \end{cases}$$

The grid $G \in \{0,1\}^{3q \times 9q}$ is defined by sticking three copies of $G' \in \{0,1\}^{3q \times 3q}$ together: $G := [G', G', G']$. Thus, $N := 3q$ and $L := 9q$. Finally, we set $z_r := 9$, $z_c := 1$, and $k := 9q^2$. It is now easy to see that G contains a z_r-z_c-zeros feasible block of size k if and only if $\mathcal{C}$ contains an exact cover for X.

An exact cover $\mathcal{C}' = \{C_{i_1}, \ldots, C_{i_q}\} \subseteq \mathcal{C}$ of X has $|\mathcal{C}'| = q$ sets. Each set contains three elements. Thus, the rows $i_1, \ldots, i_q$ in G' have 3 zeros per row. We set $B' := \{G'[i_j] : 1 \leq j \leq q\}$. Because every $x \in X$ occurs in exactly one set of $\mathcal{C}'$, each column of B' has 1 zero. It follows that $B := \{G[i_j] : 1 \leq j \leq q\}$ is a feasible 9-1-zeros block in G of size $3qL' = qL = 9q^2 = k$.

Let now be B a feasible 9-1-zeros block of size k of greater. If B has q rows $G[b_1], \ldots, G[b_q]$ it has width L (otherwise its size is smaller than k) which gives us an exact cover $\{C_{b_1}, \ldots, C_{b_q}\}$ of X. Now assume that B has $q+1$ or more rows. A block of size k has at least $3q = k/N$ columns (min{width}=area/max{height}), which means that it contains all columns of G' (maybe a cyclic shift). But each selection of $q+1$ (or more) rows of all columns of G' contains at least one column with 2 zeros (or more). Thus, B is not 9-1-zeros feasible, a contradiction. □

Theorem 2. *Deciding whether p_r-p_c-percent has a solution B with $\mathrm{size}(B) \geq k$ is $\mathcal{NP}$-complete.*

Proof. Given an arbitrary instance of RXC3 we construct in polynomial time an instance (G, p_r, p_c) of the corresponding decision problem of p_r-p_c-percent as follows. Let G' be defined as in the proof of Theorem 1. We set $G := \begin{bmatrix} G' & G' & G' \\ 1 & 1 & 1 \end{bmatrix} \in \{0,1\}^{6q \times 9q}$, where $\mathbf{1}$ is a $3q \times 3q$ matrix of 1's. Thus, $N := 6q$ and $L := 9q$. Finally, we set $p_r := \frac{1}{q}$, $p_c := \frac{1}{4q}$, and $k := 36q^2$. It is now easy to see that G contains a p_r-p_c-percent feasible block of size k if and only if $\mathcal{C}$ contains an exact cover for X.

Let $C_{i_1}, \ldots, C_{i_q}$ be an exact cover of X. As in the proof of Theorem 1, the rows $G[i_1], \ldots, G[i_q], G[3q+1], \ldots, G[6q]$ are p_r-p_c-percent feasible: Each row has width $9q$ and contains at most 9 zeros. Each column has height $4q$ and contains 1 zero. This gives us a feasible $\frac{1}{q}$-$\frac{1}{4q}$-percent block of size $4q9q = 36q^2$.

Let now be B a feasible $\frac{1}{q}$-$\frac{1}{4q}$-percent block of size k of greater. Such a block has at least $6q = k/N$ columns, so it contains all columns of G'. Assume that B has more than $4q$ rows. Because it contains all columns of G' there is a column with at least 2 zeros, which is only feasible for height $8q > N$, a contradiction. Thus, B has $4q$ rows and $9q$ columns. B contains exactly q of the rows $G[1]$, $\ldots, G[3q]$ (otherwise a column contains 2 or more zeros). Thus $B = \{G[b_1], \ldots, G[b_q], G[3q+1], \ldots, G[6q]\}$, which gives us an exact cover $C_{b_1}, \ldots, C_{b_q}$ of X. □

Theorem 3. *Deciding whether m_r-m_c-mean has a solution B with $\mathrm{size}(B) \geq k$ is $\mathcal{NP}$-complete.*

Proof. Given an arbitrary instance of RXC3 we construct in polynomial time an instance (M, m_r, m_c) of the corresponding decision problem of m_r-m_c-mean as follows. We set $m_r := \frac{q-1}{q}$, $m_c := \frac{4q-1}{4q}$, $k := 36q^2$ and use G as defined in the proof of Theorem 2 as our M. It is now easy to see that M contains a m_r-m_c-mean feasible block of size k if and only if $\mathcal{C}$ contains an exact cover for X.

Let $C_{i_1}, \ldots, C_{i_q}$ be an exact cover of X. The rows $M[i_1], \ldots, M[i_q], M[3q+1]$, $\ldots, M[6q]$ have mean value $(9q-9)/9q$ or higher and the selected columns have mean value $(4q-1)/4q$. So we have a feasible block of size $4q9q = 36q^2$.

Let now be B a feasible $\frac{q-1}{q}$-$\frac{4q-1}{4q}$-mean block of size k or greater. As in the proof of Theorem 2, B contains all columns of G'. If B has more than $4q$ rows it contains at least one column with 2 (or more) zeros. Such a column with height h has mean value $\frac{h-2}{h}$ and is therefore only feasible if $\frac{h-2}{h} \geq \frac{4q-1}{4q}$, i.e., $h \geq 8q > N$, a contradiction. The rest follows as in the proof of Theorem 2. $\square$

4 Polynomial-Time Algorithms for Relaxations of the Problems

The hardness of the three problems z_r-z_c-zeros, p_r-p_c-percent, and m_r-m_c-mean originates from the request of *simultaneously* satisfying the column and row constraints. In this section we investigate the three relaxed problems in which the column constraints are omitted, and present polynomial-time algorithms for solving them.

We define the relaxed problems z-zeros as z-∞-zeros, p-percent as p-100-percent, and m-mean as m-0-mean. These problems have in common that, if ℓ_B and r_B are known, the row set R can be constructed with a linear sweep over the grid in $\mathcal{O}(LN)$ time by checking whether the columns $\ell_B, \ldots, r_B$ of the current row fulfill the row constraint. Hence, we only focus on how to find ℓ_B and r_B.

Lemma 4. *Let $c_P(\ell, r)$ be the number of rows g of the grid such that $g[\ell..r]$ fulfills the given row constraint of problem P. If $c_P(\ell, r)$ is given for all ℓ and r, we can compute* size(B), ℓ_B *and* r_B *in* $\mathcal{O}(L^2)$ *time.*

Proof. It is easy to see that size$(B) = \max_{\ell \leq r}[(r - \ell + 1)c_P(\ell, r)]$. The borders ℓ_B and r_B can be computed as the arguments of the equation above. $\square$

Notation. A row-block without zeros is called 1-*row-block* or 0-*zeros feasible*. A row-block with at most z zeros is called 1-*row-block with at most z zeros* or z-*zeros feasible*. We use corresponding notions for p-percent and m-mean.

Solving 0-zeros and z-zeros in $\mathcal{O}(LN + L^2)$. The following simple, but crucial observation makes it possible to solve 0-zeros and z-zeros in linear time plus a post-processing step.

Observation 1. i) *Let $g[\ell..r]$ be a 1-row-block. For $\ell \leq \ell' \leq r' \leq r$ the subset $g[\ell'..r']$ is a 1-block, too.* ii) *Let $g[\ell..r]$ be z-zeros feasible. For $\ell \leq \ell' \leq r' \leq r$ the subset $g[\ell'..r']$ is z-zeros feasible, too.*

We set $c := c_{0\text{-zeros}}$ as defined in Lemma 4. Hence, $c(\ell, r)$ is the number of rows g in G such that $g[\ell..r]$ is a 1-row-block. Updating c (for each 1-row-block $g[x..y]$: $c(x, y)++$) for a given row g needs $\mathcal{O}(L^2)$ time in the worst-case (namely if $g = 11..1$). Therefore, we use an auxiliary matrix cT to count inclusion maximal 1-row-blocks in $\mathcal{O}(L)$ time per row. Later we use Observation 1 to compute c from cT in $\mathcal{O}(L^2)$ time. The whole process is given in Alg. 1.

The start and end positions of all inclusion maximal 1-row-block in a row g can be identified in linear time with two pointers that sweep over g. For each pair of start and end positions s and e we update $cT(s, e) := cT(s, e) + 1$.

To compute c from cT we use $c(\ell, r) = \sum_{x \leq \ell, r \leq y} cT(x, y)$ by definition of cT. It follows that $c(\ell, r) = c(\ell, r + 1) + \sum_{x \leq \ell} cT(x, r)$. So we can compute c from right to left and from top to bottom in $\mathcal{O}(L^2)$ time and $\mathcal{O}(L)$ space with Alg. 2.

Algorithm 1. 0-zeros	**Algorithm 2.** compute c from cT

foreach row g in G **do** $//\mathcal{O}(N)$
  foreach start s and end e of an incl.
  max. 1-row-block in g **do** $//\mathcal{O}(L)$
    $cT(s, e)++$
compute c with Alg. 2 $//\mathcal{O}(L^2)$
compute ℓ_B, r_B of B via Lem. 4 $//\mathcal{O}(L^2)$
construct R as described above $//\mathcal{O}(LN)$
	$colSum := [0, \ldots, 0]$
	for $i = 1, \ldots, L$ **do** $//\mathcal{O}(L)$
	  $colSum(L) += cT(i, L)$
	  $c(i, L) = colSum(L)$
	  for $j = L - 1, \ldots, i$ **do** $//\mathcal{O}(L)$
	    $colSum(j) += cT(i, j)$
	    $c(i, j) = colSum(j) + c(i, j + 1)$

Theorem 5. *0-zeros can be solved in* $\mathcal{O}(LN + L^2)$ *time and* $\mathcal{O}(L^2)$ *space.* □

We set $c_z := c_{z\text{-zeros}}$ as defined in Lemma 4. Hence, $c_z(\ell, r)$ is the number of feasible z-zeros row-blocks $g[\ell..r]$. The same approach as above can *not* be used to solve z-zeros: If we would define cT_z as the number of inclusion maximal feasible z-zeros row-blocks, Alg. 2 fails to compute c_z. Example (with $z = 1$):

$$G = \boxed{0\,|\,1\,|\,0} \quad cT_z = \begin{bmatrix} 0 & 1 & 0 \\ & 0 & 1 \\ & & 0 \end{bmatrix} \quad \text{correct } c_z = \begin{bmatrix} 1 & 1 & 0 \\ & 1 & 1 \\ & & 1 \end{bmatrix} \quad c_z \text{ from Alg. 2} = \begin{bmatrix} 1 & 1 & 0 \\ & 2 & 1 \\ & & 1 \end{bmatrix}$$

Instead we do as follows: We define $cC_z(\ell, r)$ as the number of rows g in G such that $g[\ell..r]$ is the largest feasible z-zeros row-block with fixed right border r, i.e., for all rows count in $cC_z(\ell, r)$ the block $g[(\ell-1)..r]$ is not z-zeros feasible. In other words, for fixed r we search the leftmost ℓ such that $g[\ell..r]$ is feasible. In the example above we counted $g[2..2]$ two times, because it is a subset of the inclusion maximal blocks $g[1..2]$ and $g[2..3]$. With the new definition, $g[2..2]$ is only counted in *one* entry $cC_z(\cdot, 2)$. This solves the problem of overlapping feasible z-zeros row-blocks. It follows from the definition of cC_z that $c_z(\ell, r) = \sum_{x \leq \ell} cC_z(x, r)$. Hence, we can compute c_z from cC_z via partial column sums in $\mathcal{O}(L^2)$ time.

Next we address the question how to compute cC_z. Assume we know the borders $(\ell_1, r_1), (\ell_2, r_2), \ldots$ of all inclusion maximal z-zeros row-blocks in the current row g. For a given right border $1 \leq y \leq L$ we search for the smallest r_i such that $y \in \{\ell_i, \ldots, r_i\}$. It follows that $g[\ell_i..y]$ is feasible and $g[(\ell_i-1)..y]$ is infeasible (otherwise there would be a $r_j < r_i$ such that $g[(\ell_i-1)..y] \subseteq g[\ell_j..r_j]$ and $y \in \{\ell_j, \ldots, r_j\}$, a contradiction to the choice of r_i). Hence, we can update

$cC_z(\ell_i, y)++$. This can be done with a linear sweep over all y, so we get a runtime of $\mathcal{O}(L)$ per row of G.

Finally, we address how to compute $(\ell_1, r_1), (\ell_2, r_2), \ldots$ With the auxiliary arrays $posZ$, $leftB$ and $rightB$ we store the positions of all zeros, the left border and the right border of each 1-row-block in the current row g. Example (left):

<div align="center"><small>1 2 3 4 5 6 7 8 9</small> $g = [1, 1, 0, 1, 1, 0, 1, 0, 0]$ $leftB = [1, 1,\ \ , 4, 4,\ \ , 7,\ \ ,\ \ ]$ $rightB = [2, 2,\ \ , 5, 5,\ \ , 7,\ \ ,\ \ ]$ $posZ = [3, 6, 8, 9]$</div>	**Algorithm 3.** ℓ_i and r_i for z-zeros <hr>**for** $i = 1, \ldots, \|posZ\| - z + 1$ **do** $//\mathcal{O}(L)$ $\quad$ $\ell_i := posZ[i]$, $r_i := posZ[i + z - 1]$ $\quad$ // Are there 1's left or right of $g[\ell_i..r_i]$? $\quad$ **if** $g[\ell_i - 1] == 1$ **then** $\ell_i := leftB[\ell_i - 1]$ **if** $\quad$ $g[r_i + 1] == 1$ **then** $r_i := rightB[r_i + 1]$

If $|posZ| \leq z$ we can select the whole row: $cC_z(1, L)++$. Otherwise, $\{posZ_i, \ldots, posZ_{i+z-1}\}$ for $i = 1, \ldots, |posZ| - z + 1$ are the possible positions for z zeros in such a row-block. For each set of zeros $g[posZ_i..posZ_{i+z-1}]$ is z-zeros feasible. Two of the zeros occur at the borders $g[posZ_i]$ and $g[posZ_{i+z-1}]$. If $g[posZ_i-1] = 1$, there is an adjoined 1-row-block that we can add. Its left border is $leftB[posZ_i-1]$. We do the same for the rightmost zero $g[posZ_{i+z-1}]$. The process is shown in Alg. 3.

The example below (left side) illustrates the definition of cC_z.

<div align="center">$G := \boxed{1\|1\|0\|1\|1\|0\|1\|0\|0}$, $z := 1$ $cC_z =$ $\qquad\qquad$ $c_z =$ $\begin{bmatrix}1\,1\,1\,1\,1\,0\,0\,0\,0\\0\,0\,0\,0\,0\,0\,0\,0\\0\,0\,0\,0\,0\,0\,0\\0\,0\,1\,1\,0\,0\\0\,0\,0\,0\,0\\0\,0\,0\,0\\0\,1\,0\\0\,0\\1\end{bmatrix} \begin{bmatrix}1\,1\,1\,1\,1\,0\,0\,0\,0\\1\,1\,1\,1\,0\,0\,0\,0\\1\,1\,1\,0\,0\,0\,0\\1\,1\,1\,1\,0\,0\\1\,1\,1\,0\,0\\1\,1\,0\,0\\1\,1\,0\\1\,0\\1\end{bmatrix}$</div>	**Algorithm 4.** z-zeros <hr>**foreach** row in G **do** $//\mathcal{O}(N)$ $\quad$ compute $posZ$, $leftB$, $rightB$ $//\mathcal{O}(L)$ $\quad$ compute all (ℓ_i, r_i) with Alg. 3 $//\mathcal{O}(L)$ $\quad$ **foreach** $y = 1, \ldots, L$ **do** $//\mathcal{O}(L)$ $\quad\quad$ as described above: $cC_z(\cdot, y)++$ compute c_z via partial column sums $//\mathcal{O}(L^2)$ compute ℓ_B, r_B, B and R as in Alg. 1

The whole process to solve the z-zeros problem is given in Alg. 4.

Theorem 6. *z-zeros can be solved in $\mathcal{O}(LN + L^2)$ time and $\mathcal{O}(L^2)$ space.* $\quad\square$

Solving p-percent and m-mean in $\mathcal{O}(L^2N)$. Note that Observation 1 only holds for the z-zeros problem. We set $c_p := c_{p\text{-percent}}$ as defined in Lemma 4. Thus, $c_p(\ell, r)$ is the number of feasible p-percent row-blocks $g[\ell..r]$. In the same way we define $c_m := c_{m\text{-mean}}$.

Lemma 7. *p-percent and m-mean can be solved in $\mathcal{O}(L^2N)$ time and $\mathcal{O}(L^2)$ space.*

Proof. Use Alg. 5 resp. Alg. 6 to compute c_p resp. c_m. Compute ℓ_B and r_B via Lemma 4 and construct R as described at the beginning of the section. $\quad\square$

In the sequel, we refer to the approach of this section as *optimized block trimming*. All algorithms have been implemented in C++. The code can be freely obtained from http://github.com/hedtke/SequenceTrimming for scientific use. In practice, the code achieves excellent runtimes close to the speed of reading data from disk for z-zeros, and still very good ones for the two other problems.

Algorithm 5. compute c_p	
foreach row g in G **do**	$//\mathcal{O}(N)$
for $i = 1, \ldots, L$ **do**	$//\mathcal{O}(L)$
$zeros := 0$	
for $j = i, \ldots, L$ **do**	$//\mathcal{O}(L)$
if $g[j] == 0$ **then** $zeros{+}{+}$ **if**	
$\frac{zeros}{j-i+1} \leq p$ **then** $c_p(i,j){+}{+}$	

Algorithm 6. compute c_m	
foreach row g in M **do**	$//\mathcal{O}(N)$
for $i = 1, \ldots, L$ **do**	$//\mathcal{O}(L)$
$mean := 0$	
for $j = i, \ldots, L$ **do**	$//\mathcal{O}(L)$
$mean := \frac{(j-i)mean+g[j]}{j-i+1}$	
if $mean \geq m$ **then** $c_m(i,j){+}{+}$	

5   Case Studies

Data. For comparing optimized block trimming to traditional stepwise trimming, we use data sets SRR1030717 and SRR985867 from NCBI GEO [4] with sample accession number GSM1267149 and GSM1231194, respectively. Data set SRR1030717 contains about 87 million Illumina single-end reads with a length of 97 bp, and data set SRR985867 published in [2] contains about 21 million Illumina single-end reads with a length of 50 bp. In both cases, we convert the SRA files into FASTQ files using fastq-dump.2.3.4 from the SRA Toolkit [10].

Per position quality after trimming. The $\mathcal{NP}$-hard problems z_r-z_c-zeros, p_r-p_c-percent, and m_r-m_c-mean presented in Section 2 impose both column and row constraints, whereas the easy problems z-zeros, p-percent, and m-mean presented in Section 4 omit the column constraints. The quality distributions of NGS data sets intuitively suggests that the column constraints could be satisfied automatically for many quality measures Q and many thresholds T, without imposing them, by satisfying the row constraints. To test this intuition, we perform trimming of data sets SRR1030717 and SRR985867 by the p-percent algorithm using different quality measures Q and thresholds T. Specifically, we choose five percentiles ranging from 75% to 95% as quality measures Q and six thresholds T ranging from 20 to 30, and we plot the column quality for each of the columns of the trimmed data set for each of the 30 combinations of Q and T.

Figure 1 shows the results for the two data sets and the two percentiles 75% and 95%. The plots for the other percentiles (80%, 85%, and 90%) are similar. In the 75% percentile case, the column quality after trimming is way above 75% for both data sets, all positions, and all quality thresholds T. In the more stringent 95% percentile case, the column quality after trimming is above 95% for both data sets and most – but not all – positions and quality thresholds. Here, we find that for thresholds 20, 22, 24, and 26 the column constraint is satisfied for all positions, but for thresholds 28 and 30 there are a few columns, at the ends of the trimmed reads, for which the column constraint is violated.

Comparison of stepwise and optimized trimming. We compare the traditional stepwise trimming method and the optimal block trimming by calculating the number of remaining bases after trimming. As example we use the p-percent algorithm. In Figure 2, we show the number of bases that remain after trimming for different quality measures Q and different quality thresholds T. Of course,

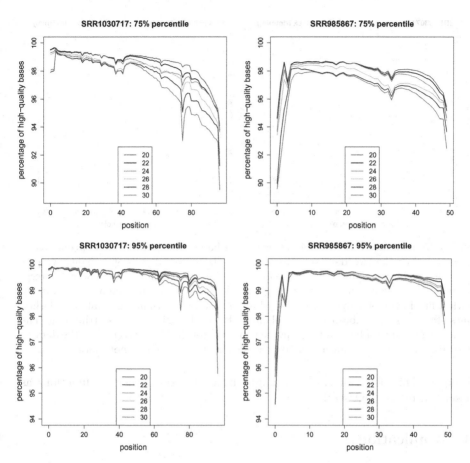

Fig. 1. Quality after optimal block trimming using p-percent. For the two data sets SRR1030717 (left) and SRR985867 (right) we plot the percentage of high-quality bases per position, i.e., column in the grid, after trimming. As parameters we use $p=0.25$ resp. $p=0.05$, which corresponds to the 75% resp. 95% percentile, and quality thresholds from 20 to 30. A base with quality score greater than or equal to the quality threshold is called high-quality base.

optimal trimming can never be worse than the traditional approach. We find that the number of remaining bases after optimal trimming is often greater than that of traditional stepwise trimming for all combinations of Q and T. Especially for high percentile values and high quality thresholds, the difference in the number of retained bases becomes surprisingly high. This indicates that previous studies using traditional stepwise trimming have discarded a considerable amount of bases that could have been retained.

The increase in the number of untrimmed bases stems from the increase in retained positions. This is a consequence of the used objective function "max[size(B)]", where adding a column is more favorable due to the fact that

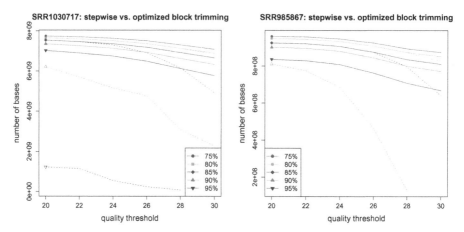

Fig. 2. Comparison of the number of remaining bases after stepwise and optimal block trimming. For the two data sets SRR1030717 (left) and SRR985867 (right) we plot the number of remaining bases depending on the quality threshold. The color resp. symbols of the lines encode the used percentile values. The continuous lines starting with a filled symbol show the results using optimal block trimming, while the dashed lines starting with a non-filled symbol show the results using stepwise trimming. The remaining number of bases after stepwise trimming for the 95% percentile of the dataset SRR985867 equals zero for all quality thresholds and is therefore not plotted.

$L \ll N$. This behavior can be changed by using other objective functions as described in the next section.

6   Applications

Trade-off between optimal solution and number of lost reads. The optimal solution for trimming as computed by p-percent maximizes the number of bases remaining after trimming, but this does not ensure that a high percentage of reads is kept. One can imagine examples where only 50% or less are kept in the optimal solution, which might be unacceptable for practical applications.

We solved the simplified problems by computing the matrices c and c_z resp. c_p and c_m. Given these matrices, we can ask for an optimal solution that keeps a given percentage of reads. That is, we allow the user to impose as constraint the minimal percentage of reads to be kept after trimming. Figure 3 shows the number of bases in the optimal solution for a given minimal percentage of reads to be kept in the solution. The plot was made with quality threshold 30 and for percentage values ranging from 75% to 95%. A single scan over the input suffices to compute the auxiliary matrices from which the trade-off curves are derived. This enables the user to play with the input parameters of each of the optimized block trimming methods and to get immediate feedback about the properties of the optimal solution without rereading a single bit of the input file. Hence, our methods can also serve as a decision support tool.

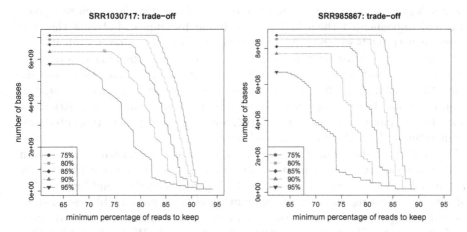

Fig. 3. Trade-off between the remaining number of bases and the minimal percentage of reads that have to be kept after trimming. For both data sets we have used p-percent for different percentile values and a quality threshold of 30.

Choosing an objective function. The calculation of optimal solutions is based on the auxiliary matrices c, c_z, c_p, and c_m, respectively. So far, an *optimal solution* was simply a block fulfilling the desired quality properties with maximal number of bases, i. e., the objective was "width" $\times$ "# selected rows" or expressed in terms of the matrix $(r-\ell+1) \cdot c(\ell, r)$. However, as soon as the auxiliary matrix, say c, is available, it is possible to optimize any objective function f of the form $f(\ell, r, c_{\ell r})$ with $f: \{1..L\} \times \{1..L\} \times \{1..N\} \to \mathbb{R}$. Even additional constraints are possible. The choice of an appropriate objective function is non-trivial and clearly depends on the downstream analysis, the read length, the read coverage, etc. Our algorithms allow the user to change the objective function and play with different parameters without expensive recomputations. For example, if we want a minimal read length after trimming of 50 bases, keep at least 80% of the reads, and weigh the number of selected rows logarithmicaly, this can be expressed as

$$\max \ (r - \ell + 1) \cdot \log c(\ell, r)$$
$$\text{s.t.} \ \ r - \ell + 1 \geq 50 \text{ and } c(\ell, r) \geq 0.8 \cdot N.$$

7   Conclusions

Read trimming is a fundamental first step of the analysis of NGS data, but there is surprisingly little literature on best practices or algorithmic aspects of read trimming. In this paper, we formalized the problem of block-based trimming, a popular trimming approach by which (i) low-quality positions at both ends of each read are truncated and (ii) truncated reads with low quality are removed. We analyzed the computational complexity of the three resulting constrained optimization problems, and we found all of them to be $\mathcal{NP}$-hard.

The distribution of quality scores in NGS data sets suggested the idea of relaxing the three optimization problems by omitting the column constraint.

We developed polynomial-time algorithms for solving the three relaxed problems and implemented them in C++. We found by empirical studies that the column constraint is satisfied at all positions, without imposing it explicitly, by satisfying the row constraints for almost all of the tested combinations of quality measures Q and quality thresholds T. Only for a combination of high-quality percentiles and high-quality thresholds the column constraint was not satisfied at all positions. However, even in these cases there were only few positions at which the column constraint was violated, and these positions were located predominantly at the ends of the trimmed reads.

We found that the number of bases resulting from optimal trimming is often greater than that resulting from stepwise trimming. The difference in the number of retained bases became quite large for high percentile values and high-quality thresholds, indicating that a significant number of bases was possibly lost by stepwise trimming in previous studies. Our methods enable the user to evaluate a whole range of possible trimming settings simultaneously. We finally remark that the related problems of paired-end trimming can be solved with similar methods, e.g., the corresponding z-zeros problem can be solved in $\mathcal{O}(L^2N+L^4)$ time.

Acknowledgements. We thank Claus Weinholdt for valuable discussions and DFG (grant no. GR 3526/2) for financial support.

References

1. Bardet, A.F., He, Q., Zeitlinger, J., Stark, A.: A computational pipeline for comparative ChIP-seq analyses. Nature Protocols 7(1), 45–61 (2012)
2. Bhargava, V., Head, S.R., Ordoukhanian, P., Mercola, M., Subramaniam, S.: Technical variations in low-input RNA-seq methodologies. Scientific Reports 4(3678) (2014)
3. Del Fabbro, C., Scalabrin, S., Morgante, M., Giorgi, F.: An extensive evaluation of read trimming effects on Illumina NGS data analysis. PLoS ONE 8(12), e85024 (2013)
4. Edgar, R., Domrachev, M., Lash, A.: Gene Expression Omnibus: NCBI gene expression and hybridization array data repository. Nucleic Acids Res. 30(1), 207–210 (2002), http://www.ncbi.nlm.nih.gov/geo
5. Ewing, B., Hillier, L., Wendl, M., Green, P.: Base-calling of automated sequencer traces using phred. I. Accuracy assessment. Genome Research 8(3), 175–185 (1998)
6. Ewing, B., Green, P.: Base-calling of automated sequencer traces using phred. II. Error probabilities. Genome Research 8(3), 186–194 (1998)
7. Gonzalez, T.F.: Clustering to minimize the maximum intercluster distance. Theor. Comput. Sci. 38, 293–306 (1985)
8. Hannon Lab: FASTX Toolkit, http://hannonlab.cshl.edu/fastx_toolkit/
9. Koboldt, D., Steinberg, K., Larson, D., Wilson, R., Mardis, E.R.: The next-generation sequencing revolution and its impact on genomics. Cell 155(1), 27–38 (2013), http://www.sciencedirect.com/science/article/pii/S0092867413011410
10. NCBI – SRA Toolkit, http://eutils.ncbi.nih.gov/Traces/sra/?view=software
11. Patel, R.K., Jain, M.: NGS QC Toolkit: A toolkit for quality control of next generation sequencing data. PLoS ONE 7(2), e30619+ (2012)
12. Schmieder, R., Edwards, R.: Quality control and preprocessing of metagenomic datasets. Bioinformatics 27(6), 863–864 (2011)
13. UC Davis Bioinformatics Core: sickle - Windowed Adaptive Trimming for fastq files using quality, http://hannonlab.cshl.edu/fastx_toolkit/

On the Implementation
of Quantitative Model Refinement

Bogdan Iancu[1,2,3], Diana-Elena Gratie[1,2,3], Sepinoud Azimi[1,2,3], and Ion Petre[1,2,3]

[1] Computational Biomodeling Laboratory
[2] Turku Centre for Computer Science
[3] Department of IT, Åbo Akademi University
Joukahainengatan 3-5, 20520 Åbo, Finland
{biancu,dgratie,sazimi,ipetre}@abo.fi

Abstract. The iterative process of adding details to a model while preserving its numerical behavior is called *quantitative model refinement*, and it has been previously discussed for ODE-based models and for *kappa*-based models. In this paper, we investigate and compare this approach in three different modeling frameworks: rule-based modeling, Petri nets and guarded command languages. As case study we use a model for the eukaryotic heat shock response that we refine to include the acetylation of the heat shock factor. We discuss how to perform the refinement in each of these frameworks in order to avoid the combinatorial state explosion of the refined model. We conclude that Bionetgen (and rule-based modeling in general) is well-suited for a compact representation of the refined model, Petri nets offer a good solution through the use of colors, while the PRISM refined model may be much larger than the basic model.

Keywords: Quantitative model refinement, heat shock response, acetylation, rule-based modeling, Petri nets, model checking.

1 Introduction

Systems biology aims to holistically characterize highly complex biological systems. A hierarchical system-level representation is very adequate in this context. Formal frameworks turn out to be fundamental in the effort of understanding the behavior of such complex systems, see [21,12]. The abstractions that lie at the core of these formalisms need to be refined to incorporate more details.

We focus in this paper on the implementation of model refinement. Within the model development process, we examine *data refinement* through three different frameworks – *rule-based modeling, Petri nets* and *guarded command languages* – and discuss their capabilities for the efficient construction of a refined model. For rule-based modeling we used the Bionetgen framework and RuleBender, for Petri nets we chose Snoopy and Charlie as modeling tools, while for modeling with guarded command languages we used PRISM. Data refinement, as described in [3] and [10], assumes the replacement of one species in the model with several of its variants, called subspecies. This type of refinement is adequate for representing post-translational modifications of proteins, e.g., acetylation, phosphorylation, etc. Given a protein P, one can indicate its state regarding

A.-H. Dediu, C. Martín-Vide, and B. Truthe (Eds.): AlCoB 2014, LNBI 8542, pp. 95–106, 2014.

post-translational modifications by replacing it with its variants. This substitution also implies a refinement of all complexes involving protein P and of all reactions involving either P or any such complex, see [10]. This might induce a combinatorial state explosion of the refined model, as in the case of ODE-based models, see [10]. The main question we are answering is whether one can avoid this problem in the other three frameworks we investigate in this paper and build a compact representation of the refined model.

We consider as a case study for our analysis the heat shock response mechanism, as described in [20] and [10]. Throughout the paper, the model in [20] will be referred to as the *basic* heat shock response model, while the model in [10] will be referred to as the *refined model*.

All models developed in this paper are available for download at [11]. Due to space restrictions, some of the details of this work were omitted. For full details, we refer the reader to [6].

2   The Heat Shock Response (HSR)

The eukaryotic *heat shock response* is a highly conserved bio-regulatory network that controls cellular function impairment produced by protein misfolding as a result of high temperatures. Elevated temperatures have proteotoxic effects on proteins, inducing protein misfolding and leading to the formation of large aggregates that thereafter trigger apoptosis (controlled cell death). Cell survival is promoted by a defense mechanism, which consists in restoring protein homeostasis by augmenting the level of molecular chaperones, see [22].

We consider the basic molecular model for the eukaryotic heat shock response proposed in [20]. *Heat shock proteins* (hsp's) play a key role in the heat shock response mechanism by chaperoning the *misfolded proteins* (mfp's). Due to their affinity to mfp's, hsp's form hsp:mfp complexes and help the misfolded proteins refold. The heat shock response is regulated by the transactivation of the hsp-encoding genes. In eukaryotes, some specific proteins, called *heat shock factors* (hsf's), promote gene transcription. In the absence of environmental stressors, heat shock factors are predominantly found in a monomeric state, extensively bound to heat shock proteins. Raising the temperature causes the correctly folded proteins (prot) to misfold and hsp:hsf complexes to break down. This switches on the heat shock response by releasing hsf's, which quickly reach a DNA binding competent state, see [20,23].

Heat stress induces dimerization (hsf$_2$) and, subsequently, trimerization (hsf$_3$) of hsf's, enabling the binding of the hsf trimers to the promoter site of the hsp-encoding gene, called *heat shock element* (hse). Subsequently, DNA binding triggers the transcription and translation of the hsp-encoding gene, inducing hsp synthesis, see [20,22]. Once the level of heat shock proteins is sufficiently elevated for the cell to withstand thermal stress, hsp synthesis is turned off. Heat shock proteins sequestrate heat shock factors and break hsf dimers and trimers, constituting hsp:hsf complexes. The explicit molecular reactions constituting the model can be found in [20].

The numerical setup of the basic model (in terms of initial concentrations and kinetic constants) can be found in [20]. Acetylation has been shown to have an extensive influence in regulating the heat shock response, we refer the reader to [24]. To this end, we consider the acetylation of heat shock factors implemented through data refinement.

3 Quantitative Model Refinement

Quantitative model refinement was investigated in [19,4] regarding rule based modeling and applied to two ampler ODE-based models in [18,10].

3.1 Quantitative Model Refinement

A reaction-based model can be refined to incorporate more information regarding its reactants and/or reactions. There are two types of refinement, either of the *data* (data refinement) or of the *reactions* (process refinement). In this study, we focus on the first refinement type. Considering that one's interest lies especially on data, a species in a model could be refined by replacing it with several of its subspecies, a routine called *data refinement*. When the interest is focused on reactions, the model can be refined by replacing a collective reaction, accounting for a specific process, by a set of reactions depicting the transitional steps of the process. The last type of refinement is called *process refinement*, see [10].

The notion of quantitative model refinement has been previously addressed in systems biology in the context of rule based modelling, see [19,4,7,5]. The rule based modelling framework embodies the concept of *data refinement*, as previously introduced, implementing agent resolution as a fundamental constituent, [7]. The key refinement method in this context is rule refinement, an approach that requires the refinement of the set of rules ensuring the preservation of the dynamic behavior of the system, see [19].

We present here the *quantitative model refinement* of reaction models following the discussion in [10]. Consider a model M, comprising a number m of species $\Sigma = \{A_1, A_2, \ldots, A_m\}$ and n of reactions r_i, $1 \leq i \leq n$, as follows:

$$r_i \; : \; S_{i,1}A_1 + S_{i,2}A_2 + \ldots + S_{i,m}A_m \xrightarrow{k_i} S'_{i,1}A_1 + S'_{i,2}A_2 + \ldots + S'_{i,m}A_m,$$

where $S_{i,1}, \ldots, S_{i,m}, S'_{i,1}, \ldots, S'_{i,m} \geq 0$ are the stoichiometric coefficients of r_i and $k_i \geq 0$ is the kinetic rate constant of r_i. We discuss here a continuous, mass-action formulation of the model based on ODEs. For some details on this approach we refer to [14].

Model M can be refined to distinguish between various subspecies of any species in the model, for example, A_1. The distinction between the subspecies is very often drawn by post-translational modifications such as acetylation, phosphorylation, sumoylation, etc. All previously mentioned subspecies of A_1 take part in all reactions A_1 engaged in, conceivably obeying a different kinetic setup. Given model M and species A_1, substituting subspecies $B_1, \ldots, B_l$ for species A_1 in M leads to attaining a new model M_R, comprising species $\{A'_2, A'_3, \ldots, A'_m\} \cup \{B_1, \ldots, B_l\}$, for some $l \geq 2$, where variables A'_i, $2 \leq i \leq m$ from M_R, coincide with A_i from model M and $B_1, \ldots, B_l$ substitute for species A_1 in M_R. Furthermore, each reaction r_i of M is replaced in the new model M_R by all possible reactions $r_{i,j}$ of the following form:

$$r_{i,j} \ : \ (T_{i,1}^{j}B_1 + \ldots + T_{i,l}^{j}B_l) + S_{i,2}A'_2 + \ldots + S_{i,m}A'_m \xrightarrow{k_{i,j}}$$
$$(T'^{j}_{i,1}B_1 + \ldots + T'^{j}_{i,l}B_l) + S'_{i,2}A'_2 + \ldots + S'_{i,m}A'_m,$$

where $k_{i,j}$ is the kinetic rate constant of $r_{i,j}$ and $(T_{i,1}^{j}, \ldots, T_{i,l}^{j}, T'^{j}_{i,1}, \ldots, T'^{j}_{i,l})$ are all possible nonnegative integers so that $T_{i,1}^{j} + \ldots + T_{i,l}^{j} = S_{i,1}$ and $T'^{j}_{i,1} + \ldots + T'^{j}_{i,l} = S'_{i,1}$. Model M_R is said to be a *data refinement of model M on variable* A_1 if and only if the following conditions are fulfilled:

$$[A_i](t) = [A'_i](t), \tag{1}$$

$$[A_1](t) = [B_1](t) + \ldots + [B_l](t), \tag{2}$$

for all $2 \le i \le m$, $t \ge 0$. Fulfilling these conditions depends on the numerical setup of model M_R, i.e., on the kinetic constants of its reactions (both those adopted from the basic model, as well as those newly introduced in the construction) and on the initial concentrations of its species.

3.2   Adding the Acetylation Details to the HSR Model through Data Refinement

We start from the basic model of the heat shock response, introduced in [20], where no post-translational modification of hsf is taken into account, and we refine all species and complexes that involve hsf taking into consideration one acetylation site for every hsf molecule. We follow here the discussion in [10]. The aim is to refine the basic model and preserve its numerical properties. For hsf$_2$, hsf$_3$, hsf$_3$:hse and hsp:hsf, the refinement is performed conforming to the number of hsf constituents respectively. This leads to the following data refinements: hsf $\rightarrow$ {rhsf, rhsf$^{(1)}$}; hsf$_2$ $\rightarrow$ {rhsf$_2$, rhsf$_2$$^{(1)}$, rhsf$_2$$^{(2)}$}; hsp:hsf $\rightarrow$ {rhsp:rhsf, rhsp:rhsf$^{(1)}$}; hsf$_3$ $\rightarrow$ {rhsf$_3$, rhsf$_3$$^{(1)}$, rhsf$_3$$^{(2)}$, rhsf$_3$$^{(3)}$}; hsf$_3$: hse $\rightarrow$ {rhsf$_3$:rhse, rhsf$_3$$^{(1)}$:rhse, rhsf$_3$$^{(2)}$:rhse, rhsf$_3$$^{(3)}$:rhse}. The refinement based on the above data refinements involves substantial changes in the list of reactions. For example, the reversible reaction of dimerization 2 hsf $\rightleftarrows$ hsf$_2$ in the basic model is replaced by three reactions as follows: 2 rhsf $\rightleftarrows$ rhsf$_2$; rhsf + rhsf$^{(1)}$ $\rightleftarrows$ rhsf$_2$$^{(1)}$; 2rhsf$^{(1)}$ $\rightleftarrows$ rhsf$_2$$^{(2)}$.

The refined model of [10] consists of 20 species and 55 irreversible reactions, compared to 10 species and 17 irreversible reactions in the basic model of [20].

4   Quantitative Refinement in Rule-Based Models

4.1   A RuleBender Implementation of the Basic HSR Model

This section focuses on the RuleBender implementation of the basic heat shock response model, as introduced in Section 2. We model all reactions to follow the principle of mass action. Conforming to the implementation presented here, Bionetgen source code comprises a set of twelve rules, which generate a total number of seventeen irreversible reactions. Due to the symmetry that some of the species exhibit, the collision frequency (e.g. in our case dimerization, trimerization, etc) and the existence of multiple paths from substrates to products in some reactions (e.g. for the heat shock response model, the unbinding of trimers), kinetic rate constants for those specific reactions are multiplied in Bionetgen by diverse symmetry and/or statistical factors, see [2]. For example, the collision frequency of two different types of reactants A and B, $A + B$, is twice that of identical types of reactants $A + A$. Another example concerns the multiple reaction paths from reactants to products, which may generate statistical factors. Preserving the fit of the heat shock response model attained in [20] required a multiplication of some rate constants by the inverse of the aforementioned factors respectively.

RuleBender generates during the process of model development a contact map which depicts the connectivity between the molecules. The contact map for the basic model of the heat shock response is shown in Figure 1.

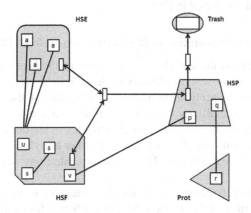

Fig. 1. The RuleBender generated contact map for the basic model of the heat shock response. It depicts the possible interconnections among the model's species.

One can notice in Figure 1 that hsf's have been represented as having 4 sites (s, s, u, v). The two s sites are involved in the generation of dimers and trimers. The other two sites, u and v, are used to illustrate the process of DNA binding/unbinding and hsf sequestratation/dimer (trimer) dissipation. Trimers are considered to be circular structures, each of the 's' site of one hsf being bound to the 's' sites of the consequent hsf's, no two hsf's having both sites 's' bound to the same partner. The promoter, hse, has been represented as having three identical sites (a, a, a), so as to be connected to the trimer in such a way that the symmetry is not affected. Heat shock proteins are modeled to have two sites 'p' and 'q', used for the modelling of unbinding of dimers and trimers and for the sequestration of misfolded proteins. The model takes into account a species called Prot, which has a site with two possible states, one of which accounts for misfolded proteins 'm' and another one 'f', that accounts for folded proteins. A "dummy" component, called Trash, has been introduced to help encode the degradation of heat shock proteins.

The contact map in Figure 1 illustrates the connectivity between the species in the model. The link between the 's' sites of the hsf molecule denotes the formation of dimers and trimers through the agency of these sites. Once trimers are formed, they can bind to the heat shock element (hse), the connection being illustrated by three links connecting hsf trimers to the heat shock element (one can notice three 'a' sites the heat shock element component exhibits). The middle connector encodes for a number of reactions, such as: DNA unbinding, HSP synthesis and breaking of dimers and trimers. The link between the site 'v' of the hsf component and the site 'p' of the hsp component illustrates hsf sequestration. The link between the hsp component and the prot component encodes the following reactions: protein misfolding, protein refolding and mfp sequestration. By linking the component *Trash* to the hsp component, we encoded for the degradation of hsp's.

We chose a deterministic simulation for the basic model. The simulation results for DNA binding for a temperature of 42°C showed that RuleBender prediction are in accordance with the results reported in [20].

4.2 A RuleBender Implementation of the Acetylation-Refined HSR Model

We focus in this section on the acetylation-refinement of the heat shock response, as described in [10]. There are several changes to do in Rulebender to refine the basic model so as to include the acetylation of hsf's. The syntax of the rules remains, in this case, unchanged, since all reactions, in this model, take place regardless of the acetylation status of the molecules. We brought changes in the definition of hsf's, by introducing one acetylation site, 'w', which can be either acetylated or not, and in the initial concentrations of the molecules. The initial concentrations were set conforming to [10].

As expected from the refinement conditions, the simulation of the refined model for a temperature of 42°C showed that the Rulebender prediction for the refined model and the one for the basic model are the same.

5 Quantitative Refinement in Petri Net Models

5.1 A Petri Net for the Basic HSR Model

A standard Petri net model for the heat shock response was previously reported in [1]. We focus here on a Snoopy continuous Petri net implementation of the basic heat shock response model, shown in Figure 2. The network has 10 places and 17 transitions, encoding the 10 species and 17 irreversible reactions in the basic model definition of [20]. Verifying the model required the analysis of several properties. For instance, the model is covered by T-invariants; also, the P-invariants reported by Charlie encode all mass conservation relations reported in the ODE-based model of [20]. Moreover, all places except HSP are covered by P-invariants, which means that they are bounded. The three mass conservation relations yield three constants (accounting for the total amount of HSF, HSE and protein molecules in the system, respectively), that have been used in the PRISM implementation of the model.

5.2 Petri Nets for the Acetylation-Refined HSR Model

For the refined heat shock response that includes two types of hsf's (acetylated and non-acetylated [10]), we chose an implementation based on colored continuous Petri nets.

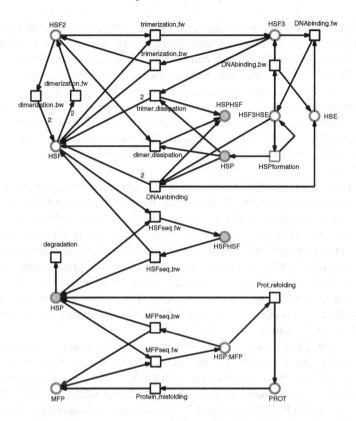

Fig. 2. Snoopy implementation of the basic heat shock response model. The text next to a place (transition) denotes the identifier of that particular place (transition). Arc multiplicities greater than 1 are included in the picture. The dashed gray circles are logical places (they may appear several times, but they represent the same species).

There are several ways of reasoning about refined species within this framework. For example, the dimer of a protein with a site that can be acetylated (1) or non-acetylated (0) can be either seen as an entity with 0, 1, or 2 acetylated sites, or as a compound where the order of the acetylated sites counts (i.e. (0,0), (0,1), (1,0), (1,1)). Depending on the approach one takes, the colored representation will have different color sets, different number of transitions and different kinetic constants.

We modeled the refined heat shock response using two approaches: one focused on keeping the structure of the basic model intact, with the same transitions and kinetic constants (we call this model *transition-focused*). This is the most compact representation. The other approach aimed to minimize the number of colors used in the model (we call this model *color-focused*). This approach uses as few colors as possible, at the cost of a complicated representation, with many conditions in a transition, and also introducing new transitions in the colored representation. Due to space limitations, we present here only the color-focused model. A more detailed description of both approaches can be found in [6].

Several choices had to be made during the modeling process. We detail the modeling options for the dimerization and trimerization of acetylated and non-acetylated hsf's. There are three types of dimers that can be formed: non-acetylated ($hsf_2^{(0)}$), single-acetylated ($hsf_2^{(1)}$) and double-acetylated dimer ($hsf_2^{(2)}$). One way of modeling the dimers is using a color set with three colors of type int (0, 1, and 2 denoting the number of acetylated sites). Another approach is using a cartesian product $\{0,1\} \times \{0,1\}$. When modeling hsf trimers, one could consider, for example, one of the following three color sets: a color set $Tri = \{0,1,2,3\}$, a compound color set $Compound = \{0,1\} \times \{0,1,2\}$, or a compound set $Trimer = \{0,1\} \times \{0,1\} \times \{0,1\}$. For the color-focused refinement, we chose the simple integer color sets.

All reactions involving the decomposition of complexes containing hsf's required additional transitions. For example, the trimer dissipation reaction $hsf_3 + hsp \rightarrow hsp:$ $hsf +2\,hsf$ is split into three transitions. One covers the case when all hsf's in the trimer have the same acetylation value (i.e. hsf_3 has color 0 or 3). In this case, there is no distinction between which hsf binds to hsp and which two hsf's become unbound, and the kinetic constant for this transition is the same as the corresponding one in the basic model. When hsf_3 has color 1 or 2, there are two binding possibilities: hsp binds to either a non-acetylated hsf, or to an acetylated hsf. For the two transitions representing these possibilities, the kinetic constant is half of the corresponding one in the basic model (following the reasoning explained in [10]).

When simulating a colored Petri net, Snoopy first unfolds it, in other words it creates an equivalent Petri net. Each place instance (each color) will correspond to a place in the unfolded net, and each transition instance (each binding) will correspond to a transition in the unfolded net; for details on colored Petri nets unfolding, see [17]. The color-focused refined model has 10 places and 25 transitions, and its corresponding flattened Petri net has 20 places and 56 transitions. This representation, although more complex than the transition-focused one, encodes a smaller flattened network. Both the transition- and color-based refinements have been compared with the basic model predictions, and they are all equivalent (data not shown).

6   Quantitative Refinement in PRISM Models

6.1   A PRISM Implementation of the Basic HSR Model

We implemented the basic heat shock response as a CTMC model that defines all possible guards (in this case reactions) within a single module. The PRISM model consists of 10 variables, each of them corresponding to one of the reactants in the model, and 17 guards representing the 17 irreversible reactions of the system. The values for upper bounds of the variables are taken from our Petri net model's P-invariants and mass-conservation relations. Upper bounds are used both for allocating memory and in the guarded commands. For example the guard corresponding to *dna binding* is expressed as follows: $hsf_3 >= 1 \wedge hse >= 1 \wedge hsf_3:hse <= N - 1 \rightarrow hsf_3 * hse * k_5 :$ $(hsf_3' = hsf_3 -1) \wedge (hse' = hse -1) \wedge (hsf_3:hse' = hsf_3:hse +1)$, where N represents the upper bound for hse in the system.

It is noteworthy to mention that the PRISM model could be obtained from the Petri net model via some format manipulations in Snoopy. However, we decided to write the model from the very beginning in order to be able to compare the modeling effort in each chosen framework.

6.2   A PRISM Implementation of the Acetylation-Refined HSR Model

The approach we took in Sections 4 and 5 to implement the acetylation-refined heat shock response model was through a compact representation of the acetylated species. Whereas colors of the places and arc expressions were employed to represent the refinement in the Petri net model, in modeling with RuleBender the solution was to introduce a new acetylation site for every hsf molecule. Both methods used structured data types for the species, thus concealing the complexity of the model in a compact representation. In PRISM this requires a method to represent the acetylation details in the definition of hsf, i.e. a composite data type. Since PRISM currently supports only simple data type (e.g. integer, boolean) variables in the model, such a definition is not possible. Alternatively, we implemented the acetylation-refined model through introducing new variables describing all possible acetylation configurations of hsf and hsf complexes. This was similar to the ODE-based approach to quantitative model refinement discussed in [10].

The refined heat shock response model is built based on the refinements given in Section 3.2 by refining all reactants and complexes involving hsf. In this approach, the strategy is to replace each guard involving any refined reactant by the guards considering all possible refined reactions.

One could also use parallel modules to implement the refinement but this approach would not help reducing the complexity of the model.

The complete PRISM implementation of the refinement is not listed here due to space limitations. The numerical setup of this model is based on [10].

6.3   Model Checking of the HSR Models

According to [15], the maximum number of states that PRISM can handle for CTMCs is 10^{10}. In both our models (basic and refined version of the heat shock response), the number of all possible states in the system exceeds this limit. This is a known problem for biological systems in PRISM, see [8]. Several studies have addressed this issue, see e.g., [9,16,8]. One of the investigated approaches is *approximate verification* of probabilistic systems, where a Monte-Carlo algorithm is used to approximate the probability of a temporal formula to be true, see [9]. We used this method to verify the desired properties of the heat shock response model. In this approach a large number of stochastic paths is sampled for the model and based on the defined properties, the result for each run is obtained. The information produced in this way gives an approximate result for the probability that the desired property holds for the model.

We are interested in verifying two properties discussed in [20]. The properties are: (i) the validity of three mass-conservation relations and (ii) the level of DNA binding eventually returns to the basal values, both at $37°C$ and at $42°C$.

In order to check the mass conservation properties, we used the G operator which checks if the property remains true at all states along the path. The three properties we were interested in are listed as follows:

- $p =? [G \text{ hsf} + 2\text{hsf}_2 + 3\text{hsf}_3 + 3\text{hsf}_3 : \text{hse} + \text{hsp} : \text{hsf} = \text{hsf}_{const}]$,
- $p =? [G \text{ hse} + \text{hsf}_3 : \text{hse} = \text{hse}_{const}]$,
- $p =? [G \text{ prot} + \text{mfp} + \text{hsp} : \text{mfp} = \text{prot}_{const}]$,

where $\text{hsf}_{const}, \text{hse}_{const}$ and prot_{const} represent the total amounts of hsf, hse and prot respectively. These properties check if the mass-conservation relations, corresponding to the level of hsf, hse and prot, are valid in all the states. In each case, the value of p was confirmed to be one, which was to be expected, with confidence level 95%, i.e. the mass conservation laws are respected in the model.

For the second property, we verified in PRISM that for time points larger than 14400, the value of $\text{hsf}_3 :$ hse reactants returns to their initial value. We formulated the following property: $p =? \quad [F >= 14400 \quad \text{hsf}_3 : \text{hse} = 3]$. The probability value calculated by PRISM was one for this property as well, with confidence level 95%.

We also checked if the model confirms the experimental data of [13] on DNA binding. One approach could be to run the simulation for many times and plot the average run. Due to the memory issues of the PRISM, we were not able to follow this approach. Since we are using a stochastic model, our second approach was to check the probability of having a data point within the interval $[0.9 \cdot d, 1.1 \cdot d]$ in the time period $[0.9 \cdot t, 1.1 \cdot t]$, where d is the experimental data point at time t. The confidence interval for all the properties and the number of simulations were 95% and 150 respectively. We interpret the high values we obtained as a result as a confirmation that the two PRISM models are in accordance with the experimental data of [13].

7   Discussion

We focused in this paper on analyzing the capability of three different frameworks to implement the concept of quantitative model refinement: rule based modelling (with Bionetgen), Petri nets (with Snoopy) and guarded command languages (with PRISM). Handling the combinatorial explosion due to accounting for a post-translational modification throughout our refinement proved to be fundamentally different in the approaches we considered. These modeling methods are not restricted to the analysis of our case study solely, but their applicability extends to other reaction-based models. Rule-based modelling tackles the complexity of refinement through a compact model representation based on a partial presentation of the details of the model species, leading to more effective model construction and analysis techniques. Colored Petri nets integrate programmability by including data types (color sets) as an intrinsic property of places. The color set assignation reflects on the structure of the network, affecting the dimensions of the corresponding flattened network. PRISM model checker promotes a low level implementation of data structures and it does not allow the modeler to introduce more complex data structures.

Our study shows that some modeling frameworks are more suitable for model refinement than others, with respect to the compactness of the representation of the refined model. A key ingredient for this is the spectrum of internal data structures supported by the modeling framework. Data structures may encapsulate a large amount of information, and their effective manipulation can substantially reduce the complexity of a model's representation. RuleBender provides data structures suitable for modeling biological systems: species, sites, links, partial description of species, rendering a straightforward refinement procedure with a very compact representation. In contrast, Petri nets are not primarily a biology-focused framework. Colored Petri nets introduce programmability in this modeling formalism, incorporating data types into the places of the network. New data types can be implemented based on primitive built-in types and composition rules. In refining a Petri net model, one has to define the appropriate data structures, and associate a biological meaning to each of them. The modeling choices affect both the compactness of the representation and the complexity of the corresponding flattened Petri net model. PRISM on the other hand only supports primitive data types. This translates into an explicit detailing of all elements of the refined model.

Our study shows that quantitative model refinement is a potentially viable approach to building a large biomodel. The approach can be used together with a multitude of modeling paradigms, allowing the modeler to increase the level of details of the model, while preserving its numerical behavior. Moreover, on any level of detail one can switch from a modeling paradigm to another, taking full advantage of the various analysis tools made possible in different model formulations, in terms of fast simulations, model checking or compact model representation. While our case-study shows the potential of the quantitative model refinement approach to model building, its scalability remains to be tested on a larger case study.

Acknowledgement. This research was partially supported by Academy of Finland under project 267915.

The authors thank Monika Heiner for her help with issues related to Snoopy and Charlie, James Faeder and Leonard Harris for advice on the Bionetgen implementation of the heat shock response, and Adam Smith for technical support regarding RuleBender.

References

1. Back, R., Ishdorj, T., Petre, I.: A petri net formalization of heat shock response model. In: Petre, I., Rozenberg, G. (eds.) Workshop on Natural Computing and Graph Transformations, pp. 19–28 (2008)
2. Blinov, M.L., Yang, J., Faeder, J.R., Hlavacek, W.S.: Graph theory for of biochemical networks. In: Priami, C., Ingólfsdóttir, A., Mishra, B., Riis Nielson, H. (eds.) Transactions on Computational Systems Biology VII. LNB, vol. 4230, pp. 89–106. Springer, Heidelberg (2006)
3. Czeizler, E., Rogojin, V., Petre, I.: The phosphorylation of the heat shock factor as a modulator for the heat shock response. In: Proceedings of the 9th International Conference on Computational Methods in Systems Biology, pp. 9–23. ACM (2011)
4. Danos, V., Feret, J., Fontana, W., Harmer, R., Krivine, J.: Rule-based modelling and model perturbation. In: Priami, C., Back, R.-J., Petre, I. (eds.) Transactions on Computational Systems Biology XI. LNCS, vol. 5750, pp. 116–137. Springer, Heidelberg (2009)
5. Faeder, J., Blinov, M., Goldstein, B., Hlavacek, W.: Rule-based modeling of biochemical networks. Complexity 10(4), 22–41 (2005)
6. Gratie, D., Iancu, B., Azimi, S., Petre, I.: Quantitative model refinement in four different frameworks, with applications to the heat shock response. Tech. Rep. 1067, TUCS (2013)

7. Harmer, R.: Rule-based modelling and tunable resolution. EPTCS 9, 65–72 (2009)
8. Heath, J.K., Kwiatkowska, M., Norman, G., Parker, D., Tymchyshyn, O.: Probabilistic model checking of complex biological pathways. In: Priami, C. (ed.) CMSB 2006. LNB, vol. 4210, pp. 32–47. Springer, Heidelberg (2006)
9. Hinton, A., Kwiatkowska, M., Norman, G., Parker, D.: Prism: A tool for automatic verification of probabilistic systems. In: Hermanns, H., Palsberg, J. (eds.) TACAS 2006. LNCS, vol. 3920, pp. 441–444. Springer, Heidelberg (2006)
10. Iancu, B., Czeizler, E., Czeizler, E., Petre, I.: Quantitative refinement of reaction models. IJUC 8(5-6), 529–550 (2012)
11. Iancu, B., Gratie, D., Azimi, S., Petre, I.: Computational modeling of the eukaryotic heat shock response: The bionetgen implementation, the petri net implementation and the prism implementation (2013), http://combio.abo.fi/research/computational-modeling-of-the-eukaryotic-heat-shock-response/
12. Kitano, H.: Systems biology: a brief overview. Science 295(5560), 1662–1664 (2002)
13. Kline, M., Morimoto, R.: Repression of the heat shock factor 1 transcriptional activation domain is modulated by constitutive phosphorylation. Molecular and cellular biology 17(4), 2107–2115 (1997)
14. Klipp, E., Herwig, R., Kowald, A., Wierling, C., Lehrach, H.: Systems biology in practice: Concepts, implementation and application. Wiley-Vch (2005)
15. Kwiatkowska, M., Norman, G., Parker, D.: Quantitative analysis with the probabilistic model checker prism. Electronic Notes in Theoretical Computer Science 153(2), 5–31 (2006)
16. Kwiatkowska, M., Norman, G., Parker, D.: PRISM 4.0: Verification of probabilistic real-time systems. In: Gopalakrishnan, G., Qadeer, S. (eds.) CAV 2011. LNCS, vol. 6806, pp. 585–591. Springer, Heidelberg (2011)
17. Liu, F., Heiner, M., Yang, M.: An efficient method for unfolding colored petri nets. In: Laroque, C., Himmelspach, J., Pasupathy, R., Rose, O., Uhrmacher, A. (eds.) Proceedings of the Winter Simulation Conference, vol. 295. Winter Simulation Conference (2012)
18. Mizera, A., Czeizler, E., Petre, I.: Self-assembly models of variable resolution. In: Priami, C., Petre, I., de Vink, E. (eds.) Transactions on Computational Systems Biology XIV. LNCS, vol. 7625, pp. 181–203. Springer, Heidelberg (2012)
19. Murphy, E., Danos, V., Feret, J., Krivine, J., Harmer, R.: Rule Based Modelling and Model Refinement. In: Elements of Computational Systems Biology, pp. 83–114. Wiley Book Series on Bioinformatics, John Wiley & Sons, Inc. (2010)
20. Petre, I., Mizera, A., Hyder, C., Meinander, A., Mikhailov, A., Morimoto, R., Sistonen, L., Eriksson, J., Back, R.: A simple mass-action model for the eukaryotic heat shock response and its mathematical validation. Natural Computing 10(1), 595–612 (2011)
21. Raman, K., Chandra, N.: Systems biology. Resonance 15(2), 131–153 (2010)
22. Rieger, T., Morimoto, R., Hatzimanikatis, V.: Mathematical modeling of the eukaryotic heat-shock response: Dynamics of the hsp70 promoter. Biophysical Journal 88(3), 1646–1658 (2005)
23. Shi, Y., Mosser, D., Morimoto, R.: Molecular chaperones as hsf1-specific transcriptional repressors. Genes & Development 12(5), 654–666 (1998)
24. Westerheide, S., Anckar, J., Stevens Jr., S., Sistonen, L., Morimoto, R.: Stress-inducible regulation of heat shock factor 1 by the deacetylase sirt1. Science Signalling 323(5917), 1063–1066 (2009)

HapMonster: A Statistically Unified Approach for Variant Calling and Haplotyping Based on Phase-Informative Reads

Kaname Kojima*, Naoki Nariai, Takahiro Mimori, Yumi Yamaguchi-Kabata,
Yukuto Sato, Yosuke Kawai, and Masao Nagasaki*

Department of Integrative Genomics,
Tohoku Medical Megabank Organization, Tohoku University
2-1 Seiryo-machi, Aoba-ku, Sendai-shi, Miyagi 980-8573, Japan
{kojima,nariai,mimori,yamaguchi,yuksato,kawai,nagasaki}
@megabank.tohoku.ac.jp,
http://nagasakilab.csml.org/en/

Abstract. Haplotype phasing is essential for identifying disease-causing variants with phase-dependent interactions as well as for the coalescent-based inference of demographic history. One of approaches for estimating haplotypes is to use phase-informative reads, which span multiple heterozygous variant positions. Although the quality of estimated variants is crucial in haplotype phasing, accurate variant calling is still challenging due to errors on sequencing and read mapping. Since some of such errors can be corrected by considering haplotype phasing, simultaneous estimation of variants and haplotypes is important. Thus, we propose a statistically unified approach for variant calling and haplotype phasing named HapMonster, where haplotype phasing information is used for improving the accuracy of variant calling and the improved variant calls are used for more accurate haplotype phasing. From the comparison with other existing methods on simulation and real sequencing data, we confirm the effectiveness of HapMonster in both variant calling and haplotype phasing.

Keywords: Next generation sequencing, variant call, haplotype phasing.

1 Introduction

Next generation sequencing (NGS) technologies enables the detection of novel rare variants in genome wide scale. For rare variant association studies, variants are often grouped in exon or gene level, and effects of multiple variants and phase-dependent interactions such as compound heterozygosity and cis-effect are considered for the analysis [2]. Thus, haplotype phasing is essential for identifying association between rare variants and disease phenotypes. In addition, haplotype phasing information is required for the coalescent-based inference of demographic history [7].

---

* Corresponding authors.

A.-H. Dediu, C. Martín-Vide, and B. Truthe (Eds.): AlCoB 2014, LNBI 8542, pp. 107–118, 2014.

Haplotype phasing is usually estimated by two types of approaches. One approach is based on linkage disequilibrium between variant sites and estimates haplotype phasing probabilistically [3,4,11]. Although this type of approach provides accurate phasing results for common variants, it requires genotyping results for multiple samples and its accuracy for low-frequency variants or variants around recombination hot spots tends to be low. Another approach is to use phase-informative reads that span multiple heterozygous variant positions. Due to the current length of NGS reads, the range size of the estimated haplotypes is limited. However, since length of sequence reads is growing rapidly, the rate of heterozygous sites that can be phased by phase-informative reads is increasing, and hence this type of approach is considered as a promising ways for phasing low-frequency variants.

Although the quality of estimated variants is essential for haplotype phasing, accurate variant calling is still challenging especially for regions with insufficient read coverage or regions containing errors on sequencing and read mapping. Since variant calling on such regions can be improved by using haplotype phasing information [12], considering both variant calling and haplotype phasing simultaneously is important. However, SNP sites are treated independently on most of the variant callers such as Unified Genotyper in GATK and BCFtools [5,10], and haplotype phasing is considered separately from the variant calling. Thus, we propose a new variant calling and haplotyping method based on phase-informative reads named HapMonster. HapMonster simultaneously performs variant calling and haplotype phasing based on phase-informative reads by unifying these procedures in a statistical model. Under the model, haplotype phasing information can be used for improving the accuracy of variant calling and the improved variant calls are used for more accurate haplotype phasing. In the performance evaluation, we applied HapMonster and other existing methods to simulation and real sequencing datasets with various read coverages, and confirmed that HapMonster provides the best results in both variant calling and haplotype phasing, compared to other methods.

2   Methods

HapMonster takes mapped sequence reads to a reference genome in the SAM/BAM format as input data and estimates variants and phased haplotypes. The model of our approach is comprised of allele likelihood model and haplotype selection model. Allele likelihood model calculates the likelihood of an allele given mapped reads at a position, where errors on sequencing and read mapping are considered. Haplotype selection model represents the assignment of sequence reads to one of two haplotypes. In the following sections, we describe the details of our model and procedures for parameter estimation and inference of genotypes and haplotypes.

2.1   Modeling

Allele Likelihood Model. Let R_i be the ith read in a SAM/BAM file. R_i contains its mapping quality score in Phred scale $MAPQ_i$, strings for bases aligned to position k in the reference genome r_i^k, and vectors of base quality scores in Phred scale for the aligned bases bq_i^k. Note that r_i^k is usually one nucleotide such as 'T', but it can be a string with more than one nucleotide for representing insertion, e.g., a string 'TGC' represents an insertion 'GC' right after a base 'T'. Deletion is represented by setting r_i^k to a null string. By using these notations, an allele likelihood for A at position k is given by

$$P(r_i^k|A, bq_i^k) = \sum_{b_i^k=0}^{1} \sum_{m_i^k=0}^{1} P(r_i^k|A, b_i^k)^{I(m_i^k=1)} P_{mis}(r_i^k)^{I(m_i^k=0)} P(m_i^k) P(b_i^k|bq_i^k),$$

where m_i^k is a binary variable that takes one if the alignment of r_i^k is correct and zero otherwise. b_i^k is a vector of binary variables and each element that takes one if the corresponding base in r_i^k is correct and zero otherwise. $I(\cdot)$ is an indicator function that returns one if a condition in its argument is true, and zero otherwise. As with r_i^k, A is one nucleotide or a string with nucleotides. The term $P(r_i^k|A, b_i^k)$ is the probability of read generation for the correct read alignment, and we represent the probability as:

$$P(r_i^k|A, b_i^k) = \text{Indel}(A, r_i^k) \prod_{l=1}^{\min(|A|,|r_i^k|)} P(r_i^k[l]|A[l], b_i^k[l]),$$

where $A[l]$ is the lth nucleotide of A, $r_i^k[l]$ is the lth nucleotide of r_i^k, $b_i^k[l]$ is the lth value of b_i^k, and function Indel represents read skipping errors and insertion errors. $|\cdot|$ takes a string or set as its argument and returns length for string or size for set. We represent function Indel by using read skip error rate δ and insertion error rate ι as:

$$\text{Indel}(A, r_i^k) = \delta^{I(|A|>|r_i^k|)}(1-\delta)^{I(|A|\leq|r_i^k|)}\iota^{I(|A|<|r_i^k|)}(1-\iota)^{I(|A|\geq|r_i^k|)}.$$

In our study, both δ and ι are set to 0.001. $P(r_i^k[l]|A[l], b_i^k[l])$ models base substitution error on each base and is given by:

$$P(r_i^k[l]|A[l], b_i^k[l]) = \begin{cases} 1 & r_i^k[l] = A[l] \ \& \ b_i^k[l] = 1 \\ 1/3 & r_i^k[l] \neq A[l] \ \& \ b_i^k[l] = 0 \\ 0 & \text{otherwise} \end{cases}.$$

$P_{mis}(r_i^k)$ represents the probability of read generation for misaligned reads. We consider that reads representing indels, i.e., read with 0 length or more than one nucleotides are generated more probably than reads with one nucleotide in the misalignment, and design $P_{mis}(r_i^k)$ as:

$$P_{mis}(r_i^k) = \begin{cases} 1/N_{mis} & |r_i^k| = 1 \\ p_{mis}/N_{mis} & \text{otherwise} \end{cases}, \quad p_{mis} \geq 1,$$

where N_{mis} is the normalization factor given by $\sum_{A \in \mathcal{A}_k} 1^{I(|A|=1)} p_{mis}^{I(|A|\neq1)}$. Here, $\mathcal{A}_k$ is a set of possible alleles at position k and is given by {'A', 'T', 'G', 'C'} and null string, which is for deletion. In addition, if there exist reads with nucleotides more than one aligned at position k, the corresponding sequences are added $\mathcal{A}_k$. Here, we set p_{mis} to 1.0, i.e., we assume that the read is generated from possible alleles uniformly. $P(b_i^k|bq_i^k)$ is factorized as $\prod_l P(b_i^k[l]|bq_i^k[l])$, and each term is given by a binomial distribution with parameter $1 - 10^{-bq_i^k[l]/10}$:

$$P(b_i^k[l]|bq_i^k[l]) = \left[1 - 10^{-bq_i^k[l]/10}\right]^{I(b_i^k[l]=1)} \left[10^{-bq_i^k[l]/10}\right]^{I(b_i^k[l]=0)} .$$

$P(m_i^k)$ is given by a binomial distribution with parameter $p_{m_i^k}$. We also give a Beta distribution with parameter $\alpha_m(1-10^{-MAPQ_i/10})$ and $\alpha_m 10^{-MAPQ_i/10}$ as a prior distribution of $p_{m_i^k}$. Thus, $p_{m_i^k}$ is updated by considering both probability for alignment reliability of read r_i^k from the model and mapping quality score $MAPQ_i$. We set prior strength α_m to five.

Haplotype Selection Model. Given a genotype (A_k^1, A_k^2) at position k, haplotype selection part selects an allele, from which each read is generated, by using a binary variable h_i^k in the following manner:

$$\prod_{i\in\mathcal{I}_k} P(r_i^k|A_k^1, bq_i^k)^{I(h_i^k=1)} P(r_i^k|A_k^2, bq_i^k)^{I(h_i^k=2)} P(h_i^k),$$

where $\mathcal{I}_k$ is a set of indexes of reads that span position k. In position-independent variant callers, equally probable condition for haplotype selection is considered, i.e., $P(h_i^k = 1)$ and $P(h_i^k = 2)$ are 0.5. Here, instead of $P(h_i)$, we consider a conditional probability $P(h_i^k|z_k)$ given by:

$$P(h_i^k|z_k) = \begin{cases} p_{h_i} & h_i^k = 1 \ \& \ z_k = 1 \\ 1 - p_{h_i} & h_i^k = 2 \ \& \ z_k = 1 \ , \\ 0.5 & z_k = 0 \end{cases} \tag{1}$$

where p_{h_i} is a rate for the assignment of read R_i to a haplotype and z_k is a binary variable that represents the zygosity at position k and takes value one for heterozygote and zero for homozygote. For paired-end read data, p_{h_i} is shared by each read pair. To represent zygosity with z_k, we introduce a conditional probability $P(z_k|A_k^1, A_k^2)$ that is given by:

$$P(z_k|A_k^1, A_k^2) = \begin{cases} 1.0 & \begin{array}{c} A_k^1 = A_k^2 \ \& \ z_k = 0 \\ \text{or} \\ A_k^1 \neq A_k^2 \ \& \ z_k = 1 \end{array} \ . \\ 0 & \text{otherwise} \end{cases}$$

The role of z_k is to filter out highly probably homozygous positions from data for read assignment via Equation (1), and hence only highly probably heterozygous positions are used for haplotyping. p_{h_i} represents one of two chromosomes from which read R_i comes. Since each read comes from one of two homologous chromosomes equally probably in an ideal condition, we represent this property by setting $P(\sum_{i \in \mathcal{I}_k} p_{h_i})$ to $\mathcal{N}(\sum_{i \in \mathcal{I}_k} p_{h_i}; |\mathcal{I}_k|/2, \bar{L}|\mathcal{I}_k|/4)$, where $\mathcal{N}$ represents normal distribution and $\bar{L}$ is the average read length. $\sum_{i \in \mathcal{I}_k} p_{h_i}$ is considered as the number of reads from the same chromosome at position k. Since $\sum_{i \in \mathcal{I}_k} p_{h_i}$ is a continuous value, normal approximation of a binomial distribution with parameter 0.5 is employed. Although the mean and variance of the normal distribution approximating a binomial distribution with parameter 0.5 are respectively $|\mathcal{I}_k|/2$ and $|\mathcal{I}_k|/4$, we set the variance to $\bar{L}|\mathcal{I}_k|/4$ to normalize the effect with the average read length.

Potential Functions between Positions. If two heterozygous positions are spanned by several reads, their estimated alleles and haplotype should be concordant with all the spanning reads. However, due to the errors on sequencing and mapping, the contradiction on haplotyping sometimes occurs between spanning reads, which causes the difficulty in estimating haplotypes. We consider that pair of alleles in a haplotype supported by more spanning reads is more reliable and introduce the following potential function between alleles in heterozygous positions: $\psi_{k,k'}(A_k, A_{k'}) = n_{A_k, A_{k'}} + \alpha_h$, where $n_{A_k, A_{k'}}$ is the number of reads that span k and k' and contain alleles A_k and $A_{k'}$ at the corresponding positions. α_h is a hyperparamter and set to 1.5 in our study. Potential functions are considered only for pairs of positions that are heterozygous with high probability. The criterion for adding potential functions is described in the next section.

2.2   Parameter Estimation

We use the EM algorithm for model parameters $p_{m_i^k}$ and p_{h_i}. Since the proposed model contains loop structures, it needs high time complexity to calculate the exact marginal probabilities required in E-step of the EM algorithm. Thus, we instead calculate approximated marginal probabilities with the loopy belief propagation [13]. In M-step, parameters $p_{m_i^k}$ and p_{h_i} are updated by using the marginal probabilities calculated in E-step. Before starting the EM algorithm, we calculate marginal probabilities of z_k to select positions with $P(z_k = 1) > 0.98$ as positions for adding potential functions. For each selected position, potential functions are added with at most five other neighboring selected positions in both upstream and downstream directions.

2.3   Variant Calling and Haplotype Inference

For genotype inference at position k, a configuration of latent variables A_k^1 and A_k^2 that maximizes their marginal probability is searched. We use the loopy belief propagation to calculate the approximated marginal probability of A_k^1 and A_k^2 and then obtain the configuration maximizing the marginal probability. To reduce false positive variants, we calculate a lod score given by the log ratio of the marginal probability of estimated genotype and that of homozygous genotype for reference allele, and filter out the variants with lod scores less than 10.

From variant calling results, we construct a graph structure comprised of sequence termed a sequence graph. In sequence graph, all the sequence reads are represented as vertices, and if two sequence reads are spanning the same heterozygous variant position, they are connected with an edge. Each mate pair is also connected with an edge. Connected components of the read connection graph are detected with breadth-first search, and then heterozygous variants spanned by reads in the same components are considered as the same phased region. To save the memory space, we divide genome sequences into ranges with predetermined length and call variants separately on each range although overlapping is considered for neighboring ranges. The range size and overlapping region size are set to 2,500,000 bp and 1,000 bp, respectively. Since sequence graphs from neighboring ranges can be connected by merging vertices for sequence reads spanning overlapping regions in the ranges, phased regions can span several ranges.

3   Results

3.1   Simulation Analysis

We synthetically generated diploid genome sequences of chromosome 21 for a CEU individual NA12286 according to the variant calling and phasing results released in November 23, 2010 by the 1000 Genomes Project [14]. The number of variants is 50,090. From the diploid genome sequences, we generated paired-end sequence reads and put 0.1% base substitution errors. Base quality scores for bases were set to Q30, which corresponds to 0.1% error. We generated three types of data sets with the following conditions:

- Read length is 100 bp, and insert size is normally distributed with mean 500 bp and standard deviation 50 bp.
- Read length is 300 bp, and insert size is normally distributed with mean 1,500 bp and standard deviation 50 bp.
- Read length is 500 bp, and insert size is normally distributed with mean 2,500 bp and standard deviation 50 bp.

A BAM file for the dataset was obtained by mapping the sequence reads to the reference genome (GRCh37) for chromosome 21 with BWA-MEM [8]. In order to evaluate the performance of HapMonster with various read coverages, we downsampled the BAM file to 5×, 10×, 20×, and 40× with Picard DownsampleSam (http://picard.sourceforge.net/). Downsampled BAM files were realigned with GATK Indel Realigner.

Performance on Variant Calling. For the comparison with existing methods, we applied Unified Genotyper implemented in GATK, BCFtools with SAMtools mpileup [10], and Linkage Method [12] to the datasets with their default options. Like HapMonster, Linkage Method simultaneously estimates variants and haplotype phasing. Table 1 summarizes the performance of genotype concordance on HapMonster and other three methods. Only if an estimated genotype and true genotype at a position are the same and not homozygous for the reference allele, the estimated genotype is counted as a true positive. In addition, only if the estimated genotype is not homozygous for the reference allele and is different from the true genotype, it is counted as a false positive. Recall and precision are respectively given by $\frac{TP}{\text{the number of trues}}$ and $\frac{TP}{TP+FP}$, where TP is the number of true positives and FP is the number of false positives. F-score is given by the harmonic mean of recall and precision as $\frac{2\times\text{recall}\times\text{precision}}{\text{recall}+\text{precision}}$, and provides the overall performance of recall and precision by capturing a trade-off between them. These are valued between zero and one, and the larger value is better.

In the all conditions, HapMonster gave the best results in the both recall and F-score. For precision, HapMonster was better than other methods for datasets with read coverage of 5× and 10×. BCFtools was the best in precision for the datasets with read coverage of 20× and 40×, and HapMonster was the second best.

Table 1. Comparison on genotype concordance of HapMonster, Unified Genotyper (UG), BCFtools, and Linkage Method (LM) for simulation datasets with read length of 100 bp, 300 bp, and 500 bp. The best result in each condition is in bold.

Cov.	Method	100 bp			300 bp			500 bp		
		Recall	Precision	F-score	Recall	Precision	F-score	Recall	Precision	F-score
5×	HapMonster	**0.7469**	**0.9492**	**0.8360**	**0.7513**	**0.9496**	**0.8389**	**0.7523**	**0.9517**	**0.8403**
	UG	0.6403	0.9413	0.7622	0.6460	0.9422	0.7665	0.6434	0.9438	0.7652
	BCFtools	0.7379	0.9249	0.8209	0.7491	0.9264	0.8284	0.7522	0.9276	0.8307
	LM	0.6142	0.7034	0.6558	0.6329	0.7537	0.6881	0.5192	0.7556	0.6155
10×	HapMonster	**0.9490**	0.9912	**0.9696**	**0.9540**	0.9925	**0.9729**	**0.9533**	0.9918	**0.9722**
	UG	0.9103	0.9911	0.9490	0.9141	0.9918	0.9514	0.9124	0.9908	0.9500
	BCFtools	0.9347	**0.9912**	0.9621	0.9428	0.9924	0.9669	0.9427	0.9910	0.9662
	LM	0.8528	0.9160	0.8833	0.8959	0.9416	0.9182	0.8975	0.9453	0.9208
20×	HapMonster	**0.9911**	0.9967	**0.9939**	**0.9925**	0.9974	**0.9949**	**0.9921**	0.9973	**0.9947**
	UG	0.9905	0.9963	0.9934	0.9916	0.9964	0.9940	0.9914	0.9966	0.9940
	BCFtools	0.9850	**0.9987**	0.9918	0.9867	**0.9991**	0.9929	0.9867	**0.9992**	0.9929
	LM	0.9046	0.9473	0.9254	0.9906	0.9906	0.9861	0.9786	0.9896	0.9841
40×	HapMonster	**0.9953**	0.9963	**0.9958**	**0.9958**	0.9971	**0.9965**	**0.9959**	0.9971	**0.9965**
	UG	0.9947	0.9963	0.9955	**0.9958**	0.9967	0.9963	0.9958	0.9967	0.9963
	BCFtools	0.9892	**0.9988**	0.9940	0.9900	**0.9993**	0.9947	0.9900	**0.9994**	0.9947
	LM	0.8729	0.9259	0.8986	0.9939	0.9960	0.9949	0.9939	0.9962	0.9951

Performance on Haplotype Phasing. We evaluate the performance on haplotype phasing based on concordance between estimated haplotypes and haplotyping of true heterozygous variants. Let v_i and s_i be the ith true heterozygous variant and the position for v_i, respectively. We denote an estimated genotype at position s_i by g_{s_i} and also denote a genotype whose allele ordering is reversed from g_{s_i} by $\bar{g}_{s_i}$. B_i is the ID of the estimated phased region spanning position s_i, and D_i is a tertiary variable that takes 1 if $g_{s_i} = v_i$ holds, -1 if $\bar{g}_{s_i} = v_i$ holds, and 0 otherwise. Since genotyping results to be phased are different between methods, we define the following switching cost by using the above notations:

$$\sum_{i=1}^{|\mathcal{V}|}[I(D_i = 0) + I(D_i \neq 0)I(B_i \neq B_{p_i}) + 2I(D_i \neq 0)I(D_i \neq D_{p_i})I(B_i = B_{p_i})],$$

where $\mathcal{V}$ is a set of true heterozygous variants and p_i is the index of previously correctly estimated variant, e.g., if genotype of v_{i-3} is correctly estimated and those of v_{i-2} and v_{i-1} are not, p_i is $i - 3$. The interpretation of the switching cost is as follows. If the estimated genotype is different from the true variant, the cost is increased by one with the first term. If the estimated genotype is correct but the estimated phased region is not spanned between positions s_i and s_{p_i}, the cost is increased by one with the second term as s_i and s_{p_i} are not phased. Finally, if the estimated haplotype is not correct between s_i and s_{p_i}, the cost is increased by two with the third term. Since a strategy for extending phased region even for uncertain case gets reward by chance for the penalty less than two in the third term, the penalty for the third term is set to two. No penalization is applied only if both the estimated genotype and estimated phase relationship between positions s_i and s_{p_i} are correct.

In addition to Linkage Method, we used Read Backed Phasing (RBP) approach implemented in GATK and HapCompass [1] as existing haplotype phasing methods based on phase-informative reads. Variant calls from Unified Genotyper were used as input variant data of RBP and HapCompass. Switching costs of HapMonster and other methods are summarized in Fig. 1. The haplotyping results of HapMonster are smaller switching costs than those of other methods for all the conditions. In the results of all the methods, switching costs are smaller for the results from the dataset of the higher read coverage and longer read length data.

3.2 Real Data Analysis

For real human sequencing data, we used 100 bp paired-end sequencing data of NA12878, one of samples analyzed in the 1000 Genomes Project. The data was sequenced on Illumina HiSeq 2000 with read coverage of $45\times$ and average insert size of 300 bp. Sequence reads were mapped to the reference genome (GRCh37) with BWA [9], and the mapped dataset was downsampled to $5\times$, $10\times$, $20\times$, and $40\times$ with Picard DownsampleSam. These downsampled datasets were realigned with GATK Indel Realigner, and their base quality scores were recalibrated with GATK Base Quality Score Recalibration.

Performance on Variant Calling. We evaluate the performance by assessing the concordance of estimated variants from these variant callers for datasets with various read coverages with SNP array genotyping results from Illumina OMNI 2.5 BeadChip. The number of SNP sites designed in the array for chr21 is 32,076, and 10,579 variants are contained in the sites. We also assessed the concordance with the variant calling result in the 1000 Genomes Project, which contain 52,454 genotyping results including 39,691 variants.

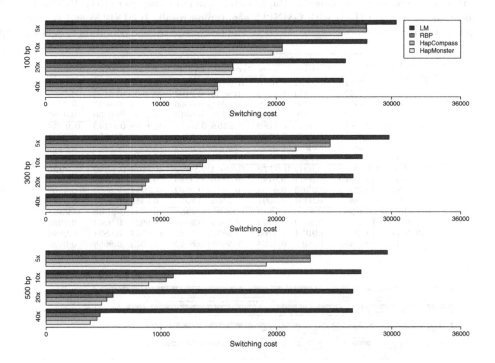

Fig. 1. Switching costs of HapMonster, HapCompass, Read Backed Phasing (RBP), and Linkage Method (LM) for simulation datasets with read length of 100 bp, 300 bp, and 500 bp. The smaller switching cost is better.

Table 2 summarizes the performance of genotype concordance on HapMonster and three existing methods for NA12878 from the real datasets with read coverages of 5×, 10×, 20×, and 40×. Accuracy is given by $\dfrac{TP+TN}{TP+FP+TN+FN}$, where TN is the number of true negatives and FN is the number of false negatives. HapMonster showed the best results in recall, F-score, and accuracy for all the read coverages. In precision, HapMonster provided the best results for the datasets with 5× on the agreement with the SNP array genotyping results and for the datasets with 5× and 10× on the agreement with the genotyping results in the 1000 Genomes Project. For other cases, the results of BCFtools were the best, and those of HapMonster were the second best.

Performance on Haplotype Phasing. We prepared phased variant calling results as follows: i) Heterozygous sites for NA12878 were phased with SHAPEIT2 using the phased haplotypes for the Phase1 integrated variant calls released in September, 17, 2013 as the reference panel. ii) Heterozygous sites not in the reference panel were phased by using variant calls of NA12878's parents, NA12891 and NA12892. Switching costs of HapMonster, HapCompass, Read Backed Phasing (RBP), and Linkage Method (LM) are summarized in Fig. 2. Variant calls

Table 2. Genotype concordance of HapMonster, Unified Genotyper (UG), BCFtools, and Linkage Method (LM) with OMNI2.5 genotyping results (OMNI2.5) and variant calls in the 1000 Genomes Project (1KGP). The best result in each condition is in bold.

Cov.	Method	OMNI2.5				1KGP			
		Recall	Precision	F-score	Accuracy	Recall	Precision	F-score	Accuracy
5×	HapMonster	**0.7028**	**0.9492**	**0.8076**	**0.9013**	**0.7337**	**0.9550**	**0.8298**	**0.7984**
	UG	0.6773	0.9444	0.7889	0.8930	0.7102	0.9500	0.8128	0.7806
	BCFtools	0.6717	0.9439	0.7849	0.8914	0.6960	0.9471	0.8023	0.7699
	LM	0.5636	0.7446	0.6416	0.8535	0.6014	0.7690	0.6750	0.6981
10×	HapMonster	**0.9112**	0.9891	**0.9486**	**0.9698**	**0.9412**	0.9928	**0.9663**	**0.9554**
	UG	0.8854	0.9875	0.9337	0.9612	0.9184	0.9922	0.9539	0.9381
	BCFtools	0.8754	**0.9900**	0.9292	0.9583	0.8972	0.9924	0.9424	0.9222
	LM	0.7944	0.8964	0.8423	0.9285	0.8229	0.9078	0.8633	0.8658
20×	HapMonster	**0.9618**	0.9942	**0.9778**	**0.9864**	**0.9902**	0.9978	**0.9940**	**0.9925**
	UG	0.9579	0.9917	0.9745	0.9846	0.9885	0.9977	0.9931	0.9912
	BCFtools	0.9466	**0.9963**	0.9708	0.9818	0.9692	**0.9981**	0.9834	0.9767
	LM	0.8810	0.9391	0.9091	0.9560	0.9480	0.9222	0.9225	
40×	HapMonster	**0.9663**	0.9956	**0.9808**	**0.9880**	**0.9951**	0.9981	**0.9966**	**0.9961**
	UG	0.9644	0.9904	0.9772	0.9859	0.9948	0.9981	0.9964	0.9959
	BCFtools	0.9569	**0.9971**	0.9766	0.9852	0.9796	**0.9987**	0.9891	0.9846
	LM	0.8553	0.9199	0.8864	0.9461	0.8764	0.9373	0.9058	0.9062

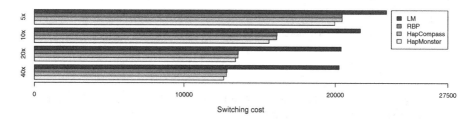

Fig. 2. Switching costs of HapMonster, HapCompass, Read Backed Phasing (RBP), and Linkage Method (LM) for real sequencing datasets. The smaller switching cost is better.

from Unified Genotyper were used for input variant data of HapCompass and RBP. For all the dataset, switching costs of HapMonster are smaller than those of other methods. In the results of all the methods, switching costs are smaller for the results from the dataset of the higher read coverage.

3.3   Required Computational Resource

The computational time and memory usage of HapMonster, Unified Genotyper (UG), BCFtools, Read Backed Phasing (RBP), HapCompass, and Linkage Method (LM) for the simulation datasets with read length of 500 bp and real datasets are summarized in Table 3. HapMonster is implemented in Java. All the computation was performed on Intel Xeon CPU E5-2670 processors with a single thread. The computational time of HapMonster is similar to that of BCFtools, and slightly more than that of Unified Genotyper. Although Read Backed Phasing and HapCompass are very fast on the simulation datasets, they are slower than HapMonster on the real datasets. HapMonster requires more memory space

Table 3. Running time and memory usage for HapMonster, Unified Genotyper (UG), BCFtools, Read Backed Phasing (RBP), HapCompass, and Linkage Method (LM) for (a) simulation datasets with read length of 500 bp and (b) real sequencing datasets.

(a) Simulation datasets					(b) Real datasets			
Cov.	Method	Running Time	Memory Usage		Cov.	Method	Running Time	Memory Usage
5×	HapMonster	2.6 [min]	2.1 [GB]		5×	HapMonster	5.6 [min]	1.5 [GB]
	UG	4.0 [min]	0.9 [GB]			UG	4.6 [min]	0.9 [GB]
	BCFtools	3.5 [min]	0.1 [GB]			BCFtools	4.0 [min]	0.1 [GB]
	RBP	0.5 [min]	1.0 [GB]			RBP	18.7 [min]	1.0 [GB]
	HapCompass	0.1 [min]	1.4 [GB]			HapCompass	27.9 [min]	1.8 [GB]
	LM	29.8 [min]	1.6 [GB]			LM	21.6 [min]	5.6 [GB]
10×	HapMonster	3.1 [min]	2.3 [GB]		10×	HapMonster	8.4 [min]	2.1 [GB]
	UG	5.0 [min]	0.9 [GB]			UG	5.9 [min]	0.9 [GB]
	BCFtools	5.9 [min]	0.1 [GB]			BCFtools	7.4 [min]	0.1 [GB]
	RBP	1.0 [min]	1.0 [GB]			RBP	40.7 [min]	1.0 [GB]
	HapCompass	0.3 [min]	1.7 [GB]			HapCompass	281.3 [min]	1.8 [GB]
	LM	56.8 [min]	5.6 [GB]			LM	108.3 [min]	6.2 [GB]
20×	HapMonster	4.8 [min]	2.8 [GB]		20×	HapMonster	13.2 [min]	4.0 [GB]
	UG	6.8 [min]	0.9 [GB]			UG	8.1 [min]	0.9 [GB]
	BCFtools	10.8 [min]	0.1 [GB]			BCFtools	14.4 [min]	0.1 [GB]
	RBP	2.3 [min]	1.0 [GB]			RBP	72.4 [min]	1.0 [GB]
	HapCompass	0.6 [min]	1.8 [GB]			HapCompass	419.5 [min]	2.6 [GB]
	LM	129.7 [min]	9.6 [GB]			LM	154.9 [min]	20.8 [GB]
40×	HapMonster	13.2 [min]	4.5 [GB]		40×	HapMonster	26.0 [min]	5.8 [GB]
	UG	10.0 [min]	0.9 [GB]			UG	12.8 [min]	0.9 [GB]
	BCFtools	20.5 [min]	0.1 [GB]			BCFtools	30.4 [min]	0.1 [GB]
	RBP	3.9 [min]	1.0 [GB]			RBP	130.4 [min]	1.0 [GB]
	HapCompass	1.0 [min]	2.0 [GB]			HapCompass	529.1 [min]	3.1 [GB]
	LM	396.6 [min]	27.9 [GB]			LM	633.6 [min]	85.5 [GB]

than other methods other than Linkage Method. However, the required memory space is less than 6GB, and hence it can work currently available laptop PCs.

4   Conclusions

HapMonster is a statistically unified model for variant calling and haplotyping by using phase-informative reads. By considering variant calling and haplotyping simultaneously, phased information improved the accuracy of variant calling, and the accurately estimated variants support the reliable haplotyping synergistically. The model structure of HapMonster is similar to that of our previous work, a variant caller considering pedigree information and phase-informative reads [6]. However, due to the consideration of potential functions between heterozygous positions, the performance of HapMonster on haplotype phasing becomes much higher than that of the previous work.

By using the simulation sequencing datasets with various read length and real human NGS datasets, we confirmed the effectiveness of our method from the comparison with Unified Genotyper, BCFtools, and Linkage Method for variant calling and with Linkage Method, Read Backed Phasing, and HapCompass for haplotype phasing. We also showed that HapMonster was more effective for datasets with longer reads, especially in haplotype phasing. Although HapMonster considers only phase-informative reads for haplotype phasing, use of linkage disequilibrium is a promising way for the further improvement on accurate haplotype phasing. We would like to consider such an extension in future work.

Acknowledgements. This work was supported (in part) by MEXT Tohoku Medical Megabank Project. All computational resources were provided by the ToMMo phase0 cluster computer.

References

1. Aguiar, D., Istrail, S.: Haplotype assembly in polyploid genomes and identical by descent shared tracts. Bioinformatics 29(13), i352–i360 (2013)
2. Bansal, V., Libiger, O., Torkamani, A., Schork, N.J.: Statistical analysis strategies for association studies involving rare variants. Nature Reviews Genetics 11, 773–785 (2010)
3. Browning, R., Browning, B.L.: Rapid and accurate haplotype phasing and missing data inference for whole genome association studies using localized haplotype clustering. Ametican Journal of Human Genetics 81, 1084–1097 (2007)
4. Delaneau, O., Marchini, J., Zagury, J.F.: A linear complexity phasing method for thousands of genomes. Nature Methods 9(2), 179–181 (2011)
5. DePristo, M.A., et al.: A framework for variation discovery and genotyping using next-generation DNA sequencing data. Nature Genetics 43, 491–498 (2011)
6. Kojima, K., Nariai, N., Mimori, T., Takahashi, M., Yamaguchi-Kabata, Y., Sato, Y., Nagasaki, M.: A statistical variant calling approach from pedigree information and local haplotyping with phase informative reads. Bioinformatics 29(22), 2835–2843 (2013)
7. Kuhner, M.K.: Coalescent genealogy samplers: Windows into population history. Trends in Ecology and Evolution 24(2), 86–93 (2009)
8. Li, H.: Aligning sequence reads, clone sequences and assembly contigs with BWA-MEM. arXiv:1303.3997 (2013)
9. Li, H., Durbin, R.: Fast and accurate short-read alignment with Burrows-Wheeler Transform. Bioinformatics 25(14), 1754–1760 (2009)
10. Li, H., Ruan, J., Durbin, R.: Mapping short DNA sequencing reads and calling variants using mapping quality scores. Genome Research 18(11), 1851–1858 (2008)
11. Li, Y., Willer, C.J., Ding, J., Scheet, P., Abecasis, G.R.: MaCH: Using sequence and genotype data to estimate haplotypes and unobserved genotypes. Genetic Epidemiology 34(8), 816–834 (2010)
12. Sasaki, E., Sugino, R.P., Innan, H.: The linkage method: a novel approach for SNP detection and haplotype reconstruction from a single diploid individual using next generation sequence data. Molecular Biology and Evolution (9), 2187–2196 (2013)
13. Yedidia, J.S., Freeman, W.T., Weiss, Y.: Constructing free-energy approximations and generalized belief propagation algorithms. IEEE Transactions on Information Theory 51(7), 2282–2312 (2005)
14. 1000 Genomes Project Consortium, Abecasis, G.R., Altshuler, D., Auton, A., Brooks, L.D., Durbin, R.M., Gibbs, R.A., Hurles, M.E., McVean, G.A.: A map of human genome variation from population-scale sequencing. Nature, 467(7319), 1061–1073 (2010)

Mapping-Free and Assembly-Free Discovery of Inversion Breakpoints from Raw NGS Reads

Claire Lemaitre[1], Liviu Ciortuz[1,2], and Pierre Peterlongo[1]

[1] INRIA/IRISA/GenScale, Campus de Beaulieu, 35042 Rennes cedex, France
{claire.lemaitre,pierre.peterlongo}@inria.fr
[2] Faculty of Computer Science Iasi, Romania
ciortuz@info.uaic.ro

Abstract. We propose a formal model and an algorithm for detecting inversion breakpoints without a reference genome, directly from raw NGS data. This model is characterized by a fixed size topological pattern in the *de Bruijn Graph*. We describe precisely the possible sources of false positives and false negatives and we additionally propose a sequence-based filter giving a good trade-off between precision and recall of the method. We implemented these ideas in a prototype called TAKEABREAK. Applied on simulated inversions in genomes of various complexity (from *E. coli* to a human chromosome dataset), TAKEABREAK provided promising results with a low memory footprint and a small computational time.

Keywords: structural variant, NGS, reference-free, de Bruijn graph.

1 Introduction

Structural variation is an important source of variations in genomes, that can be involved in phenotypic variations, inherited diseases, evolution and speciation. The extent of structural variations in populations has been only recently acknowledged, thanks mainly to next generation sequencing (NGS). In fact, by sequencing the genomes of several human individuals, one can find more DNA involved in structural variations than in single nucleotide polymorphism (SNP) [8]. However, due to the small size of the reads these variants are much more difficult to identify than SNPs. Most methods proposed so far rely on mapping the reads on a reference genome. The main approach calls structural variant breakpoints when mapped read pairs show discordant mappings with respect to expected insert-size and orientation of the reads [7]. Due mainly to repetitions in complex genomes and mapping errors, these methods suffer from high false positive rates and a small overlap between predictions obtained by different methods [1]. Noteworthy, copy number variations seem to have focussed most attention and efforts, whereas balanced structural variants such as inversions have been less investigated [8], suggesting that the latter are even more difficult to detect in short read data.

All these approaches rely on a reference genome and on a first mapping step. This is a strong limitation when dealing with organisms with no available reference genome or one of poor quality or too distantly related. On the other hand,

A.-H. Dediu, C. Martín-Vide, and B. Truthe (Eds.): AlCoB 2014, LNBI 8542, pp. 119–130, 2014.

one can also perform full *de novo* assembly of re-sequenced genomes and compare the resulting assemblies [6], however *de novo* assembly remains a difficult and resource intensive task and this could be reduced by targeting directly inversion variants. The problem we address is therefore: can we identify inversion signatures directly in raw NGS reads without the need of any reference genome nor full assembly of the reads? Several methods have been developed recently for calling biological events of interest directly from raw unassembled reads, by targeting specific patterns in assembly graphs. Some of them are dedicated to detect SNPs or small indels [10,9,13] or alternative slicing events in RNA-seq data [11]. Cortex_var [4] claims to detect any variant generating a *bubble* pattern, but does not target specifically inversions or other balanced structural variants.

The main contribution of this paper is an analysis and a formal modeling of topological patterns generated by inversions in the *de Bruijn Graph*. Additionally, we propose an algorithm detecting such inversion patterns. This algorithm was implemented in a tool called TAKEABREAK that was used as a proof of concept and that can be downloaded from `http://colibread.inria.fr/TakeABreak/`. Applying this tool on simulated datasets showed that i) the described model detects with high recall and precision inversion breakpoints ii) approximate repeats present in complex genomes only slightly decrease performances iii) time and memory are very limited.

2   Inversion Pattern in the de Bruijn Graph

2.1   Preliminaries

Given a sequence s on the DNA alphabet $\Sigma = \{A, C, G, T\}$, we use the concept of k-*mers* that are words of length k in s. We denote by $\overleftarrow{s}$ the reverse-complement of sequence s, for instance with $s = TTGC$, $\overleftarrow{s} = GCAA$.

de Bruijn Graph. The approach we propose is based on the use of a *de Bruijn Graph*. Given a set of sequences such as reads in the assembly framework, a *de Bruijn Graph* is a directed graph where the set of vertices corresponds to the k-mers from the sequences, and a directed edge connects a source k-mer a to a target k-mer b if the $k-1$ suffix of a is equal to the $k-1$ prefix of b. Usually, in the genome assembly framework [14], a *de Bruijn Graph* node stores explicitly a k-mer and implicitly its reverse complement. Thus there are two ways of traversing a node: either reading the explicit k-mer or reading the implicit one; we denote this notion by the *state* of the node: *explicit* or *implicit*. In this context, each edge is labeled by the states of its source and target nodes. In the following $n_{\omega}^{\rightarrow}$ denotes the node storing explicitly or implicitly ω, in the state such that reading $n_{\omega}^{\rightarrow}$ provides the k-mer ω. Respectively, $n_{\omega}^{\leftarrow}$ denotes the same node in the state such that reading $n_{\omega}^{\leftarrow}$ provides the k-mer $\overleftarrow{\omega}$.

Given two nodes $n_{start}^{\rightarrow}$ and $n_{stop}^{\rightarrow}$, we say that a path of length l exists from node $n_{start}^{\rightarrow}$ to node $n_{stop}^{\rightarrow}$, *iif* node $n_{stop}^{\rightarrow}$ can be reached using l nodes from node $n_{start}^{\rightarrow}$ and for any traversed node, it should be entered and left in the same state (i.e. explicit or implicit). Let k-path denote a path of length k.

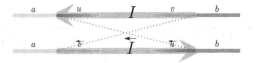

Fig. 1. Sequences aIb and $a\overleftarrow{I}b$, showing the four particular k-mers a, u, v and b at the breakpoints

Inversion. Given a fixed k value and a set of input sequences, we define an inversion as a sequence I of length larger or equal to k such that aIb and $a\overleftarrow{I}b$ occur both at least once, each in any of the input sequences, with a and b being two k-mers. We call u (resp. v) the prefix (resp. suffix) of length k of I. Our inversion model imposes $a \neq \overleftarrow{b}$ and $u \neq \overleftarrow{v}$. Figure 1 proposes a graphical representation of an inversion. We call the *breakpoints* of the inversion, the junctions between the inverted segment and the non-inverted parts. Such a rearrangement generates therefore two breakpoints in each sequence.

2.2   Inversion Pattern

An inversion, such as shown in Figure 1, generates a particular motif in the *de Bruijn Graph*. The only differences in terms of k-mers between both sequences, with and without the inversion, involve the breakpoints of the inversion: only the $k - 1$ k-mers spanning each breakpoints differentiate the two sequences. The breakpoints at the left of the inverted segments are then characterized by a *fork* in the *de Bruijn Graph*, which is defined by a common node $n_a^\rightarrow$ that branches to two distinct k-paths that end respectively in $n_u^\rightarrow$ and $n_v^\leftarrow$. Similarly, the other two breakpoints (at the right) are characterized by two k-paths starting from $n_u^\leftarrow$ and $n_v^\rightarrow$ that join in $n_b^\rightarrow$. These two *forks*, being connected by two common nodes (corresponding to the k-mers u and v respectively, and their reverse complements), lead to a particular motif in the *de Bruijn Graph*, that we call the *inversion pattern*, as exemplified in Figure 2.

It is important to note that the definition of the inversion pattern imposes conditions on the four k-mers a, u, v, b. First, $a \neq \overleftarrow{b}$ and $u \neq \overleftarrow{v}$ for the two distinct forks to exist. Second the node $n_a^\rightarrow$ must be branching, that is the first nucleotide of u must be different from the first nucleotide of $\overleftarrow{v}$.

One major advantage of this motif is that it can be traversed by 4 k-paths in the *de Bruijn Graph*: one from $n_a^\rightarrow$ to $n_u^\rightarrow$; one from $n_u^\leftarrow$ to $n_b^\rightarrow$, one from $n_b^\leftarrow$ to $n_v^\rightarrow$; one from $n_v^\rightarrow$ to $n_a^\leftarrow$. Being composed of only fixed length paths, finding the presence of such motif in a *de Bruijn Graph* is rapid and rather simple.

Notice that this motif presents some drawbacks. First, it detects the presence of inversion breakpoints but it does not provide the inversion itself. As second drawback, the motif is perfectly symmetrical: starting from node $n_b^\leftarrow$, or $n_u^\leftarrow$ or $n_v^\rightarrow$ leads to the discovery of the same inversion. As presented Section 3.2, we propose a way to output only once each inversion breakpoints. Finally, such a motif may witness approximate repeats instead of inversions (see Section 3.4).

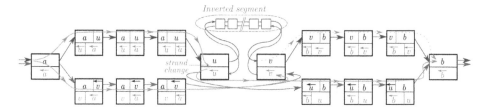

Fig. 2. Schematic example of the inversion pattern generated by sequences aIb (the blue path) and $a\overleftarrow{I}b$ (the red path) in a *de Bruijn Graph* with $k = 4$. Nodes are represented as two-stage boxes, with the upper part in black showing the explicit k-mer and the lower part, in grey, the implicit one. DNA k-mers are not represented, instead the node content shows proportion of full or junction of the four main k-mers a, u, v, b and their reverse complements. For sake of simplicity and without loss of generality, we consider that all k-mers of au, vb, $a\overleftarrow{v}$ and $\overleftarrow{u}b$ are explicitly stored. The state of a node traversed by edges entering and leaving its upper (resp. lower) part is explicit (resp. implicit). The green paths represent the paths enumerated by TAKEABREAK algorithm. The dashed green path is only checked, once the nodes n_v and n_b are found.

3   Algorithm for Inversion Pattern Detection

3.1   Main Algorithm

This section describes an algorithm for efficient detection of the inversion pattern from an already constructed *de Bruijn Graph*.

A "naive" algorithm for detecting the inversion pattern would be to check for each possible starting k-mer a a k-path from $n_a^\rightarrow$ to $n_u^\rightarrow$, then from $n_u^\leftarrow$ to $n_b^\rightarrow$, then from $n_b^\leftarrow$ to $n_v^\leftarrow$ and then from $n_v^\rightarrow$ to $n_a^\leftarrow$ and finally checking that $a = \alpha$. This approach would lead to the construction of $4k$-paths from $n_a^\rightarrow$ leading to a combinatorial explosion in complex genomes.

We propose an algorithm whose longest walked paths are of length $2k$, then strongly limiting the search space. The main idea is to start from any branching node (a node having more than one outgoing edge) $n_a^\rightarrow$, to detect all nodes reachable by a k-path, storing them in several sets N_α depending on the first letter α of these nodes. The second main step is then to detect any node $n_b^\rightarrow$ ($\overleftarrow{b} \neq a$) such that there exist a k-path from $n_u^\leftarrow$ to $n_b^\rightarrow$ and a k-path from $n_b^\leftarrow$ to $n_v^\leftarrow$, with $n_u^\leftarrow \in N_\alpha$ and $n_v^\leftarrow \in N_\beta$ and $\alpha \neq \beta$. In such case the pair of sequences au and vb is output. Algorithm 1 proposes a high level presentation of our algorithm.

3.2   Canonical Representation of Occurrences

The inversion pattern presents some symmetries. In most cases (see Section 3.3), the inversion pattern generated by an inversion will be detected by our algorithm as four distinct occurrences each starting from one of the four main nodes: $n_a^\rightarrow$, $n_u^\leftarrow$, $n_b^\leftarrow$ and $n_v^\rightarrow$. The output of the algorithm 1 is a pair of words au and vb

1. **Input:** A list of branching nodes and a *de Bruijn Graph* of all input reads.
2. **Provides:** A set of pairs of inversion breakpoint sequences
3. **for** each branching node $n_a^{\rightarrow}$ **do**
4.   Compute all paths of length k starting from $n_a^{\rightarrow}$
5.   Store all reached nodes starting with letter α in N_α ($\alpha \in \{A, C, G, T\}$)
6.   **for** each $\alpha \in \{A, C, G\}$ **do**
7.   **for** each $n_u^{\rightarrow} \in N_\alpha$ **do**
8.   Compute all paths of length k from $n_u^{\leftarrow}$
9.   Store all reached nodes in B
10.   **for** each $n_b^{\rightarrow} \in B$ **do**
11.   **for** each $n_v^{\leftarrow} \in \cup N_{\beta > \alpha}$ **do**
12.   **if** a path of length k exists from $n_b^{\leftarrow}$ to node $n_v^{\leftarrow}$ **then**
13.   Output (au, vb)

Algorithm 1. Main algorithm to detect the inversion pattern

depending both on the starting node $n_a^{\rightarrow}$ and the order of detection between $n_u^{\rightarrow}$ and $n_v^{\leftarrow}$. To avoid outputting several times the same inversion, we define its *canonical representation* as the smallest 2-words output in lexicographical order among the eight possible rearrangements: (au, vb), $(a\overleftarrow{v}, \overleftarrow{u}b)$, $(\overleftarrow{u}\,\overleftarrow{a}, \overleftarrow{b}\,\overleftarrow{v})$, $(\overleftarrow{u}b, a\overleftarrow{v})$, (vb, au), $(v\overleftarrow{a}, \overleftarrow{b}\,u)$, $(\overleftarrow{b}\,\overleftarrow{v}, \overleftarrow{u}\,\overleftarrow{a})$, $(\overleftarrow{b}\,u, v\overleftarrow{a})$. Only the canonical representation is reported and only once.

3.3   Presence of Small Inverted Repeats at the Breakpoints

If an inversion contains an inverted repeat of size larger or equal to $k - 1$ at its breakpoints, this inversion will not generate the inversion pattern since it does not generate new k-mers nor new paths in the *de Bruijn Graph* with respect to the non inverted sequence. This is the case for instance if $a = \overleftarrow{b}$ or $u = \overleftarrow{v}$.

In the case of an inverted repeat whose length is smaller than $k - 1$, such inversion still generates the inversion pattern, however the latter is not be fully symmetrical. Suppose there is an inverted repeat of size $x < k - 1$ at the breakpoints or overlapping the breakpoints. As the first node $n_a^{\rightarrow}$ must be branching, it imposes that the repeated sequence is included in k-mer a and considered outside the inverted segment (note that even with the full sequences at hand, we can not decide if the inversion includes the repetition entirely, partially or not at all). The suffix of size x of a is then equal to the prefix of size x of $\overleftarrow{b}$. It implies also that there are no more $k - 1$ distinct k-mers at each breakpoint that differentiate the two sequences, but $k - 1 - x$ k-mers. Therefore the two forks of the inversion pattern, represented in Figure 2, are shortened. In this case, the nodes $n_u^{\leftarrow}$ and $n_v^{\rightarrow}$ reached after k-paths are not necessarily branching and can not constitute starting k-mers in other occurrences of the inversion pattern. Instead k-mers at the end of $(k - x)$-paths in the fork constitute the other starting k-mers.

In fact, such an inversion will still be detected as 4 occurrences but the sets of k-mers a, u, v and b will be different depending on the starting k-mer. Starting from inside ($n_u^{\leftarrow}$ or $n_v^{\rightarrow}$) or outside ($n_a^{\rightarrow}$ or $n_b^{\leftarrow}$) the inverted segment I will

Fig. 3. Example of an inversion with small inverted repeats (red arrows) at the breakpoints. Breakpoint sequences au, vb (resp. $a'u', v'b'$) are obtained starting from nodes $n_a^{\rightarrow}$ or $n_b^{\leftarrow}$ (resp. $n_{u'}^{\leftarrow}$ or $n_v^{\rightarrow}$). The unique canonical representative is represented by the two grey bottom lines.

generate two distinct sets of $2k$ words overlapping on $2k - x$ characters. To avoid duplicating once again artificially the number of occurrences, the output of the algorithm truncates the k-mers u and $\overleftarrow{v}$ such that all starting k-mers give the same sets of words (here of size $2k - x$) and a unique canonical representative can be computed for each of the four occurrences (Figure 3).

3.4   Distinguishing Inversions from Approximate Repeats

Some approximate repeats may generate the inversion pattern in the *de Bruijn Graph* and are thus a source of false positives. Consider for instance that a given sequence au has at least four approximate copies in the sequence, such that au, au', $a'u$ and $a'u'$ with $u' \simeq u$ (at least the first letter is different) and $a' \simeq a$ (at least one substitution or indel anywhere in a). In this situation, without loss of generality, calling $\overleftarrow{b} = a'$ and $\overleftarrow{v} = u'$, the four paths au, $vb(= \overleftarrow{a'u'})$, $a\overleftarrow{v}(= au')$, and $\overleftarrow{u}b(= \overleftarrow{a'u})$ exist and mimics the inversion pattern. More generally, high similarity between a and $\overleftarrow{b}$ and between u and $\overleftarrow{v}$ is characteristic of an approximate repeat.

In order to distinguish inversions from false positives due to approximate repeats, we filter out occurrences of the inversion model where a and $\overleftarrow{b}$ and where u and $\overleftarrow{v}$ have a Longest Common Subsequence (LCS) size higher than a given threshold. As an optimization, we try to detect earlier cases where a and $\overleftarrow{b}$ are too similar during the k-path search from $n_u^{\leftarrow}$ to $n_b^{\rightarrow}$. During this step, we forbid paths that go back on the previous path towards first node $n_a^{\rightarrow}$, since the longer we take the former path, the more similar will be k-mers a and $\overleftarrow{b}$. However, to permit the detection of inversions with small inverted repeats at the breakpoints, we tolerate to go back on the former path for a given maximum number of nodes (this parameter is usually fixed to 8).

Additionally, it is well known that high copy number repeats with approximate copies are an important source of complexity generating highly branching subparts in the *de Bruijn Graph*. Searching for inversions in such complex part of the graph presents two main drawbacks. First, as previously mentioned, it is a source of false positives, and second, it generates a possible huge number of k-paths whose enumeration can be highly time consuming. To overcome these two drawbacks we stop the inversion pattern detection from a node n_a as soon as the product of the cardinality of the two largest sets N_α is bigger than a limit (called

LCT for *Local Complexity Threshold*). This product is a lower approximation of the minimal number of couples of k-paths that are to be enumerated from the starting n_a. Similarly, we apply the same strategy once a set of nodes B (see Algorithm 1 line 9) is detected from a node $n_u^{\rightarrow}$: if the cardinality of B times the cardinality of the largest set N_α is larger than LCT, then the exploration from node $n_u^{\rightarrow}$ stops. This last product reflects another lower bound of the number of paths to be checked. Note that this approach highly limits both false positives and computational time, while having a limited impact on recall, as shown in results Section 4.2.

3.5 TAKEABREAK Implementation

We implemented the proposed algorithm in a prototype called TAKEABREAK. It takes as input one or several sets of sequences in fasta or fastq format. Its main parameters are the k-mer size k; $max\_sim \in [0, 100]$ the maximal similarity authorized between a and $\overleftarrow{b}$ and between u and $\overleftarrow{v}$, expressed as a percentage of k-mer size; and LCT: the Local Complexity Threshold (see Section 3.4). Prior to the inversion pattern detection phase, the *de Bruijn Graph* is constructed using the Minia data structure [2,12]. This graph is constructed using only k-mers having at least 3 occurrences in order to discard sequencing errors. This is a very common parameter used for *de Bruijn Graph*-based assembly. The second phase implements algorithm 1. The output is a *fasta* file containing, for each detected inversion, its breakpoint sequences. These are the $2k - 2$ (or $2k - x - 2$ in the case of an inverted repetition of size x) words centered on the canonical representation (au, vb). By removing the two extreme nucleotides, it ensures that the output paths are made of the k-mers that overlap the breakpoints and that must be specific to each sequence.

TAKEABREAK was implemented in C++ with the GATB library [3], providing notably the Minia data structure, and it can be downloaded from `http://colibread.inria.fr/TakeABreak/`.

4 Results

To evaluate the ability of TAKEABREAK to detect inversions in reads, we generated artificial read datasets. First, non-overlapping inversions of varying sizes were simulated in a copy of a real genome. Then we simulated the sequencing processing on both genomes, the original one and the one with artificial inversions. Finally both read sets were given as input to TAKEABREAK. To classify the results of TAKEABREAK as true positive or false positive, we first generated for each simulated inversion its canonical representation of breakpoints such as described in sections 3.2 and 3.3 and then called a prediction of TAKEABREAK as true positive if it is exactly present in this set of true breakpoints. Finally, recall and precision were computed as follows: recall as the number of true positives over the number of simulated inversions, and precision as the number of true positives over the number of predictions.

In more details, inversions were simulated as follows. Each inversion was put sequentially. For each inversion, its first breakpoint is chosen uniformly along the sequence, then its size is sampled uniformly in a given interval (here $[k - 1000]$), finally if it does not overlap and is sufficiently far from a formerly placed inversion (the min distance was fixed to k nucleotides) the inversion is kept and its sequence is reversed-complemented. To simulate reads, 100 bp reads are sampled uniformly along the genome, sequencing errors are put also uniformly with 1 % rate, the depth of coverage was fixed to 40x for each genome.

4.1   Results on a Bacterial Genome

TAKEABREAK was first evaluated on a simple and small dataset based on the bacterial *E. coli K12* genome, in which 1000 random inversions were simulated. TAKEABREAK was applied on this simulated dataset with default parameters ($k = 31$, $max\_sim = 80\%$, $LCT = 100$). On this simple dataset, TAKEABREAK proved to be highly efficient to detect inversion breakpoints, since it predicted the 1000 true positive inversions, leading to a 100% recall for 100% precision (see Table 1). Cortex_var bubble caller [4] was run on the same data and failed to detect any of the simulated inversions.

Table 1. Precision and recall results for TAKEABREAK on simulated datasets. The first part of the table presents results obtained with default parameters ($k = 31$, $max\_sim = 80\%$, $LCT = 100$), the second part shows the decrease of precision when relaxing filtering parameters ($k = 31$, $max\_sim = 100$, $LCT = 10000$). # FP indicates the amounts of false positives.

	Recall (%)	Precision (%)	# FP
E. coli genome - default parameters	100.00	100.00	0
C. elegans genome - default parameters	96.00	99.07	9
Human chromosome 22 - default parameters	87.60	92.50	71
C. elegans genome - relaxed parameters	99.60	0.37	271,374
Human chromosome 22 - relaxed parameters	93.50	0.06	1,442,760

4.2   Results on More Complex Genomes

Bacterial genomes are small and contain few repeats, leading to rather simple *de Bruijn Graph* and few false positives of the inversion pattern. To evaluate TAKEABREAK on more complex genomes, we simulated inversions in eukaryotic genomes and chromosomes, first in the full *C. elegans* genome ($\sim$ 100 Mbp) and second in human chromosome 22 ($\sim$ 35 Mbp without N bases). As expected (see Section 3.4), precision and recall decrease when the repeat content of the genome increases, as shown in Table 1. However, this effect is greatly limited by the use of filtering parameters max_sim and LCT, since relaxing these parameters leads to millions of false positives (see Table 1).

Note that these parameters have to be fixed carefully as they can also affect the recall, as shown in Figure 4 where precision and recall results are represented for varying values of *max_sim* and *LCT*. This figure shows that both parameters are useful to decrease the false positive rate and that the proposed default parameters offer a good trade-off between precision and recall.

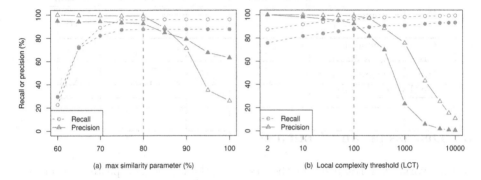

Fig. 4. Effect of the filtering parameters, *max_sim* (a) and *LCT* (b), on precision and recall values for the *C. elegans* (open symbols) and human chromosome 22 (solid symbols) datasets. Vertical dashed lines represent the default parameters.

4.3   Time and Memory Performances

These tests were performed with 2.3 GHz Intel Core i7 processors, with 8GB RAM. Table 2 shows time and memory performances of the prototype TAKEA-BREAK for the different datasets. Time and memory increase with the complexity of the datasets. Even if the human chromosome dataset is smaller than the *C. elegans* one, the computational time is much larger for human. This shows that the complexity of the graph is not solely linked to the size of the genome, but also to its repeat content, with human chromosome 22 high copy number repeats generating sub-parts of the graph with high density of branching nodes and imbricated patterns of inversions.

Nevertheless, as presented Table 2, TAKEABREAK scales up to complex and large datasets. The highest memory consumption is reached during the *de Bruijn Graph* construction and is limited to 1GB, allowing TAKEABREAK to be executed on a standard desktop (note that the full human genome would need 6GB of memory [12]). The graph construction time is limited to at most 20 minutes for the most complex dataset we used. The time needed for enumerating all inversion patterns is sensitive to genome complexity (from 1 second for *E. coli* to one hour and a half for human chromosome 22) and still remains acceptable. Moreover, in addition to dramatically improving the precision (see Section 4.2), we can notice that the default filters highly reduce the computational time (*e.g.* from 7h40 without filters to 1h30 with filters on the human dataset).

Table 2. Time and memory performances of TAKEABREAK on simulated datasets with default parameters. For each dataset we indicated the number of reads and the total number of nucleotides it contains. Time values given in parenthesis are those obtained while relaxing the filter parameters (bottom part of Table 1).

	Time (s)		Memory	
	Graph construction	Inversion detection	Graph construction	Inversion detection
E. coli genome (3.7M reads 370 Mbp)	24	1	1GB	3MB
C. elegans genome (80M reads, 8 Gbp)	78	935 (7408)	1GB	53MB
Human chromosome 22 (28M reads 2.8 Gbp)	1205	5412 (27554)	1GB	153MB

5 Discussion and Conclusion

In this work, we formalized for the first time the topological pattern generated by the inversion of a DNA segment in the *de Bruijn Graph* representing both sequences, with and without the inversion.

We also proposed a first analysis of what kind of variant or sequence feature can or can not generate this pattern. The pattern involves only the $2k$ sequences around the breakpoints of the inversion (k being the k-mer size of the *de Bruijn Graph*). Therefore the size of the inversion does not limit the existence of the pattern as long as it is greater than k. The pattern is based on four k-mers at each side of the breakpoints that must be identical between both sequences with and without the inversion. As a consequence, the breakpoint regions must not contain any substitution or indel at distance less than k from both breakpoints, that is as if the inversion was generated by perfect blunt-ended double strand breaks. Finally, another feature that can prevent an inversion from generating this pattern is the presence of an inverted repeat of size $\geq k - 1$ at each breakpoints since all breakpoint sequences will follow the same paths in the *de Bruijn Graph*.

On the other hand, we showed that the pattern can be generated by other sequence features than inversions. First, some approximate repeats with appropriate combinations of differences can easily generate this pattern, these are considered as false positive or noise since they do not differentiate the compared genomes. If in small bacterial genomes, this situation is quite rare, our tests show that in more complex genomes this can dramatically increases the number of false positive calls, explaining why we added a sequence-based filter to this topology-based pattern. Indeed, with high copy number repeats, such as transposable elements in eukaryotic genomes, such combinations of at least two differences in repeats of size $2k$ is very likely to happen. Another variant that can generate the inversion pattern is the reciprocal translocation, since it has also two breakpoints per sequence (with and without the translocation) with the same combinations of four k-mers. We consider this as another advantage of this pattern, because, in this case, this is also a structural variant that can differentiates genomes and has therefore a potential biological interest.

In this work, we also proposed and implemented an efficient algorithm to enumerate all inversion patterns in a *de Bruijn Graph*, together with powerful filtering strategies to avoid false positives due to approximate repeats. The tests we performed on simulated data prove that this approach enables to recover almost all simulated inversions quite rapidly. The power of this pattern lies mainly in its fixed size. Contrary to structural variants with only one breakpoint, such as insertions and deletions, it is not necessary to traverse in the graph the full inverted segment to detect the presence of the inversion. In fact, insertions and deletions generate *bubble* patterns that can only be detected by traversing the full inserted or deleted sequence [4,11], this strongly limits the size of detectable events (at least in complex genomes) and increases the computational time. Inversions too could be detected as *bubbles* but Cortex_var did not detect them, even the smallest ones, probably because it requires bubbles not to contain any branching node while such nodes are inherent of inversion patterns, as shown in this paper.

The tests and simulations we performed were meant to demonstrate the validity of our pattern and of our algorithm, we are aware that they can still be improved to better fit actual genome re-sequencing data. Indeed, only inversions were simulated without any other polymorphism that could impact the breakpoints. Inversions were put following a uniform distribution, whereas rearrangement distribution is likely not random and some rearrangements can be linked for instance to repeated sequences. Finally, only perfect blunt-ended breakpoints were simulated which may not reflect all molecular mechanisms of such events (for instance, non-homologous end joining is known to generate small indels at the very breakpoint). For all these reasons, recall values we obtained are likely to be over-estimated with respect to real inversions. However, our promising results on such simulated inversions open the way to further improvements of the model.

First, the model could largely be improved by additionally including SNP or small indel detection models such as [13]. Thus both SNP and inversion detection would not suffer from each other. This would improve recall for events that lie close to each other and could be used as preliminary step of the assembly process. Second, the breakpoint detection algorithm could be coupled with a third party local assembly or gap-filling tool, such as MindTheGap [5], to get the sequence of the inverted segment and not only its breakpoints. Finally, other biological variants can benefit from this approach. As already mentioned, reciprocal translocations can be detected by the proposed model as is. Additionally, the model could be extended to the detection of other rearrangements that have more than two breakpoints, such as transpositions that generate a three-fork model, thus showing high similarity with the model proposed in this paper.

Acknowledgments. The authors warmly thank Erwan Drezen and Guillaume Rizk for implementation support and Marie-France Sagot for interesting discussions. This work was supported by the Région Bretagne SAD-MIRAGE project and by the ANR (French National Research Agency), ANR-12-BS02-0008 *Colib'read* project and ANR-12-EMMA- 0019-01 *GATB* project.

References

1. Alkan, C., Coe, B.P., Eichler, E.E.: Genome structural variation discovery and genotyping. Nat Rev. Genet. 12, 363–376 (2011)
2. Chikhi, R., Rizk, G.: Space-efficient and exact de bruijn graph representation based on a bloom filter. Algorithms for Molecular Biology 8, 22 (2013)
3. Drezen, E., et al.: The Genome Assembly and Analysis Tool Box, http://gatb.inria.fr/ (Manuscript in Prep. 2014)
4. Iqbal, Z., Caccamo, M., Turner, I., Flicek, P., McVean, G.: De novo assembly and genotyping of variants using colored de bruijn graphs. Nature Genetics 44, 226–232 (2012)
5. Lemaitre, C., et al.: MindTheGap Software, http://mindthegap.genouest.org/ (Manuscript in Prep. 2014)
6. Li, Y., Zheng, H., Luo, R., Wu, H., Zhu, H., Li, R., et al.: Structural variation in two human genomes mapped at single-nucleotide resolution by whole genome de novo assembly. Nat. Biotechnol. 29, 723–730 (2011)
7. Medvedev, P., Stanciu, M., Brudno, M.: Computational methods for discovering structural variation with next-generation sequencing. Nat Methods 6, S13–S20 (2009)
8. Mills, R.E., Walter, K., Stewart, C., Handsaker, R.E.: 1000 Genomes Project: Mapping copy number variation by population-scale genome sequencing. Nature 470, 59–65 (2011)
9. Nordström, K.J.V., Albani, M.C., James, G.V., et al.: Mutation identification by direct comparison of whole-genome sequencing data from mutant and wild-type individuals using k-mers. Nature Biotechnology 31, 325–330 (2013)
10. Peterlongo, P., Schnel, N., Pisanti, N., Sagot, M.-F., Lacroix, V.: Identifying sNPs without a reference genome by comparing raw reads. In: Chavez, E., Lonardi, S. (eds.) SPIRE 2010. LNCS, vol. 6393, pp. 147–158. Springer, Heidelberg (2010)
11. Sacomoto, G.A., Kielbassa, J., Chikhi, R., Uricaru, R., et al.: Kissplice: de-novo calling alternative splicing events from rna-seq data. BMC Bioinformatics 13, S5 (2012)
12. Salikhov, K., Sacomoto, G., Kucherov, G.: Using Cascading Bloom Filters to Improve the Memory Usage for de Brujin Graphs. In: Darling, A., Stoye, J. (eds.) WABI 2013. LNCS, vol. 8126, pp. 364–376. Springer, Heidelberg (2013)
13. Uricaru, R., et al.: discoSnp Software, http://colibread.inria.fr/discosnp/ (Manuscript in Prep. 2014)
14. Zerbino, D.R., Birney, E.: Velvet: algorithms for de novo short read assembly using de bruijn graphs. Genome Research 18, 821–829 (2008)

Modeling the Geometry
of the Endoplasmic Reticulum Network

Laurent Lemarchand[1], Reinhardt Euler[1], Congping Lin[2], and Imogen Sparkes[3]

[1] Lab-STICC UMR 6285, UBO-Université Européenne de Bretagne, Brest, France
{Laurent.Lemarchand,Reinhardt.Euler}@univ-brest.fr
[2] Mathematics, University of Exeter, UK
c.lin@exeter.ac.uk
[3] Biosciences, University of Exeter, UK
I.Sparkes@exeter.ac.uk

Abstract. We have studied the network geometry of the endoplasmic reticulum by means of graph theoretical and integer programming models. The purpose is to represent this structure as close as possible by a class of finite, undirected and connected graphs the nodes of which have to be either of degree three or at most of degree three. We determine plane graphs of minimal total edge length satisfying degree and angle constraints, and we show that the optimal graphs are close to the ER network geometry. Basically, two procedures are formulated to solve the optimization problem: a binary linear program, that iteratively constructs an optimal solution, and a linear program, that iteratively exploits additional cutting planes from different families to accelerate the solution process. All formulations have been implemented and tested on a series of real-life and randomly generated cases. The cutting plane approach turns out to be particularly efficient for the real-life testcases, since it outperforms the pure integer programming approach by a factor of at least 10.

Keywords: endoplasmic reticulum, plane graph, 0-1 programming, separation procedure.

1 Problem

The endoplasmic reticulum (ER) is a membrane-bound organelle that forms a highly complicated interconnected network of tubules and flattened sacs (known as cisternae) [10,6]. As the cortical ER in plant cells occupies a very thin, almost two-dimensional, layer of cytoplasm beneath the plasma membrane, our study of the ER network will be based on 2D approximations of the ER network in Tobacco leaf epidermal cells. Figure 1 (a) shows an instance of live cell images of an ER network [13]. Transition between tubules and cisternae can be highly dynamic and tubules can also dynamically change their polygonal network [10]. The dynamic ER shape is suggested to be adaptable to the cells requirements for ER function; for example, ER cisternae may be the preferred site of protein translocation while tubules might be the preferred site for ER vesicle budding.

A.-H. Dediu, C. Martín-Vide, and B. Truthe (Eds.): AlCoB 2014, LNBI 8542, pp. 131–145, 2014.

As an expanding number of proteins have been identified that mediate the generation and shape of the ER network, the stage is now set for investigations into the mechanisms regulating ER morphology within the cell. To be able to carry out such investigations, better tools are required to quantify the morphology and dynamic rearrangements that the ER undergoes. It is important to consider that whilst the proteins and stresses on the system driving formation and changes in the remodelling may differ between eukaryotic systems, the network geometry of the ER appears to be fairly well conserved in terms of it consisting of a polygonal network of interconnected tubules and cisternae. Therefore, the problem considered here has universal appeal towards understanding the constraints placed on ER network formation in all systems.

(a)   (b)   (c)

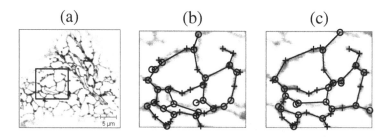

Fig. 1. Illustration of the ER network and abstracted geometric graphs. (a) shows a static ER network, where rectangle regions highlight a region with no ER cisternae. (b)-(c) show zoomed local networks in the rectangle region and corresponding abstracted geometric graphs from two images. The abstracted graphs are obtained using the image processing method introduced in [11], where markers '+' and 'o' represent persistent and non-persistent nodes, respectively, and lines represent edges (only the largest connected components in the chosen rectangle regions are considered, and tubules extending outside the chosen region are not included in the abstracted graphs). The experimental ER images are taken from [13] (www.plantcell.org, Copyright American Society of Plant Biologists).

A quantitative analysis [11] of the ER network in tobacco leaf epidermal cells suggests that it is a perturbed Euclidean Steiner network between terminals, where terminals are persistent nodes (static elements of tubules) and degree-1 nodes. A Euclidean Steiner network is a locally minimal network, i.e., a network in which any local perturbation of non-terminal nodes would increase the total length of the network [11]. This is analogous to Steiner trees in Euclidean space [8] in which the non-terminals (called Steiner points) have degree 3. Note that local optima such as Euclidean Steiner networks might not give a unique network topology, which is essential for modelling its dynamics. In this paper, we reanalyze live cell confocal microscopy data of native ER networks from [13] in an attempt to understand whether there is an optimization principle governing the network shape beyond local minimization. For quantitative analysis, we abstract ER networks into geometric graphs using the image processing method from [11]; examples of abstracted graphs are shown in Figure 1 (b,c).

The ER membrane surface, serving as a transport network, is intuitively suspected to be a minimal film [12]. Due to the existence of cycles commonly observed in networks, minimal spanning trees do not sufficiently explain the shape of the ER network. Also, ER tubules generate 3-way junctions when branching and angles at these degree-3 nodes follow a normal distribution with mean around $120°$ [11]. Here, we include a degree constraint and an angle constraint for the nodes while minimizing total edge-length, and we test whether the optimal graph under these constraints could mimic the ER network geometry. To test the optimal solutions from our model, we quantitatively compare them to the abstracted ER networks from a tobacco leaf epidermal cell.

2   Model

We construct two models: one basic model with only degree constraints and one full model with both degree and angle constraints.

Basic Modelization   Given a set of nodes V (corresponding to all the nodes in an abstracted ER network), a subset V_b of V (corresponding to the degree-3 branching nodes in the network), the problem is to find an undirected connected plane graph, whose nodes in V_b have degree 3, those outside V_b have degree at most 3, and which minimizes the sum of the Euclidean distances associated with the connecting edges.

Full Modelization   A full model is the basic model with an additional angle constraint. More precisely, for a branching node $u \in V_b$, and its neighbours v_1, v_2, v_3, we add the constraint that any two angles at the node u formed from edges $uv_i (i = 1, 2, 3)$ have a sum no less than θ where θ is a given angle around $180°$. The idea comes from the concern of force balance acting on the branching node. For each degree-3 branching node (tubule junction), as modeled in [11], each of the three ER filaments is assumed to apply a membrane tension force on this tubule junction. Thus, it is unlikely that the tension forces are in the same direction of a half plane, i.e., the sum of two angles of the branching node is less than $180°$. This angle constraint means that none of two angles of a branching node form a sum less than $\theta \approx 180°$ in the optimal solution.

3   Problem Formulation and Resolution

Given the complete edge-weighted graph $G = (V, E, w)$ and a set $V_b \subseteq V$, the basic problem $\mathcal{BP}$ can be formulated as follows:

$$\text{minimize} \sum_{x_{uv} \in E} w_{uv} x_{uv} \tag{1}$$

subject to

$$1 \leq \sum_{v \neq u} x_{uv} \leq 3 \forall u \in V \setminus V_b, \tag{2}$$

$$\sum_{v \neq u} x_{uv} = 3 \forall u \in V_b, \tag{3}$$

$$\delta(W) \geq 1 \forall W \subset V, |W| \geq 2 \tag{4}$$

$$x_{uv} + x_{wz} \leq 1 \forall u, v, w, z \in V, \text{edges } uv \text{ and } wz \text{ cross} \tag{5}$$

$$x_{uv} + x_{uw} + x_{uz} \leq 2 \forall u \in V_b, v, w, z \in V, \text{angle}(vuw) + \text{angle}(wuz) < \phi \tag{6}$$

$$x_{uv} \in \{0, 1\}, \forall uv \in E. \tag{7}$$

We are looking for a minimum-weight, connected and plane, spanning subgraph of G, where:

- (1) the objective function represents the total Euclidean distance of the connecting subgraph;
- (2) and (3) represent the constraints on nodes including degree-3 nodes;
- (4) ensures the connectivity of the resulting subgraph, where

$$\delta(W) = \sum_{i \in W, j \notin W} x_{ij}$$

- Angle equations (6) are set with a degree of $\phi = 180^o$. u is a branching node, and v, w, z are its neighbours in a solution;
- x_{uv} and x_{vu} represent the same variable, and $x_{uv} = 1$ iff edge uv is selected in the solution.

We remark that the complexity status of our problem still seems to be open. The particular case of finding a minimum-weight 3-regular connected spanning subgraph of G is known to be NP-hard [3]. Moreover and throughout our calculations, we have observed only few solutions containing a pair of crossing edges. Therefore, instead of adding constraints (5) at start, we check after each iteration whether two edges cross, in case of which the corresponding constraint (5) is added on the fly.

Similar incompatibility constraints are added for the angle restrictions of the full model.

3.1   Binary Linear Programming Resolution

Problem $\mathcal{BP}$ is solved using the CPLEX MIPS solver as follows. The initial $0-1$ programming problem is solved without connectivity constraints (4). Cuts are added iteratively in order to discard connected components which cover only a subset of the nodes. More precisely, if $W \subset V$, with $1 < |W| < |V|$, is the node set of a connected component obtained as a result, we add the following constraint to $\mathcal{BP}$:

$$\delta(W) \geq 1. \tag{8}$$

This process is repeated until the resulting graph is connected. As described above, a plane embedding is checked for at each iteration.

3.2 Linear Programming Formulation

Let $\mathcal{RP}$ denote the linear relaxation of $\mathcal{BP}$, i.e., the linear program obtained by replacing the constraints (7) by $0 \leq x_{uv} \leq 1 \; \forall uv \in E$. Relaxing $\mathcal{BP}$ to $\mathcal{RP}$ allows to replace the IP-solver by an LP-one, but we are now faced with the possibility of fractional optimal solutions.

Three types of search algorithms have been used to "cut off" such fractional solutions, leading to 4 different *separation* procedures.

The first one (*2-cut* procedure) is based on a min-cut algorithm. If an $s-t$-cut leads to a cut value < 1, the connectivity constraint is violated. This situation is detected using the Stoer-Wagner algorithm [14]. If V_1, V_2 is the corresponding partition of V, we add the constraint

$$\delta(V_1) \geq 1 \tag{9}$$

The second one relies on a p-partition of the node set V, $P = \{V_1, V_2, ..., V_p\}$. In this case we add the constraint

$$\frac{1}{2} \sum \delta(V_i) \geq (p-1) \tag{10}$$

The problem is now to find a partition of V, whose inequality is violated by the current optimal (and fractional) solution.

We have implemented 2 separation procedures for this type. The *r-cut* procedure is based on a recursive splitting of partitions using a min-cut algorithm, whereas the *k-cut* procedure is a 1-edge-like contraction procedure.

- The *r-cut* procedure for multi-cuts is inspired by [1]. Given a p-partition, find a minimum cut in each part of it. For the cut with minimum value among all these min-cuts, break the associated component to obtain a $(p+1)$-partition.
- The *k-cut* procedure for multi-cuts goes as follows. The partition is induced by the components of the graph $G' = (V, E' = \{e \in E | x_e \geq \alpha\})$, with $\alpha \in \;]0..1]$ as given by the current optimal solution. Case $\alpha = 1$ corresponds to the component search of $\mathcal{BP}$. This separation procedure is applied for $\alpha = 0.8, 0.6, 0.4$.

The third one is a *b-cut* procedure based on *blossom* inequalities arising from Edmonds' description of matching-polytopes [5]. For this we just recall that the incidence vectors of *u-capacitated b-matchings* are the solutions of the following constraints:

$$\sum_{e \in \delta(i)} x_e \leq b_i \qquad\qquad \forall i \in V, \tag{11}$$

$$0 \leq x_e \leq u_e \qquad\qquad \forall e \in E, \tag{12}$$

$$x_e \in \mathbb{Z} \qquad\qquad \forall e \in E, \tag{13}$$

and if we let $b_i = 3 \; \forall i \in V$ and $u_e = 1 \; \forall e \in E$, the following *blossom* inequalities

$$\sum_{e \in E(W)} x_e + \sum_{f \in F} x_f \leq \left\lfloor \frac{3|W| + |F|}{2} \right\rfloor, \forall W \subset V, F \subset \delta(W) \quad \text{with} \quad 3|W| + |F| \text{ odd}$$

(14)

are valid for any solution of our basic problem.

The separation procedure presented in [9] is based on cut-trees. In our implementation we use Gusfield's algorithm [7] for cut-tree computations, and we also make use of the igraph C library [4].

The initial binary algorithm is thus split into two phases :

- *Linear phase* : Solve $\mathcal{RP}$ with an LP-solver, and add constraints for all disconnected components. When the obtained solution is connected but fractional, find iteratively minimal cuts (2-cuts of value less than one, k-cuts or b-cuts, whose corresponding inequality is currently violated) and add the corresponding constraints.
- *Binary phase* : When no more efficient cuts are found, solve the resulting 0-1 programming problem with an MIPS solver.

According to the kind of linear constraints we add in the linear phase, this leads to 5 versions of the mixed linear/binary algorithms:

- *BP*: the binary formulation,
- *LP*: the linear formulation,
- *LPr*: the linear formulation with recursive cuts,
- *LPrk*: the linear formulation with both recursive and parametric cuts.
- *LPrkb*: the linear formulation with recursive, parametric and blossom cuts.

4   Tests

All formulations have been implemented using the CPLEX API in C. A set of 50 real-life testcases has been provided. Randomly generated testcases, with distance values in the range [1..100] are also used for the tests, especially with a large number of nodes. We look at the comparative results in terms of runtimes for both the binary and the linear formulation as well as the different cutting techniques for the basic problem solution. Random sets are used to evaluate the behaviour of the different algorithms with respect to problem characteristics such as nodes and percentage of branching nodes.

Finally, real-life testcases are used to evaluate the behaviour of the algorithms in view of the different pieces of the model.

4.1   Runtimes for the Solution of the Basic Problem

We examine the runtimes of 5 different algorithms for solving the basic model: (1) the binary formulation *BP*, (2) the linear formulation *LP*, (3) the linear formulation with recursive cuts *LPr*, (4) the linear formulation with both recursive

---

Algorithm 1: General Algorithm. *BP* (*solveBinary* = *True*), *LP* (*solveBinary* = *False*), *LPr* (*solveBinary* = *False*, *checkRcuts* = *True*), *LPrk* (*solveBinary* = *False*, *checkRcuts* = *True*, *checkKcuts* = *True*) and *LPrkb* (*solveBinary* = *False*, *checkRcuts* = *True*, *checkKcuts* = *True*, *checkBlossom* = *True*) are derivated from this general algorithm

Data: a set of points V with a subset of branching nodes V_b, a set I of incompatible edge pairs $\{e, e'\}$, a set of boolean variables $\{solveBinary, checkPlan, check2cuts, checkRcuts, checkKcuts, checkBlossom\}$

Result: a connected plane subgraph

1 **begin**
2   graph $G := (V, E)$, with $E := V^2$, w_{uv} the Euclidean distance between u and v
3   generate problem P with constraints (2) and (3)
4   add constraints (5) for all edge pairs in I
5   **if** *solveBinary* = *True* **then**
6 | add constraints (7) to P
7   **end**
8   f := solve (P)
9   **if** $G_f = (V, f)$ *is connected and binary* **then**
10   **if** *checkPlan* = *True* **then**
11 | **if** *checkPlanarity*(f) = *OK* **then** goto 31
12 | **else** add corresponding constraints (5)
13   **end**
14   **end**
15   **if** $G_f = (V, f)$ *is not connected* **then**
16 | compute component set $W = \{W_1, ..., W_p\}$
17 | add corresponding constraints (4)
18   **end**
19   **if** *solveBinary* = *False* **then**
20   **if** *check2cuts* = *True* **then** check and add corresponding constraints (9)
21   **if** *checkRcuts* = *True* **then** check and add corresponding constraints (10)
22   **if** *checkKcuts* = *True* **then** check and add corresponding constraints (10)
23   **if** *checkBlossom* = *True* **then** check and add corresponding constraints (14)
24   **if** *no new constraint added* **then**
25 | add constraint (7) to P
26 | *solveBinary* := *True*
27   **end**
28   **end**
29
30   goto 8
31 **end**

and parametric cuts $LPrk$ with $\alpha = \{0.4, 0.6, 0.8\}$, and finally (5) $LPrkb$, that is $LPrk$ plus the blossom-cut separation procedure.

We have generated a set of problems of increasing size (in number of nodes) with distance values in the range [1..100] and with 30 % of branching nodes. Problem sizes of Figure 2(a) are comparable to real-life testcases : the average percentage of branching nodes is 27.54 %, and the number of nodes is in the range [12..76] for the testcase set of Figure 3. We measure the average runtimes of 10 trials for the different algorithms, according to problem size. Corresponding results are presented in Figure 2 (a). Figure 2 (b) shows results for much larger problems obtained with BP and, the most efficient procedure, $LPrk$.

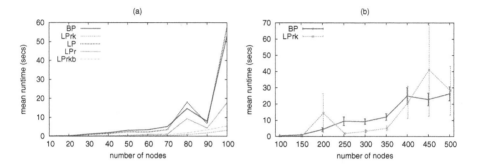

Fig. 2. (a) BP and LPx formulation runtimes for series of 10 random testcases with distances in [10..100] and 30 % branching nodes. (b) BP and $LPrk$ runtimes (and standard error) on large testcase series with 45% of branching nodes.

For random testcases of Figure 2 (a), $LPrk$ outperforms slightly the BP approach. Adding blossom inequalities in $LPrkb$ is time-consuming but does not improve the average results of $LPrk$. BP runtimes dramatically increase with problem size.

For large instances (Figure 2 (b)), the efficiency of $LPrk$ is clearly visible for problems with up to 400 nodes, number which is much greater than that of the real testcases. For more than 400 nodes this advantage disappears, because the solver swaps into Binary programming mode rather early in relation to the runtime of the whole optimization process. Time won in early steps does not lead to significant improvements in cost for later steps. For those large cases, $LPrk$ is often faster, but suffers from a lack of stability, with some instances degrading heavily the average performance, as shown by the large standard error of the runtimes for this procedure.

Figure 3 shows the runtimes for the different real-life testcases (*frames*) and the different procedures. The average runtimes are, respectively, 102.29 s, 98.96 s, 6.67 s, 0.33 s and 0.63 s for BP, LP, LPr, $LPrk$ and $LPrkb$.

Figure 3 clearly shows that $LPrk$ is very efficient for real-life testcases when compared to other approaches for solving the basic problem. Its maximum

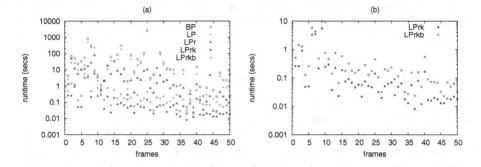

Fig. 3. (a) *BP* and all *LP* formulation runtimes for real-life testcases. (b) *LPrk* and *LPrkb* only.

runtime over the 50 cases is 5.58 s, compared to 9.68 s for *LPrkb*, 78 s for *LPr* and two non-terminated cases for *BP* and *LP*.

Runtimes according to the number of branching nodes We have generated a set of problems of increasing size (in number of nodes), and we have measured the total runtimes of *LPrk* for different percentages of branching nodes within each instance.

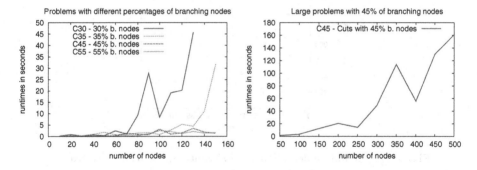

Fig. 4. Runtimes of *LPrk* according to the size of the problem and the number of branching nodes. Distances are in the range [1..100].

As shown in Figure 4, the runtimes of *LPrk* depend heavily on the number of branching nodes within the problem, that we recall to have a fixed degree of 3. For large size random problems, the algorithm behaves correctly provided the percentage of branching nodes is above 40.

Conclusion The *LPrk* approach is very efficient for solving real-life problems with the basic model, in comparison to other methods, especially *BP*. *LPrk* is also the most efficient for a large part of the random problems, i.e., those with up to 400 nodes. However, it gives bad runtimes for a few instances.

The percentage of branching nodes has a great influence on the runtimes whatever the solution algorithm is. The higher this percentage, the better the runtimes.

4.2   Runtimes of Real-Life Testcases with the Full Model

Figure 5 shows the runtimes for the 50 testcases and the full model (see section 2) solved with both the *BP* and *LPrk* algorithms. As opposed to the basic model, constraints on angles are added. Planarity problems arise only in two cases. Thus, it is not necessary to check for a plane embedding at each iteration. Instead, it suffices to check it at the end, and the solution process is restarted after the addition of appropriate constraints whenever needed.

With the *LPrk* solver, the full model is solved in 46.52 s on average compared to 0.33 s on average for the basic model. One case is very costly and represents half of the total runtime. If this case is removed, the average runtime is 21.75 s, which is 2 orders of magnitude more than the basic model runtime. With *BP*, mean runtime was 96.92 s for the full model. In this series, one case (different from the one already mentioned) was very time consuming again. If removed, the mean runtime is 64.24 s, 3 times the average runtime for *LPrk*.

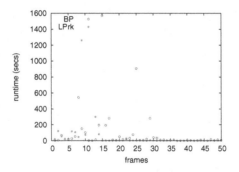

Fig. 5. For the full model, *BP* and *LPrk* runtimes for real-life testcases

These last results show that the complexity as induced by the model's advanced constraints leads to a runtime explosion for the *LPrk* algorithm, although the problem is still solvable with *LPrk* in reasonable time. *BP* runtimes remain stable when taking into account the full model. But those runtimes are still more than two times higher than those of the *LPrk* approach on average for the 50 frames testbench. Finally, notice that sometimes one of the algorithms behaves well while the other spends a lot of time for solving a particular case.

In the next section, we show, however, that those additional constraints are mandatory in order to obtain results that are close to real-life networks. That means that the runtime aspect would be crucial if larger networks are to be computed. Instead of relying on the MILP solver branch-and-bound algorithm when cutting planes are not found anymore, we could exploit these cuts further by embedding them into a branch-and-cut procedure. However, this implies the development of a specific branching strategy. This could help improving the runtimes, especially for those instances where the MILP solver is called early.

4.3   Real-Life Testcase Results and a Comparison with Actual Topologies

We have implemented the proposed technique to find optimal graphs for $N = 50$ problem instances both for the basic model (with degree-3 constraints on nodes $V_b \subset V$) and the full model (with angle constraints as illustrated in Section 2 in addition to the degree-3 constraints). The nodes are abstracted from native ER networks in chosen regions with no cisternae using the image processing method given in [11]. Figure 6 shows different optimal graphs for given sets of nodes and degree-3 nodes together with the abstracted ER network. To quantify the difference between these optimal graphs and the abstracted ER network we may use the concepts of *effective similarity, error correcting matching* and *angle distribution*.

We define an effective similarity $s(G_1, G_2)$ between two graphs G_1, G_2 as the average of two percentages: percentage of edges in G_1 that do appear in G_2 and percentage of edges in G_2 that do appear in G_1. This measurement can be calculated via the adjacency matrices of the two graphs. The measurement $s(G_1, G_2)$ ranges from 0 to 1; it equals 0 if none of the edges coincides in the two graphs and it equals 1 if the connecting structures between the two graphs are the same. Note that the adjacency matrix Adj_G is symmetric whenever G is.

As a complement, and in analogy to the notion of error correcting graph matching [2], we define a normalized error correcting matching $m(G_1, G_2)$ between two graphs G_1, G_2 over the same set of nodes as the ratio of the minimum number of edit operations (edge addition and edge deletion) necessary to transform one graph into the other, to the number of edges of the complete graph with the same set of nodes. This measurement can be calculated via the adjacency matrix as well. It also ranges from 0 (no correction needed) to 1.

Figure 6 illustrates an example of an ER network in comparison to different optimal graphs (including minimal spanning tree, optimal graph from the basic model and optimal graph from the full model). The optimal graphs from the full model show a higher similarity and a lower error correcting matching. Moreover, we show in Table 1 that overall the ER network is closer to an optimal graph from the full model than that with/without degree constraints, and the θ in the angle constraint would not lead to a significant difference in terms of similarity and error correcting matching when compared to the ER network. In addition, Figure 7 shows that the distribution of angles of degree-3 nodes in the model with both degree and angle constraints is much closer to that of the abstracted ER networks than that in the minimal spanning trees and in the optimal graphs with only degree constraints. This suggests that beyond degree constraints, angle constraints are necessary for understanding the principles governing the ER network geometry.

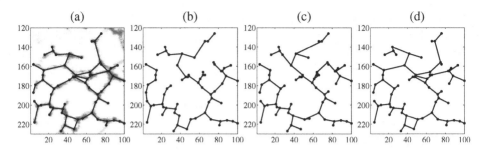

Fig. 6. A comparison between: an abstracted ER network (a), a minimal spanning tree (b), an optimal graph from the basic model (c), an optimal graph from the full model with $\theta = 180°$ (d). Their total lengths are 627.8, 508.3, 558.0 and 602.3, respectively, in units of pixel. The effective similarities of (b-d) with respect to the abstracted network in (a) are $0.726, 0.812$ and 0.938, respectively, while the normalized error correcting matchings with respect to the abstracted network in (a) are $0.120, 0.012$ and 0.004, respectively. The underlying ER image in (a) is from the imaging data in [13].

Table 1. A comparison of optimal solutions G_1 from different models with the abstracted ER network G in terms of similarity $s(G, G_1)$ and matching $m(G, G_1)$. Data indicate the mean±error of the mean (N=50). T tests show significant differences on $s(G, G_1), m(G, G_1)$ between the case where G_1 is a minimal spanning tree and the case where G_1 is an optimal graph from the basic model ($p < 0.0001$ for both measurements), and slight differences between the basic and the full model with $\theta = 160$ ($p = 0.0058, p = 0.0136$ for similarity and matching, respectively). One-way ANOVE tests show no significant differences between graphs from the full model ($p = 0.1083, p = 0.0869$ for similarity and matching, respectively).

N=50	MSP model	basic model	full model ($\theta = 160°$)	full model ($\theta = 170°$)	full model ($\theta = 180°$)
$s(G, G_1)$	0.7376 ± 0.0015	0.9065 ± 0.0103	0.9185 ± 0.0092	0.9270 ± 0.0081	0.9271 ± 0.0084
$m(G, G_1)$	0.2930 ± 0.0195	0.0106 ± 0.0011	0.0096 ± 0.0011	0.0087 ± 0.0010	0.0087 ± 0.0011

In addition, considering that the nodes abstracted from the ER network using the image processing method in [11] may have errors in their position, we have analyzed the sensitivity of optimal solutions with respect to perturbations of the node positions by randomly modifying one pixel among the 8 connected neighbours. Note, that in the presence of angle constraints, optimal solutions should change when a pair of nodes forming a sum of angles close to the critical θ-value are perturbed. Figure 8 shows examples of optimal graphs with perturbed nodes together with those without such a perturbation. However, statistical analysis on similarity and error correcting matching with the ER networks for 10% perturbed nodes gives $m(G, G_1) = 0.9200 \pm 0.0029 (N = 500), s(G, G_1) = 0.0096 \pm 000039 (N = 500)$. This indicates that there is no significant difference in these measurements between optimal graphs with/without node perturbation.

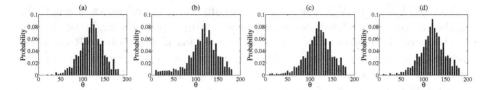

Fig. 7. A comparison of angle distributions from an abstracted ER network (a), an optimal graph from the basic model (b) and an optimal graph from the full model with $\theta = 160°$ (c) and with $\theta = 180°$ (d). Observe, that the distributions in (c) and (d) are much closer (with p-values of 0.863 and 0.975, respectively) to that in (b) (with a p-value of 0.111) when compared to the distribution in (a). The p-values are the asymptotic p-values for the null hypothesis that distribution in (b) ((c) and (d)) and that in (a) are from the same distribution. Angles are taken from all 3 angles of degree-3 nodes in all 50 problem instances. The sample size is $N = 1428$ in each distribution.

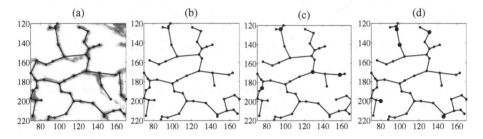

Fig. 8. A comparison of graphs with perturbed node positions. (a) shows an abstracted ER network. (b) shows the optimal graph from the full model with $\theta = 180°$. (c) and (d) show optimal graphs with perturbed node positions (shown as larger bold dots) from the full model with $\theta = 180°$. The underlying ER image in (a) is from the imaging data in [13].

5   Discussion

In this article, a quantitative comparison between optimal graphs and the ER network geometry suggests that the ER tubule network in the native state minimizes the total tubule length between branching nodes and non-branching nodes. The difference between an optimal graph and the corresponding ER network might be that there are other principles behind the ER network geometry. For instance, Figure 9 shows that the ER network contains a cycle while the optimal graph for this instance does not have a cycle. We suspect this ER network to be more robust in structure. Indeed, we have compared the number of cycles and the natural connectivity for all 50 times points in the data set, and we show in Figure 10 that both the number of cycles and the natural connectivity in the abstracted ER networks are overall larger than in corresponding optimal graphs. The natural connectivity $\lambda(G)$ [15] is a measurement of structural robustness of a graph G in terms of a weighted sum of numbers of closed walks. More precisely,

$\lambda(G) = \log(\sum_{k=0}^{\infty} \frac{n_k}{k!}/|V|)$, where n_k is the number of closed walks of length k. Another reason for cycles disappearing in an optimal solution might be due to the choice of the region, where branching junctions in the global ER network might only have one or two edges (branching junctions connecting with tubules outside the chosen regions are not included in the abstracted graph) and thus are not in the set of degree-3 nodes. To avoid this, one may wish to choose a non-rectangle region where branching junctions are all of degree 3 in the abstracted graph. However, these branching junctions might be connected with ER cisternae and thus techniques need to be developed to distinguish ER cisternae and ER tubules.

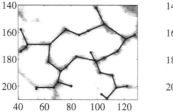 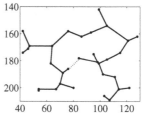

Fig. 9. Left: an abstracted ER network. Right: the optimal graph from the full model ($\theta = 180°$). They only differ in one edge shown as a dotted line in the right panel. The underlying ER image in the left panel is from the imaging data in [13].

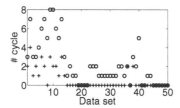 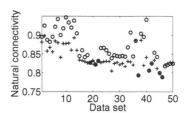

Fig. 10. A comparison of the number of cycles (left) and natural connectivity (right) between abstracted ER networks (circles) and optimal graphs (crossings) from the full model with $\theta = 180°$ for each data set

The study of the optimal principles behind ER network geometry could, in the future, allow predictions of network dynamics as a consequence of node movement. This would require image processing methods to track the node movements and one challenge for this is the dramatic dynamics of ER networks themselves where nodes may appear and disappear. We leave these aspects for future study.

As to algorithmic efficiency, there should be place for improvement. Setting $V = V_b$ and omitting crossing and angle constraints leads to the minimum-weight d-regular connected spanning subgraph problem with $d = 3$. For $d = 2$, we encounter the traveling salesman problem, well-studied both for its polyhedral structure and algorithmic aspects. A suitable transfer of this knowledge to our case, combined with the modifications indicated at the end of section 4.2, could considerably improve the efficiency of our method.

Acknowledgments. CL thanks grant BB/J009903/1 for support of the time.

References

1. Barahona, F.: On the k-cut problem. Oper. Res. Lett. 26, 99–105 (2000)
2. Bunke, H.: Error correcting graph matching: on the influence of the underlying cost function. IEEE Trans. Pattern Anal. Mach. Intell. 21, 917–922 (1999)
3. Cheah, F., Corneil, D.: The complexity of regular subgraph recognition. Discrete Appl. Math. 27, 59–68 (1990)
4. Csardi, G., Nepusz, T.: The igraph software package for complex network research. Inter. Journal, Complex Systems, 1695 (2006), http://igraph.sf.net
5. Edmonds, J.: Maximum matching and a polyhedron with 0-1 vertices. J. Res. Nat. Bur. Standards 69B, 125–130 (1965)
6. Goyal, U., Blackstone, C.: Untangling the web: Mechanisms underlying ER network formation. Biochim. Biophys. Acta 1833, 2492–2498 (2013)
7. Gusfield, D.: Very simple methods for all pairs network flow analysis. SIAM J. Comput. 19(1), 143–155 (1990)
8. Hwang, F.K., Richards, D.S., Winter, P.: The steiner tree problem. Ann. Discrete Math., 53 (1992)
9. Letchford, A., Reinelt, G., Theis, D.: Odd minimum cut sets and b-matchings revisited. SIAM J. Discrete Math. 22(4), 1480–1487 (2008)
10. Levine, T., Rabouille, C.: Endoplasmic reticulum: one continuous network compartmentalized by extrinsic cues. Curr. Opin. Cell. Biol. 17, 362–368 (2005)
11. Lin, C., Ashwin, P., Sparkes, I.A., Zhang, Y.: Structure and dynamics of ER networks and perturbed euclidean steiner networks (2014) (in revision)
12. Sparkes, I.A., Hawes, C., Frigerio, L.: FrontiERs: movers and shapers of the higher plant cortical endoplasmic reticulum. Curr. Opin. Plant. Biol. 14(6), 658–665 (2011)
13. Sparkes, I.A., Runions, J., Hawes, C., Griffing, L.: Movement and remodeling of the endoplasmic reticulum in nondividing cells of tobacco leaves. Plant Cell 21, 3937–3949 (2009)
14. Stoer, M., Wagner, F.: A simple min-cut algorithm. J. ACM 44, 585–591 (1997)
15. Wu, J., Barahona, M., Tan, Y.J., Deng, H.Z.: Spectral measure of structural robustness in complex networks. IEEE Trans. Syst., Man, Cybern.A, Syst., Humans 41, 1244–1252 (2011)

On Sorting of Signed Permutations
by Prefix and Suffix Reversals and Transpositions

Carla Negri Lintzmayer and Zanoni Dias

Institute of Computing, University of Campinas (Unicamp)
Av. Albert Einstein, 1251, Campinas, São Paulo, Brazil
{carlanl,zanoni}@ic.unicamp.br

Abstract. A reversal inverts a segment and the signs of the elements of this segment in a permutation. A transposition exchanges the position of two consecutive segments. These are the most common kinds of genome rearrangements. In this paper, we introduce the study of prefix and suffix versions of these operations, that is, when only segments of the beginning or of the end are involved, when considering signed permutations. We gave asymptotic approximation algorithms of factor two for three new problems: when prefix and suffix reversals are allowed, when prefix reversals and prefix transpositions are allowed, and when prefix and suffix reversals and prefix and suffix transpositions are allowed.

Keywords: sorting signed permutations, reversals, transpositions, prefix operations, suffix operations.

1 Introduction

In Genome Rearrangements, chromosomes are usually represented as sequences of segments which are shared by the genomes we are comparing. Generally, due to the maximum parsimony, the minimum distance between two genomes is considered to be a reasonable representation of the evolution distance between them [3]. When we represent one of them as the identity, the comparison is resumed in sorting the other and finding the events that occurred during the transformation. In this paper we will consider the orientation of the segments in the genomes, that is, we will represent them as signed permutations.

A reversal occurs when a segment of the permutation is inverted as well as the signs of its elements. The problem of Sorting by Signed Reversals is polinomial, as shown by Hannenhalli and Pevzner [7] when they presented a $O(n^4)$ algorithm. Improvements were made and a $O(n^{3/2}\sqrt{\lg n})$ algorithm was introduced [13].

Another well studied problem is Sorting by Signed Prefix Reversals, or the Burnt Pancake Flipping problem. It was introduced by Gates and Papadimitriou [5], whose concerns were in finding the maximum number needed to sort any stack of pancakes of a given size n. They showed that this value lies between $\frac{3n}{2} - 1$ and $2n + 3$ but the best known lower and upper bounds are, respectively, $\frac{3n+3}{2}$ and $2n - 6$ [9,1]. The best known algorithm for this problem was given by Cohen and Blum [2] and it is a 2-approximation.

A.-H. Dediu, C. Martín-Vide, and B. Truthe (Eds.): AlCoB 2014, LNBI 8542, pp. 146–157, 2014.

Walter *et al.* [14] considered a variation in which transpositions are allowed along with signed reversals and gave a 2-approximation algorithm. A transposition occurs when two consecutive segments of the permutation exchange position. Gu *et al.* [6] considered a third event, the transreversal, which is a transposition in which one of the segments are reversed, and also gave a 2-approximation algorithm when the three are allowed. Lin and Xue [10] yet considered a fourth event that they called revrev, which consists in a transposition where the two segments are reversed. When considering the four events, they gave a 1.75-approximation algorithm but the best approximation factor is 1.5 [8].

In this paper, we present approximation algorithms for three sorting problems of signed permutations for which prefix and suffix versions of reversals and transpositions are allowed. Despite the Burnt Pancake Problem, to our knowledge there are not considerations of prefix or suffix events for signed permutations in the literature, specially when more than one rearrangement is allowed. We note that for unsigned permutations, prefix reversals and prefix transpositions were already considered together by Sharmin *et al.* [12], who gave a 3-approximation algorithm, and prefix reversals, suffix reversals, prefix transpositions and suffix transpositions were considered by Lintzmayer and Dias [11].

The rest of this paper is divided as follows. Section 2 presents some important definitions. Sections 3 to 6 describe the algorithms we considered in this paper while Section 7 shows the results.

2   Basic Definitions

We represent a genome as a signed permutation $\pi = (\pi_1 \ \pi_2 \ \ldots \ \pi_n)$ where $\pi_i \in \{-n, -(n-1), \ldots, -1, +1, +2, \ldots, +n\}$, for $1 \le i \le n$, represents a syntenic block (a gene or a block of genes) and $|\pi_i| \ne |\pi_j|$ for all $i \ne j$. It is common to omit the plus sign. Two special permutations are the *identity*, $\iota = (1 \ 2 \ \ldots \ n)$, and the *reverse*, $\eta = (-n \ -(n-1) \ \ldots \ -1)$. The *composition* between two permutations π and σ is the permutation $\pi \cdot \sigma = (\pi_{\sigma_1} \ \pi_{\sigma_2} \ \ldots \ \pi_{\sigma_n})$.

The extended permutation, which is also denoted as π, has two fixed elements $\pi_0 = 0$ and $\pi_{n+1} = n + 1$, always positive. From now on, a permutation will always refer to its extended version, even if the extra elements are omitted.

A *signed reversal* $\bar{\rho}(i,j)$, $1 \le i \le j \le n$, is a rearrangement that transforms π into $\pi \cdot \bar{\rho}(i,j) = (\pi_1 \ \ldots \ \pi_{i-1} \ \overline{-\pi_j \ -\pi_{j-1} \ \ldots \ -\pi_{i+1} \ -\pi_i} \ \pi_{j+1} \ \ldots \ \pi_n)$. A *signed prefix reversal* $\bar{\rho}_p(j)$ is a signed reversal $\bar{\rho}(1,j)$, $1 \le j \le n$, while a *signed suffix reversal* $\bar{\rho}_s(i)$ is a signed reversal $\bar{\rho}(i,n)$, $1 \le i \le n$.

A *transposition* $\tau(i,j,k)$, $1 \le i < j < k \le n + 1$, is a rearrangement that transforms π into $\pi \cdot \tau(i,j,k) = (\pi_1 \ \ldots \ \pi_{i-1} \ \pi_j \ \pi_{j-1} \ \ldots \ \pi_{k-1} \ \pi_i \ \pi_{i+1} \ \ldots \ \pi_{j-1} \ \pi_k \ \ldots \ \pi_n)$. A *prefix transposition* $\tau_p(j,k)$ is a transposition $\tau(1,j,k)$, $2 \le j < k \le n + 1$, while a *suffix transposition* $\tau_s(i,j)$ is a transposition $\tau(i,j,n+1)$, $1 \le i < j \le n$. Note that transpositions do not change the signs of the elements. Therefore, when signed permutations are being used, transpositions can only be considered in problems where reversals are also allowed.

A *breakpoint* for signed problems is a pair of consecutive elements (π_i, π_{i+1}) if $\pi_{i+1} - \pi_i \ne 1$. When prefix rearrangements are considered, $1 \le i \le n$ and (π_0, π_1)

is never a breakpoint. In this case, ι is the only permutation without breakpoints. When prefix and suffix rearrangements are considered, $1 \leq i \leq n - 1$ and both (π_0, π_1) and (π_n, π_{n+1}) are never breakpoints. In this case, ι and η are the only permutations without breakpoints.

A *strip* is a subsequence of elements of π without breakpoints, except for its extremities. For example, $\pi = (3\ 4\ -1\ -2\ -6\ -5)$ has four strips: $[3\ 4]$, $[-1]$, $[-2]$ and $[-6\ -5]$. A strip is *positive* if its elements are positive and *negative* otherwise. A *singleton* is a strip composed by one element.

Let β be the set of rearrangements allowed in a sorting problem. We denote the number of breakpoints of a permutation π by $b_\beta(\pi)$. A sequence or rearrangements $\beta_1, \ldots, \beta_k$, for which $\beta_i \in \beta$, $1 \leq i \leq k$, is called *sorting sequence* if $\pi \cdot \beta_1 \cdots \beta_k = \iota$. The minimum value of k needed to satisfy this equation is called the *sorting distance* of π and it is represented by $d_\beta(\pi)$.

We can now define the lower bounds for the sorting distance of the problems we consider in this paper. Since the identity has the smallest number of breakpoints, one can see that sorting a permutation is the same of reducing its number of breakpoints. A prefix reversal $\bar{\rho}_p(i)$ (resp. a suffix reversal $\bar{\rho}_s(j)$) can reduce the number of breakpoints by at most one unit, because it separates elements $[\pi_0, \pi_1]$ and $[\pi_i, \pi_{i+1}]$ (resp. $[\pi_{j-1}, \pi_j]$ and $[\pi_n, \pi_{n+1}]$) but (π_0, π_1) (resp. (π_n, π_{n+1})) is never a breakpoint. This leads directly to the next lemma.

Lemma 1. *For any permutation π, $d_{\bar{\rho}_p}(\pi) \geq b_{\bar{\rho}_p}(\pi)$ and $d_{\bar{\rho}_p\bar{\rho}_s}(\pi) \geq b_{\bar{\rho}_p\bar{\rho}_s}(\pi)$.*

Moreover, a prefix transposition $\tau_p(i,j)$ (resp. a suffix transposition $\tau_s(i,j)$) can reduce the number of breakpoints by at most two units, because it separates $[\pi_0, \pi_1]$, $[\pi_{i-1}, \pi_i]$ and $[\pi_{j-1}, \pi_j]$ (resp. $[\pi_{i-1}, \pi_i]$, $[\pi_{j-1}, \pi_j]$ and $[\pi_n, \pi_{n+1}]$) but (π_0, π_1) (resp. (π_n, π_{n+1})) is never a breakpoint. This leads to the next lemma.

Lemma 2. *For any permutation π, $d_{\bar{\rho}_p\tau_p}(\pi) \geq b_{\bar{\rho}_p\tau_p}(\pi)/2$ and $d_{\bar{\rho}_p\tau_p\bar{\rho}_s\tau_s}(\pi) \geq b_{\bar{\rho}_p\tau_p\bar{\rho}_s\tau_s}(\pi)/2$.*

3   Sorting by Signed Prefix Reversals

Cohen and Blum [2] described a 2-approximation algorithm for Sorting by Signed Prefix Reversals (SBSPR), which is the best approximation so far for this problem and will be called here 2-SPR. To achieve this, they showed how is always possible to remove one breakpoint with at most two operations. In this section we describe their algorithm somewhat different from their paper because we explicitly use breakpoints and strips. Also, the algorithm that we describe is extended and greedy in a sense, since it first tries to remove one breakpoint with one prefix reversal. Therefore, 2-SPR considers three cases (only the first one is not explicitly considered by Cohen and Blum [2]):

1. If there is a $\pi_j = -\pi_1 + 1$ in π, $2 \leq j \leq n + 1$, then the reversal $\bar{\rho}_p(j-1)$ removes one breakpoint;
2. If π has a positive element out of order, let $\pi_i = k$ be the greatest such element:

    a. If there is a $\pi_j = -(k+1)$ such that $i < j \leq n$, then $\pi = (\ldots\ldots k \ldots\ldots$
$-(k+1) \ldots\ldots)$ and the sequence $\bar{\rho}_p(j) \cdot \bar{\rho}_p(j-i)$ removes one breakpoint;

    b. If there is a $\pi_j = -(k + 1)$ such that $j \leq i$, then $\pi = (\ldots\ldots - (k + 1)$
$\ldots\ldots k \ldots\ldots)$ and the sequence $\bar{\rho}_p(i) \cdot \bar{\rho}_p(i-j)$ removes one breakpoint;

    c. If there is not a $\pi_j = -(k+1)$, then $\pi = (\ldots\ldots k \ldots\ldots k+1 \, k+2 \ldots n)$
and the sequence $\bar{\rho}_p(i) \cdot \bar{\rho}_p(k)$ removes one breakpoint.

3. If π does not have a positive element out of order:

    a. If there is a $\pi_i = -(k + 1)$ and a $\pi_j = -k$ for some $k \geq 1$ such that
$i+1 < j$, then $\pi = (\ldots\ldots - (k+1) \ldots\ldots - k \ldots\ldots)$ and the sequence
$\bar{\rho}_p(i) \cdot \bar{\rho}_p(j-1)$ removes one breakpoint;

    b. If there are not elements such as the described in the previous item, then π
is of the form $(\underbrace{-p_1 \ldots -1}_{\ell_1} \underbrace{-p_2 \ldots -(p_1+1)}_{\ell_2} \ldots\ldots \underbrace{-t \ldots -(p_{x-1}+1)}_{\ell_x}$
$t+1 \, t+2 \ldots n)$ where $x = b_{\bar{\rho}_p}(\pi) \geq 2$, and the sequence $\bar{\rho}_p(t) \cdot \bar{\rho}_p(t - \ell_1)$
$\cdot \bar{\rho}_p(t) \cdot \bar{\rho}_p(t - \ell_2) \cdot \ldots \cdot \bar{\rho}_p(t) \cdot \bar{\rho}_p(t - \ell_{b_{\bar{\rho}_p}(\pi)})$ of $2b_{\bar{\rho}_p}(\pi)$ prefix reversals
sorts it [4].

4   Sorting by Signed Prefix Reversals and Signed Suffix Reversals

We developed an algorithm for Sorting by Signed Prefix Reversals and Signed
Suffix Reversals (SBSPRSSR), which will be called 2-SPRSSR and it is based on
greedy removal of breakpoints. The idea of this algorithm is somewhat similar to
the idea of 2-SPR: it tries to remove one breakpoint with one operation; if this is
not possible, it tries to remove one breakpoint with two operations; if this is not
possible either, the permutation has a special form and there is a sequence to sort
it. The main difference is that we do not consider the greatest element because
there is no sense in talking about "elements that are out of order", since one
suffix reversal would always remove them from their place. Therefore, 2-SPRSSR
considers five main cases:

1. If there is a $\pi_j = -\pi_1 + 1$ in π, $2 \leq j \leq n$, then $\bar{\rho}_p(j-1)$ removes one breakpoint;
2. If there is a $\pi_i = -\pi_n - 1$ in π, $1 \leq i \leq n-1$, then $\bar{\rho}_s(i+1)$ removes one
breakpoint;
3. If there are π_i and π_j such that:

    a. $\pi_j = -\pi_i - 1$, $1 \leq i < j \leq n$, then $\bar{\rho}_p(j) \cdot \bar{\rho}_p(j-i)$ removes one breakpoint;

    b. $\pi_j = \pi_i + 1$, $0 \leq i+1 < j \leq n$, then $\bar{\rho}_p(i) \cdot \bar{\rho}_p(j-1)$ removes one breakpoint.

4. If there are π_i and π_j such that:

    a. $\pi_i = -\pi_j + 1$, $1 \leq i < j \leq n$, then $\bar{\rho}_s(i) \cdot \bar{\rho}_s(n + 1 - (j - i))$ removes one
breakpoint;

    b. $\pi_i = \pi_j - 1$, $0 \leq i+1 < j \leq n$, then $\bar{\rho}_s(j) \cdot \bar{\rho}_s(i+1)$ removes one breakpoint.
This item is equivalent to item 3.b, therefore, only one of them needs to
be considered.

5. If it is not possible to remove one breakpoint as described in the previous
items, them π is of one of the three forms described in Lemma 3 and one
sequence of at most $b_{\bar{\rho}_p \bar{\rho}_s}(\pi) + 2$ reversals, as described in Lemma 4, sorts π.

Lemma 3. *Let π be a signed permutation for which neither one nor two operations can remove one breakpoint. Then π is of one of the three forms:*

1. η;
2. $\sigma^1 = (\underbrace{p_{b_{\bar\rho_p\bar\rho_s}(\pi)}+1 \ \ldots \ n}_{\ell_{b_{\bar\rho_p\bar\rho_s}(\pi)+1}} \ \underbrace{p_{b_{\bar\rho_p\bar\rho_s}(\pi)-1}+1 \ \ldots \ p_{b_{\bar\rho_p\bar\rho_s}(\pi)}}_{\ell_{b_{\bar\rho_p\bar\rho_s}(\pi)}} \ \ldots\ldots \ \underbrace{1 \ \ldots \ p_1}_{\ell_1});$
3. $\sigma^2 = (\underbrace{-p_1 \ \ldots \ -1}_{\ell_1} \ \underbrace{-p_2 \ \ldots \ -(p_1+1)}_{\ell_2} \ \ldots\ldots \ \underbrace{-n \ \ldots \ -(p_{b_{\bar\rho_p\bar\rho_s}(\pi)}+1)}_{\ell_{b_{\bar\rho_p\bar\rho_s}(\pi)+1}}).$

where $\ell_i \geq 1$ for all $1 \leq i \leq b_{\bar\rho_p\bar\rho_s}(\pi) + 1$ and $b_{\bar\rho_p\bar\rho_s}(\pi) \geq 1$.

Proof. It is easy to see that when $\pi = \eta$ the first four main cases of the algorithm cannot turn it into the identity, since a reversal $\bar\rho_p(n)$ is necessary and it is not considered as a reversal that would remove a breakpoint. In fact, it does not remove a breakpoint, because $b_{\bar\rho_p\bar\rho_s}(\eta)$ is already zero.

Let $\pi_i = k$ be any element of π. If $\pi_j = -(k+1)$ exists and $i < j$, then $\bar\rho_p(j) \cdot \bar\rho_p(j - i)$ remove one breakpoint. If $j < i$, then $\bar\rho_s(j) \cdot \bar\rho_s(n + 1 - (i - j))$ remove one breakpoint. Something similar happens when $-(k - 1)$ exists in π. Therefore, the elements of π must all have the same sign.

Suppose that π only has positive elements. If $b_{\bar\rho_p\bar\rho_s}(\pi) = 1$, is trivial to see that π must be of the form of σ^1. Now assume that every permutation with $b - 1$ positive strips for which is not possible to remove one breakpoint with one or two operations is of the form of σ^1.

Let π be a permutation with b positive strips for which is not possible to remove one breakpoint with one or two operations and let π_j, $1 < j \leq n$, be the first element of the last strip of π. Note that we must have $\pi_j = 1$, otherwise, we would have a $\pi_i = \pi_j - 1$ for some $i + 1 < j$, a contradiction, since $\bar\rho_p(i) \cdot \bar\rho_p(j - 1)$ could remove a breakpoint. Let π' be a permutation built from π such that $\pi'_i = \pi_i - \pi_n$ for all $1 \leq i \leq j$.

It is easy to see that π' has $b - 1$ positive strips and it is almost immediate to see that if it would be possible to remove one breakpoint from π', then it would also be possible to remove it from π. Therefore, π' is of the form of σ^1. Since π' is π relabeled without the last strip, which has the element 1, it follows that π is also of the form of σ^1.

A similar demonstration can be made when π is formed only by negative strips: it must be of the form of σ^2. □

Lemma 4. *Let π be one of the signed permutations described in Lemma 3. Then at most $b_{\bar\rho_p\bar\rho_s}(\pi) + 2$ operations sort π.*

Proof. If $\pi = \eta$, one prefix reversal $\bar\rho_p(n)$ or one suffix reversal $\bar\rho_s(1)$ suffices.

If $\pi = \sigma^1$ and $b_{\bar\rho_p\bar\rho_s}(\pi)$ is an even number, then the $b_{\bar\rho_p\bar\rho_s}(\pi) + 1$ reversals $\bar\rho_p(n-\ell_1) \cdot \bar\rho_s(\ell_2+1) \cdot \bar\rho_p(n-\ell_3) \cdot \bar\rho_s(\ell_4+1) \cdot \ldots \cdot \bar\rho_p(n-\ell_{b_{\bar\rho_p\bar\rho_s}(\pi)-1}) \cdot \bar\rho_s(\ell_{b_{\bar\rho_p\bar\rho_s}(\pi)}+1) \cdot \bar\rho_p(n - \ell_{b_{\bar\rho_p\bar\rho_s}(\pi)+1})$ sort π, as we show next.

Let π^k, $1 \leq k \leq b_{\bar\rho_p\bar\rho_s}(\pi)/2$, be the permutation we obtain after applying the first $2k$ reversals of the sequence given above: $\bar\rho_p(n - \ell_1) \cdot \bar\rho_s(\ell_2 + 1) \cdot \ldots \cdot \bar\rho_p(n -$

$\ell_{2k-1}) \cdot \bar{\rho}_s(\ell_{2k}+1)$. Let $b = b_{\bar{\rho}_p \bar{\rho}_s}(\pi)$ and $p_{b+1} = n$, for simplicity. We will show by induction on k that π^k is of the form

$$
(\underbrace{-p_{2k} \ \ldots \ - (p_{2k-1}+1) \ -p_{2k-1} \ \ldots - (p_{2k-2}+1) \ \ldots \ldots \ - p_1 \ \ldots \ -1}_{\ell_{2k}+\ell_{2k-1}+\ldots+\ell_1}
$$
$$
\underbrace{p_b+1 \ \ldots \ n}_{\ell_{b+1}} \ \ldots \ldots \ \underbrace{p_{2k+1}+1 \ \ldots \ p_{2k+2}}_{\ell_{2k+2}} \ \underbrace{p_{2k}+1 \ \ldots \ p_{2k+1}}_{\ell_{2k+1}})
$$

It is easy to see that π^k has this form when $k = 1$. Now, assume that π^{k-1} is of this form, that is, π^{k-1} is equal to

$$
(\underbrace{-p_{2k-2} \ \ldots \ - (p_{2k-3}+1) \ -p_{2k-3} \ \ldots - (p_{2k-4}+1) \ \ldots \ldots \ - p_1 \ldots -1}_{\ell_{2k-2}+\ell_{2k-3}+\ldots+\ell_1}
$$
$$
\underbrace{p_b+1 \ \ldots \ n}_{\ell_{b+1}} \ \ldots \ldots \ \underbrace{p_{2k-1}+1 \ \ldots \ p_{2k}}_{\ell_{2k}} \ \underbrace{p_{2k-2}+1 \ \ldots \ p_{2k-1}}_{\ell_{2k-1}})
$$

Since $\pi^k = \pi^{k-1} \cdot \bar{\rho}_p(n - \ell_{2k-1}) \cdot \bar{\rho}_s(\ell_{2k}+1)$, the result follows.

Now, when $k = \frac{b}{2}$, $\pi^{b/2} = (-p_b \ldots - (p_{b-1}+1) \ldots - p_1 \ldots - 1 \ p_b+1 \ldots \ n)$ and the last reversal of the sequence, $\bar{\rho}_p(n - \ell_{b+1})$, sorts $\pi^{b/2}$.

If $\pi = \sigma^2$ and $b_{\bar{\rho}_p \bar{\rho}_s}(\pi)$ is an even number, one must apply $\bar{\rho}_p(n)$ to transform it into σ^1 and then apply the $b_{\bar{\rho}_p \bar{\rho}_s}(\pi) + 1$ reversals given above.

If $\pi = \sigma^2$ and $b_{\bar{\rho}_p \bar{\rho}_s}(\pi)$ is an odd number, then the $b_{\bar{\rho}_p \bar{\rho}_s}(\pi) + 1$ reversals $\bar{\rho}_s(\ell_1+1) \cdot \bar{\rho}_p(n-\ell_2) \cdot \bar{\rho}_s(\ell_3+1) \cdot \bar{\rho}_p(n-\ell_4) \cdot \ldots \cdot \bar{\rho}_s(\ell_{b_{\bar{\rho}_p \bar{\rho}_s}(\pi)}+1) \cdot \bar{\rho}_p(n-\ell_{b_{\bar{\rho}_p \bar{\rho}_s}(\pi)+1})$ sort π. This can also be shown by a similar induction as the one done above.

If $\pi = \sigma^1$ and $b_{\bar{\rho}_p \bar{\rho}_s}(\pi)$ is odd, one must apply $\bar{\rho}_p(n)$ to transform it into σ^2 and then apply the reversals above. $\qquad \square$

Although the sequences used to handle the permutations shown in Lemma 3 are half of the size used in SBSPR, sometimes the algorithm needs two reversals to remove one breakpoint, which makes it an asymptotic 2-approximation.

Theorem 5. 2-SPRSSR *is an approximation algorithm of asymptotic factor 2.*

Proof. In the worst case, two operations will always be used to remove one breakpoint and the reverse permutation will be reached. Therefore, the number of prefix or suffix reversals used is at most $2b_{\bar{\rho}_p \bar{\rho}_s}(\pi)+1$. It follows from the lower bound that 2-SPRSSR is an asymptotic 2-approximation algorithm. $\qquad \square$

5 Sorting by Signed Prefix Reversals and Prefix Transpositions

We also developed an algorithm for Sorting by Signed Prefix Reversals and Prefix Transpositions problem (SBSPRPT). It will be called 2-SPRPT and it also has a greedy idea: first, it tries to remove two breakpoints with one prefix transposition; if it is not possible, then it tries to remove one breakpoint with either one prefix transposition or one prefix reversal.

Let $\pi \neq \iota$ and suppose that $\tau_p(i,j)$ is a prefix transposition that removes two breakpoints from π. Since $\pi \cdot \tau_p(i,j) = (\pi_i \ \dots \ \pi_{j-1} \ \overline{\pi_1 \ \dots \ \pi_{i-1}} \ \dots \ \pi_n)$, one must have that $\pi_{j-1} = \pi_1 - 1$ and $\pi_{i-1} = \pi_j - 1$. In order to maintain our approximation, we cannot have $\pi_i = 1$.

Let $\pi = (k{+}1 \ k{+}2 \ \dots \ k{+}(i{-}1) \ \pi_i \ \dots \dots)$ with $i \geq 2$ and $\pi_i \neq k{+}i$. There are two possibilities to remove one breakpoint with one prefix transposition $\tau_p(i,j)$ by increasing the first strip: (i) if its next element $\pi_j = k + i = \pi_{i-1} + 1$ exists in π; or (ii) if its previous element $\pi_{j-1} = k = \pi_1 - 1$ exists in π.

Finally, to remove one breakpoint with one prefix reversal is identical to SBSPR: if exists $\pi_{j+1} = -\pi_1 + 1$, then $\bar{\rho}_p(j)$ for $1 \leq j \leq n$ suffices.

Note that it is always possible to remove one breakpoint from π with these last steps we described, because they always manage to increase the first strip with its previous or next element, positive or negative.

The three main cases described above will be considered if and only if $\pi_1 \neq 1$. Whenever $\pi_1 = 1$, the algorithm must send the first strip to the end of the permutation, from where it will be removed when the element n is sent there, as Lemma 6 shows. Because of this, π_1 will be 1 again at most one more time, which will allow the algorithm to keep removing one or two breakpoints until the end of the sorting, as Lemmas 7 and 8 show. Theorem 9 shows how this behavior guarantees the desired approximation factor.

Lemma 6. *Let π be a signed permutation of the form $\pi = (\dots \dots \pi_{n-i} \ 1 \ 2 \ \dots \ i)$, $1 \leq i < n$. The last strip of π will be removed from there only when the element n is sent to the end.*

Proof. Since $\pi_1 \neq 1$, the algorithm will first try to remove two breakpoints from π with a transposition $\tau_p(i,j)$. Then, one must have $\pi_i = \pi_j - 1$. To remove the last strip from the end with this transposition, one must have $j = n + 1$. Therefore, $\pi_{i-1} = n$ will be sent to the end. Next, it will try to remove one breakpoint from π with a transposition $\tau_p(i,j)$ where π_i is the last element of the first strip of π. If it increases the first strip with its next element, then $\pi_j = \pi_{i-1} + 1$. To remove the last strip from the end, one must have $j = n + 1$, in which case $\pi_{i-1} = n$ will be sent to the end. If it increases the first strip with its previous element, then $\pi_{j-1} = \pi_1 - 1$. To remove the last strip from the end, π_{j-1} should be the last element of the last strip, that is, $j = n + 1$. However, in this case, the last strip would only be increasing its length. Finally, if a reversal $\bar{\rho}_p(j)$ removes one breakpoint, one must have $\pi_{j+1} = -\pi_1 + 1$. To remove the last strip from the end with this reversal, one must have $j = n$, in which case $\pi_1 = -n$ and n is sent to the end of the permutation. $\square$

Lemma 7. *Let π be a signed permutation of the form $\pi = (\dots \dots n \ 1 \ 2 \ \dots \ i)$, $1 \leq i < n$. Then, the elements n and 1 will not be separated until π is sorted. Therefore, it is always possible to keep removing one or two breakpoints.*

Proof. We have that $\pi_1 \neq 1$, so the algorithm will first try to remove two breakpoints with $\tau_p(i,j)$. Since we explicitly deny that $\pi_i = 1$ in this step, consider $\pi_j = 1$. We would have that $\pi_{j-1} = n$, but then π_1 would have to be $n{+}1$, which is impossible. Now the algorithm will try to increase the first strip, which ends at π_{i-1} with $\tau_p(i,j)$. If $\pi_i = 1$, then the first strip ends at n and this transposition would sort π and remove two breakpoints at once. If $\pi_j = 1$, then either π_{i-1} would have to be 0 or π_1 would have to be $n+1$. Both cases are impossible. Finally, the algorithm will try to perform $\bar{\rho}_p(j)$ to remove one breakpoint. If $\pi_{j+1} = 1$, then π_1 would have to be 0. Hence, the algorithm will not separate elements n and 1, unless it is one step from sorting the permutation. Except for this case, π_1 will always be different from 1 and the three main cases of the algorithm can be used. $\qquad\square$

Lemma 8. *During the sorting, π_1 will be 1 at most twice.*

Proof. When $\pi_1 = 1$ and $\pi_n \neq n$, the first strip is sent to the end of the permutation and it does not change the number of breakpoints. As Lemma 6 shows, it will only be removed from there when n goes to the end. Since the main steps of the algorithm are always trying to remove breakpoints, n will not be removed from the end until $\pi_1 = 1$ again. In this case, the number of breakpoints is increased in one unit. However, the elements n and 1 are put together and the algorithm does not separate them, as shown by Lemma 7. $\qquad\square$

Theorem 9. 2-SPRST *is an asymptotic approximation algorithm of factor 2.*

Proof. While $\pi_1 \neq 1$, the algorithm always remove one or two breakpoints. In the worst case, $\pi_1 = 1$ twice and two extra operations are needed to move the first strip as Lemma 8 shows (one of them creates one breakpoint but puts n and 1 together, which will guarantee that the last operation of the sorting removes two breakpoints as Lemma 7 shows). Therefore, at most $b_{\bar{\rho}_p\tau_p}(\pi) + 2$ operations sort any permutation. It follows from the lower bound that 2-SPRPT is an asymptotic 2-approximation algorithm. $\qquad\square$

6 Sorting by Signed Prefix Reversals, Prefix Transpositions, Signed Suffix Reversals and Suffix Transpositions

Finally, we developed an algorithm for the problem that allows all variations: Sorting by Signed Prefix Reversals, Prefix Transpositions, Signed Suffix Reversals and Suffix Transpositions (SBSPRPTSSRST), which we called 2-SPRPTSSRST. It follows the same greedy idea of 2-SPRPT with the same restriction regard the separation of elements n and 1, but it also does not allow that -1 and $-n$ be separated when they are together in this order. Thus, the main steps of this algorithm are: (i) try to remove two breakpoints with one prefix or suffix transposition; (ii) try to remove one breakpoint with one prefix or suffix transposition; and (iii) try to remove one breakpoint with one prefix or suffix reversal.

To remove two breakpoints with one prefix transposition $\tau_p(i,j)$, we also must find $\pi_{j-1} = \pi_1 - 1$ and $\pi_{i-1} = \pi_j - 1$, but this time considering that $2 \leq i < j \leq n$. To remove two breakpoints with one suffix transposition $\tau_s(i,j)$, since

$\pi \cdot \tau_s(i,j) = (\pi_1 \ \ldots \ \pi_{i-1} \ \pi_j \ \ldots \ \pi_n \ \pi_i \ \ldots \ \pi_{j-1})$, one must have that $\pi_i = \pi_n + 1$ and $\pi_j = \pi_{i-1} + 1$ also with $2 \leq i < j \leq n$. For both cases, neither $\pi_{i-1} = n$ and $\pi_i = 1$ nor $\pi_{j-1} = -1$ and $\pi_j = -n$ can happen.

The removal of one breakpoint with one prefix transposition $\tau_p(i,j)$ is almost similar to what we explained in the previous section. Let $\pi = (k+1 \ k+2 \ \ldots \ k+(i-1) \ \pi_i \ \ldots \ldots)$ with $i \geq 2$ and $\pi_i \neq k+i$, there are two possibilities to increase the first strip: (i) if its next element $\pi_j = k+i = \pi_{i-1} + 1$, with $j \leq n$, exists in π; or (ii) if its previous element $\pi_{j-1} = k = \pi_1 - 1$ exists in π. Neither $\pi_{i-1} = n$ and $\pi_i = 1$ nor $\pi_{j-1} = -1$ and $\pi_j = -n$ can happen.

Let $\pi = (\ldots \ldots \ \pi_{j-1} \ k+1 \ k+2 \ \ldots \ k+x)$ with $x \geq 1$, $j \leq n$ and $\pi_{j-1} \neq k$. There is also two possibilities to increase the last strip and remove one breakpoint with one suffix transposition $\tau_s(i,j)$: (i) if its previous element $\pi_{i-1} = \pi_j - 1 = k$, with $i \geq 2$, exists in π; or (ii) if its next element $\pi_i = \pi_n + 1 = k+x+1$ exists in π. Neither $\pi_{i-1} = n$ and $\pi_i = 1$ nor $\pi_{j-1} = -1$ and $\pi_j = -n$ can happen.

Finally, to remove one breakpoint with one prefix reversal $\bar\rho_p(j)$, it suffices to find a $\pi_{j+1} = -\pi_1 + 1$ where $1 \leq j \leq n - 1$. To remove one breakpoint with one suffix reversal $\bar\rho_s(i)$, one must find a $\pi_{i-1} = -\pi_n - 1$, with $2 \leq i \leq n$.

When none of the options given above are available, the permutation is of one of the forms described in Lemma 10 and we must perform either a reversal to sort η or a prefix transposition to concatenate the first strip with the last one. The later operation is the reason why the algorithm cannot separate the elements n and 1 or -1 and $-n$, which will guarantee the approximation factor of the algorithm, as Theorem 12 shows.

Lemma 10. *Let π be a signed permutation for which is not possible to remove at least one breakpoint with one operation. Then π is of one of the five forms:*

1. *η;*
2. *$\mu^1 = (1 \ 2 \ \ldots \ k \ \ldots \ -(k+1) \ \ldots \ -(i-1) \ \ldots \ i \ i+1 \ \ldots \ n)$, that is, $\pi_1 = 1$, $\pi_n = n$, and the elements that can increase the first and last strips are negative (the relative position between them is irrelevant);*
3. *$\mu^2 = (\underline{-n \ -(n-1) \ldots -i} \ \ldots \ (i-1) \ \ldots \ (k+1) \ \ldots \ \underline{-k \ -(k-1)\ldots -1})$, that is, $\pi_1 = -n$, $\pi_n = -1$, and the elements that can increase the first and last strips are positive (the relative position between them is irrelevant);*
4. *$\mu^3 = (k+1 \ k+2 \ \ldots \ n \ 1 \ 2 \ \ldots \ k)$;*
5. *$\mu^4 = (\underline{-k \ -(k-1)} \ \ldots \ -1 \ \underline{-n \ -(n-1) \ \ldots \ -(k+1)})$.*

where $1 \leq k$ and $k+1 < i \leq n$.

Proof. It is easy to see that when $\pi = \eta$, the first three main steps of the algorithm cannot be performed. We must explicitly perform $\bar\rho_p(n)$ to sort it.

When $\pi = \mu^1$, one must have $\pi_1 = 1$ and $\pi_n = n$. Thus, suppose $\pi_1 = \ell \neq 1$, disregard the value of π_n. Either $+(\ell-1)$ or $-(\ell-1)$ must exist in π. Therefore, a transposition or a reversal can be performed to remove one breakpoint, a contradiction. Now, suppose $\pi_n = \ell \neq n$, disregard the value of π_1. Either $-(\ell+1)$ or $+(\ell+1)$ must exist in π. Therefore, a reversal or a transposition can be performed to remove one breakpoint, a contradiction. A very similar analysis can be done when $\pi = \mu^2$. Since π_1 can only be 1 or $-n$ and π_n can only be n or -1, μ^1 and μ^2 are the only possible forms that allow this. Also, if the elements that can increase the first and the last strip were not of opposite signs regarding the strips, a transposition that removes one breakpoint could be performed.

Is easy to see that, since the algorithm does not separate n and 1 or -1 and $-n$, the cases when $\pi = \mu^3$ or $\pi = \mu^4$ cannot be dealt by the first three steps. □

Lemma 11. *Let π be a signed permutation as described in Lemma 10 such that $\pi \neq \eta$. Then, one transposition that concatenates the first strip with the last one does not create new breakpoints and guarantees that the next operations will always be able to remove at least one breakpoint.*

Proof. Let $\pi = \mu^1$. To concatenate the strips, the transposition $\tau_p(k+1, n+1)$ suffices. Then, $\pi' = \pi \cdot \tau_p(k+1, n+1) = (\ldots -(k+1)\ldots -(i-1)\ldots i\ i+1 \ldots n\ 1\ 2\ \ldots\ k)$. When $\pi = \mu^2$, the transposition $\tau_p(n-i+2, n+1)$ suffices and $\pi' = \pi \cdot \tau_p(n-i+2, n+1) = (\ldots k+1\ldots j-1\ldots -k-(k-1)\ldots -1 -n-(n-1)\ldots -i)$. In both cases, π' is not of any form given in Lemma 10. Therefore, is possible to remove at least one breakpoint with one operation.

Note that μ^1 and μ^2 will happen again when $\pi_1 = 1$ and $\pi_n = n$ or when $\pi_1 = -n$ and $\pi_n = -1$. However, this only can happen if the algorithm separate the elements n and 1 or -1 and $-n$, which is not allowed. On the other hand, this separation will have to happen when $\pi = \mu^3$ and $\pi = \mu^4$. But at this point, the algorithm will be performing its last or last but one operation. When $\pi = \mu^3$, the concatenation of the strips will lead to ι. When $\pi = \mu^4$, the concatenation will lead to η, which will be followed directly by a reversal that will sort it. □

Theorem 12. 2-SPRPTSSRST *is an asymptotic 2-approximation algorithm.*

Proof. While π is not of the forms described in Lemma 10, it can remove at least one breakpoint with one operation. Two extra operations that do not remove nor create breakpoints may be needed: one to handle μ^1 or μ^2 and one to handle η (after μ^4 is reached or simply when η itself is reached). Therefore, 2-SPRPTSSRST always sorts π with less than $b_{\bar{\rho}_p \tau_p \bar{\rho}_s \tau_s}(\pi) + 2$ operations. It follows from the lower bound that it is an asymptotic algorithm of factor 2. □

7 Results

We implemented all the algorithms we described in C language. They all have complexity of $O(n^2)$ because they run while the permutation is not sorted (the distance is proportional to the number of breakpoints, which is $O(n)$) and, in each step, they search for a next operation to apply (the search and the operations are $O(n)$). Nevertheless, we highlight that our goals in these tests are regarding the approximation factors. The algorithms were executed under a set of 1,990,000 permutations: for each value of n, $10 \leq n \leq 1,000$ in intervals of 5, 10,000 arbitrary signed permutations only with singletons were generated. For each permutation, an approximation factor was calculated by dividing the distance given by the algorithm for the theoretical lower bound.

Since all algorithms we developed are asymptotic, it is expected that sometimes the approximation factor is above 2. This happened only once for 2-SPRSSR when $n = 10$. For 2-SPRPT, this happened on 0.41% of the permutations, only when $n \leq 100$. Besides, from this amount, 72.13% happened when $n \leq 20$ while 98.65% happened when $n \leq 50$. For 2-SPRPTSSRST, the factor was above 2 on 0.44% of the permutations, which happened only when $n \leq 105$. From this amount, 76.62% happened when $n \leq 20$ while 99.35% happened when $n \leq 50$. Figure 1 shows how the average approximation factor of the 10,000 permutations of size n changes when n grows for each problem.

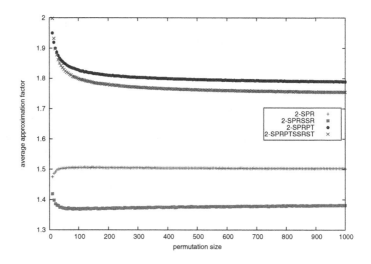

Fig. 1. Average approximation factor of all algorithms when the permutation size grows

We notice that 2-SPR presents a stable average approximation factor that is close to 1.5. In fact, for $n \geq 100$, this factor is always below 1.507. The maximum factor over all permutations tested is 2.000 and it happens when $n = 10$. When $n \geq 100$, this maximum factor does not exceed 1.700. On the other hand, 2-SPRSSR shows a slight growth when n grows. Nevertheless, the average approximation factor is very slow and does not exceed 1.381 for $n \geq 100$. For this algorithm, the maximum factor obtained is 2.111 and happens also when $n = 10$. For $n \geq 100$, the maximum factor does not exceed 1.661.

For 2-SPRPT and 2-SPRPTSSRST, the curves are similar in some sense and are always decreasing. For the former, the average factor for $n \geq 100$ is always below 1.828 while the maximum factor over all permutations is 2.400, also happening when $n = 10$, followed by 2.267 when $n = 15$. For 2-SPRPTSSRST, the average factor is below 1.799 when $n \geq 100$. The maximum factor of all permutations is 2.444 and it happens when $n = 10$ followed by 2.286 when $n = 15$.

Acknowledgements. This work was partially supported by São Paulo Research Foundation - FAPESP (grants 2013/01172-0 and 2013/08293-7) and National Counsel of Technological and Scientific Development - CNPq (grants 477692/2012-5 and 483370/2013-4).

References

1. Cibulka, J.: On Average and Highest Number of Flips in Pancake Sorting. Theoretical Computer Science 412(8-10), 822–834 (2011)
2. Cohen, D.S., Blum, M.: On the Problem of Sorting Burnt Pancakes. Discrete Applied Mathematics 61(2), 105–120 (1995)
3. Fertin, G., Labarre, A., Rusu, I., Tannier, É., Vialette, S.: Combinatorics of Genome Rearrangements. The MIT Press (2009)
4. Galvão, G.R.: Uma Ferramenta de Auditoria para Algoritmos de Rearranjo de Genomas. Master's thesis, University of Campinas, Institute of Computing, Campinas, São Paulo, Brazil (2012) (in Portuguese)
5. Gates, W.H., Papadimitriou, C.H.: Bounds for Sorting by Prefix Reversal. Discrete Mathematics 27(1), 47–57 (1979)
6. Gu, Q.P., Peng, S., Sudborough, I.H.: A 2-Approximation Algorithm for Genome Rearrangements by Reversals and Transpositions. Theoretical Computer Science 210(2), 327–339 (1999)
7. Hannenhalli, S., Pevzner, P.A.: Transforming Cabbage into Turnip: Polynomial Algorithm for Sorting Signed Permutations by Reversals. Journal of the ACM 46(1), 1–27 (1999)
8. Hartman, T., Sharan, R.: A 1. 5-Approximation Algorithm for Sorting by Transpositions and Transreversals. Journal of Computer and System Sciences 70(3), 300–320 (2005)
9. Heydari, M.H., Sudborough, I.H.: On the Diameter of the Pancake Network. Journal of Algorithms 25(1), 67–94 (1997)
10. Lin, G.H., Xue, G.: Signed Genome Rearrangement by Reversals and Transpositions: Models and Approximations. Theoretical Computer Science 259(1-2), 513–531 (2001)
11. Lintzmayer, C.N., Dias, Z.: Sorting Permutations by Prefix and Suffix Versions of Reversals and Transpositions. In: Pardo, A., Viola, A. (eds.) LATIN 2014. LNCS, vol. 8392, pp. 671–682. Springer, Heidelberg (2014)
12. Sharmin, M., Yeasmin, R., Hasan, M., Rahman, A., Rahman, M.S.: Pancake Flipping with Two Spatulas. In: International Symposium on Combinatorial Optimization (ISCO 2010). Electronic Notes in Discrete Mathematics, vol. 36, pp. 231–238 (2010)
13. Tannier, E., Bergeron, A., Sagot, M.F.: Advances on Sorting by Reversals. Discrete Applied Mathematics 155(6-7), 881–888 (2007)
14. Walter, M.E.M.T., Dias, Z., Meidanis, J.: Reversal and Transposition Distance of Linear Chromosomes. In: Proceedings of the 5th International Symposium on String Processing and Information Retrieval (SPIRE 1998), Santa Cruz, Bolivia, pp. 96–102. IEEE Computer Society (1998)

On the Diameter of Rearrangement Problems

Carla Negri Lintzmayer and Zanoni Dias

Institute of Computing, University of Campinas (Unicamp)
Av. Albert Einstein, 1251, Campinas, São Paulo, Brazil
{carlanl,zanoni}@ic.unicamp.br

Abstract. When we consider the Genome Rearrangements area, the problems of finding the distance of a permutation and finding the diameter of all permutations of the same size are the most common studied. In this paper, we considered problems for which no known results were presented regarding their diameters. We present some families of permutations whose distance is identical to the diameter for small sizes. They allowed us to gave bounds for the diameters of the problems we considered, as well as conjectures regarding the exact value.

Keywords: Sorting permutations, diameter, reversals, transpositions, prefix operations, suffix operations.

1 Introduction

Rearrangements are mutations that happen in big portions of the genomes, transforming them. It is assumed that the minimum distance between two genomes, that is, the minimum number of rearrangements needed to transform one into another, represents the evolution distance between them. We can represent them as permutations and assume that one is the identity, so that it is only needed to sort the other to find the events that occurred during the transformation.

Therefore, interesting combinatorial problems have raised from the Genome Rearrangements: given a permutation one wants to find its *distance*, that is, the minimum number of rearrangements needed to sort it. In general, approximation algorithms are developed. However, it is also common to find studies regarding the *diameter*: what is the greatest distance between the distances of all permutations of a size n? The last one is our interest in this paper.

Sorting by Reversals and Sorting by Transpositions are well studied NP-hard [4,3] problems. For the former, Bafna and Pevzner [1] showed that the diameter is $n-1$. The latter still has an unknown diameter and the best lower and upper bounds are, respectively, $\lfloor \frac{n+1}{2} \rfloor + 1$, for $n \geq 1$, and $\lfloor \frac{2n-2}{3} \rfloor$, for $n \geq 9$ [9,10]. On the contrary, Sorting by Signed Reversals is polinomial, as shown by Hannenhalli and Pevzner [14]. It is known, therefore, that its diameter is $n+1$ [11].

Walter *et al.* [21] considered the Sorting by Reversals and Transpositions and the Sorting by Signed Reversals and Transpositions problems. They gave a lower bound for the diameter of the latter, which is $\lfloor \frac{n}{2} \rfloor + 2$, and conjectured that it

A.-H. Dediu, C. Martín-Vide, and B. Truthe (Eds.): AlCoB 2014, LNBI 8542, pp. 158–170, 2014.

is the exact value. We show that this is not true. To our knowledge, no results regarding the diameter of the unsigned version is known.

Rearrangements that affect segments from the beginning of the genome are prefix rearrangements. The Pancake Flipping problem, or Sorting by Prefix Reversals, was proved to be NP-hard recently [2]. It is interesting to notice that the first computational results regarding this problem, given by Gates and Papadimitrou [13], were about its diameter. They showed that it lies between $\frac{17n}{16}$ and $\frac{5n+5}{3}$. The best known lower bound was improved eighteen years later to $\frac{15n}{14}$ [15]. The best upper bound is $\frac{18n}{11} + O(1)$ [5]. Sorting by Prefix Transpositions remains open and its diameter, also unknown, lies between $\lfloor \frac{3n+1}{4} \rfloor$, for $n \geq 2$, and $n - \log_{\frac{9}{2}} n$ [16,6]. Dias and Meidanis [8] conjectured that its exact value is $n - \lfloor \frac{n}{4} \rfloor$ for $n \geq 4$. Sharmin et $al.$ [20] introduced Sorting by Prefix Reversals and Prefix Transpositions but no results regarding the diameter is known.

Another well studied problem is Sorting by Signed Prefix Reversals, introduced by Gates and Papadimitriou [13]. They showed that the diameter lies between $\frac{3n}{2} - 1$ and $2n + 3$ but the best known lower and upper bounds are, respectively, $\frac{3n+3}{2}$ and $2n - 6$ [15,7].

Lintzmayer and Dias [17,18] introduced new problems for which signed and unsigned suffix rearrangements are allowed along with their prefix versions and presented approximation algorithms but without results regarding the diameters.

In this paper we are interested in the problems of Sorting by Reversals and Transpositions, Sorting by Prefix Reversals and Prefix Transpositions, Sorting by Prefix Reversals and Suffix Reversals, Sorting by Prefix Reversals, Prefix Transpositions, Suffix Reversals and Suffix Transpositions, and Sorting by Prefix Transpositions and Suffix Transpositions. We considered both signed and unsigned versions of all these problems, except for the last one.

This paper is divided as follows. Section 2 presents some definitions. Section 3 present some families of permutations and their distances, which will be used in Section 4 to derive some bounds on the diameter of the cited problems. Section 5 concludes our work.

2   Definitions

An unsigned permutation $\pi = (\pi_1 \ \pi_2 \ \ldots \ \pi_n)$ is a function over $\{1, 2, \ldots, n\}$. A signed permutation, also denoted as π, is a function over $\{-n, -(n-1), \ldots, -1, 1, 2, \ldots, n\}$ for which $|\pi_i| \neq |\pi_j|$ for all $i \neq j$. We will always consider their $extended$ version, in which they have two fixed elements $\pi_0 = 0$ and $\pi_{n+1} = n + 1$.

We have four special permutations when considering signed and unsigned permutations: the $identity$ $\iota_n = (1 \ 2 \ \ldots \ n)$, the $reverse$ $\eta_n = (n \ n-1 \ \ldots \ 1)$, the $signed$ $reverse$ $\bar{\eta}_n = (-n \ -(n-1) \ \ldots \ -1)$, and the $signed$ $identity$ $\bar{\iota}_n = (-1 \ -2 \ \ldots \ -n)$. Also, we have the $inverse$ permutation π^{-1}, for which $\pi^{-1}_{\pi_i} = i$. It satisfies $\pi^{-1} \cdot \pi = \iota_n$ where '·' represents a $composition$ between permutations. The permutation $\pi \cdot \sigma$ is $(\pi_{\sigma_1} \ \pi_{\sigma_2} \ \ldots \ \pi_{\sigma_n})$.

A *reversal* $\rho(i,j)$, $1 \leq i < j \leq n$, is a rearrangement that transforms π into $\pi \cdot \rho(i,j) = (\pi_1 \ldots \pi_{i-1} \; \pi_j \; \pi_{j-1} \ldots \pi_{i+1} \; \pi_i \; \pi_{j+1} \ldots \pi_n)$. A *prefix reversal* $\rho_p(j)$ is a reversal $\rho(1,j)$, $1 < j \leq n$, while a *suffix reversal* $\rho_s(i)$ is $\rho(i,n)$, $1 \leq i < n$.

A *signed reversal* $\bar{\rho}(i,j)$, $1 \leq i \leq j \leq n$, is a rearrangement that transforms π into $\pi \cdot \bar{\rho}(i,j) = (\pi_1 \ldots \pi_{i-1} \; -\pi_j \; -\pi_{j-1} \ldots \; -\pi_{i+1} \; -\pi_i \; \pi_{j+1} \ldots \pi_n)$. A *signed prefix reversal* $\bar{\rho}_p(j)$ is a signed reversal $\bar{\rho}(1,j)$, $1 \leq j \leq n$, while a *signed suffix reversal* $\bar{\rho}_s(i)$ is a signed reversal $\bar{\rho}(i,n)$, $1 \leq i \leq n$.

A *transposition* $\tau(i,j,k)$, $1 \leq i < j < k \leq n+1$, is a rearrangement that transforms π into $\pi \cdot \tau(i,j,k) = (\pi_1 \ldots \pi_{i-1} \; \pi_j \; \pi_{j-1} \; \cdots \; \pi_{k-1} \; \pi_i \; \pi_{i+1} \; \cdots \; \pi_{j-1} \; \pi_k \ldots \pi_n)$. A *prefix transposition* $\tau_p(j,k)$ is a transposition $\tau(1,j,k)$, $2 \leq j < k \leq n+1$, while a *suffix transposition* $\tau_s(i,j)$ is a transposition $\tau(i,j,n+1)$, $1 \leq i < j \leq n$. Note that transpositions do not change the signs of the elements. Therefore, when signed permutations are being used, transpositions can only be considered in problems where reversals are also allowed.

A *reversal breakpoint* is a pair of elements (π_i, π_{i+1}) of π such that $|\pi_{i+1} - \pi_i| \neq 1$. Normally, $0 \leq i \leq n$. For *prefix reversal breakpoints*, $1 \leq i \leq n$ and (π_0, π_1) is never a breakpoint. For *suffix reversal breakpoints*, $0 \leq i < n$ and (π_n, π_{n+1}) is never a breakpoint. For *prefix and suffix reversal breakpoints*, $1 \leq i < n$ and neither (π_0, π_1) nor (π_n, π_{n+1}) are breakpoints. The identity is the only permutation without breakpoints of the first three kinds. The identity and the reverse are the only ones that do not have breakpoints of the last kind.

A *transposition breakpoint* or a *signed reversal breakpoint* is a pair of elements (π_i, π_{i+1}) of π such that $\pi_{i+1} - \pi_i \neq 1$. Normally, $0 \leq i \leq n$. For *prefix transposition breakpoints* or *signed prefix reversal breakpoints*, $1 \leq i \leq n$ and (π_0, π_1) is never a breakpoint. For *suffix transposition breakpoints* or *signed suffix reversal breakpoints*, $0 \leq i < n$ and (π_n, π_{n+1}) is never a breakpoint. For *prefix and suffix transposition breakpoints* or *signed prefix and signed suffix reversal breakpoints*, $1 \leq i < n$ and neither (π_0, π_1) nor (π_n, π_{n+1}) are breakpoints. The identity permutation is the only permutation without any kind of transposition breakpoints and without any of the first three kinds of signed reversal breakpoints. The identity and the signed reverse are the only permutations without signed prefix and signed suffix reversal breakpoints.

If variations of reversals are allowed, breakpoints of that kinds are considered. For example, in Sorting by Prefix Reversals, Prefix Transpositions, Suffix Reversals and Suffix Transpositions we use prefix and suffix reversal breakpoints. We denote the number of breakpoints of a permutation π by $b_t(\pi)$, where t is the type of breakpoint being considered.

Let β be the rearrangement or rearrangements allowed in a sorting problem. We denote the minimum number of operations $\beta_i \in \beta$ needed to transform π into ι_n, that is, the minimum k for which $\pi \cdot \beta_1 \cdots \beta_k = \iota_n$, by $d_\beta(\pi)$. This number is called the *sorting distance* of π. The permutations between π and ι_n that were generated during this transformation form the *sorting sequence*. Finally, the greatest sorting distance between all the permutations of the same size n regarding β is called *diameter* and it is denoted by $D_\beta(n)$.

3 Families

Next we present families of permutations which will help us with our main results and give some lemmas regarding their distance when considering the problems we are interested in this paper.

$$\pi_n^1 = \begin{cases} (2\ 4\ 6\ \ldots\ n-4\ n\ n-2\ n-1\ n-3\ n-5\ \ldots\ 1) & \text{if } n \text{ is even} \\ (2\ 4\ 6\ \ldots\ n-5\ n-1\ n-3\ n\ n-2\ n-4\ \ldots\ 1) & \text{if } n \text{ is odd} \end{cases} \tag{1}$$

Lemma 1. *For $n \geq 4$, $d_{\rho\tau}(\pi_n^1) = \lceil \frac{n}{2} \rceil$.*

Proof. First note that $d_{\rho\tau}(\pi_n^1) \leq \lceil \frac{n}{2} \rceil$ because of Alg. 1. Now, when n is even there is not a transposition that removes three reversal breakpoints at once from π_n^1. This happens because the odd numbers are completely separated from the even numbers and one can demonstrate this by simple contradiction. However, there are four ways to remove two breakpoints: placing any even i, $2 \leq i \leq n-4$, between $i+1$ and $i-1$, placing any odd i, $1 \leq i \leq n-5$, between $i-1$ and $i+1$, reverting the segment $n-2, n-1$, or placing the segment $2, 4, 6, \ldots, n-4, n$ at the end of the permutation, between 1 and $n+1$.

In any case, note that the separation mentioned above is kept and, again, there is not a transposition that removes three breakpoints at once. If we follow this idea, we can see that the maximum number of breakpoints that can be removed at once is always two. Since removing three is never possible and removing one would only increase the size of the sorting sequence, the algorithm is optimum. A similar analysis can be done when n is odd. The only difference is that it is possible to remove three breakpoints in the first step of the sorting, and it will happen again at most one more time. $\square$

$$\pi_n^2 = \begin{cases} (n-1\ n-2\ n\ n-4\ n-6\ \ldots\ 2\ n-3\ n-5\ \ldots\ 1) & \text{if } n \text{ is even} \\ (n\ n-3\ n-1\ n-5\ n-7\ \ldots\ 2\ n-2\ n-4\ \ldots\ 1) & \text{if } n \text{ is odd} \end{cases} \tag{2}$$

Lemma 2. *For $n \geq 7$, $\lceil \frac{n}{2} \rceil \leq d_{\rho_p \tau_p}(\pi_n^2) \leq \lceil \frac{n}{2} \rceil + 1$.*

Proof. The lower bound is true because $b_{\rho_p}(\pi_n^2) = n-1$ when n is even, $b_{\rho_p}(\pi_n^2) = n$ when n is odd and $d_{\rho_p \tau_p}(\pi) \geq \lceil \frac{b_{\rho_p}(\pi)}{2} \rceil$ for any π [20]. The upper bound is given by Alg. 2. $\square$

$$\pi_n^3 = (+4\ +3\ +2\ +1\ -5\ -6\ -7\ \ldots\ -(n-1)\ -n) \tag{3}$$

Lemma 3. *For $n \geq 5$, $\lceil \frac{n}{2} \rceil + 1 \leq d_{\bar{\rho}_p \tau_p}(\pi_n^3) \leq n+1$.*

Proof. First note that $\lceil \frac{n}{2} \rceil$ is a valid lower bound, since $b_{\bar{\rho}_p}(\pi_n^3) = n$ and $d_{\bar{\rho}_p \tau_p}(\pi) \geq \lceil \frac{b_{\bar{\rho}_p}(\pi)}{2} \rceil$ for any π [17]. However, this lower bound is not tight. If it was, when

n is even the sorting should only have operations that remove two breakpoints. But it is easy to see that this is not possible. When n is odd, the sorting would need to have $\lceil \frac{n}{2} \rceil - 1$ operations that remove two breakpoints at once and one operation that removes one breakpoint. Removing two breakpoints at the first operation is not possible and the only form of removing one is by placing 4 after 3. However, now it is not possible to remove two breakpoints either. Therefore, the distance is at least $\lceil \frac{n}{2} \rceil + 1$.

The upper bound is given by a simple algorithm that reverts the segment $+4, +3, +2, +1$, reverts the whole permutation and then places each element in its correct position. $\qquad \square$

$$\pi_n^4 = \begin{cases} (n \ 1 \ n{-}2 \ n{-}4 \ n{-}6 \ \ldots \ 4 \ 2 \ n{-}3 \ n{-}5 \ n{-}7 \ \ldots \ 3 \ n{-}1) & \text{if } n \text{ is even} \\ (n \ 1 \ n{-}2 \ n{-}4 \ n{-}6 \ \ldots \ 5 \ 3 \ n{-}3 \ n{-}5 \ n{-}7 \ \ldots \ 2 \ n{-}1) & \text{if } n \text{ is odd} \end{cases} \tag{4}$$

Lemma 4. *For* $n \geq 8$, $n - 1 \leq d_{\rho_p \rho_s}(\pi_n^4) \leq n$.

Proof. The lower bound is true because $b_{\rho_p \rho_s}(\pi_n^4) = n{-}1$ and $d_{\rho_p \rho_s}(\pi) \geq b_{\rho_p \rho_s}(\pi)$ for any π [18]. The upper bound is true because of Alg. 3. $\qquad \square$

$$\pi_n^5 = (-1^n n \ -1^{n-1}(n{-}1) \ -1^{n-2}(n{-}2) \ \ldots \ +2 \ -1) \tag{5}$$

Lemma 5. *For* $n \geq 5$, $n \leq d_{\bar{\rho}_p \bar{\rho}_s}(\pi_n^5) \leq n + \lfloor \frac{n-1}{2} \rfloor$.

Proof. First note that $n - 1$ is a valid lower bound since $b_{\bar{\rho}_p \bar{\rho}_s}(\pi_n^5) = n - 1$ and $d_{\bar{\rho}_p \bar{\rho}_s}(\pi) \geq b_{\bar{\rho}_p \bar{\rho}_s}(\pi)$ for any π [17]. However, this lower bound is not tight, because if it was, it would always be necessary to remove one signed prefix and signed suffix reversal breakpoint. If n is even, then $\pi_n^5 = (+n \ -(n-1) \ +(n-2) \ -(n-3) \ \ldots \ +2 \ -1)$ and it is easy to see that there is only one possibility to remove one breakpoint at once, which will lead to $\pi' = (-n \ -(n-1) \ +(n-2) \ -(n-3) \ \ldots \ +2 \ -1)$. On the other hand, it is not possible to remove one breakpoint with one operation from π'. If n is odd, then $\pi_n^5 = (-n \ +(n-1) \ -(n-2) \ +(n-3) \ \ldots \ +2 \ -1)$, for which is not possible to remove one breakpoint of this kind with one operation. Therefore, the distance is at least n.

The upper bound is given by Alg. 4. $\qquad \square$

$$\pi_n^6 = \eta_n = (n \ n{-}1 \ n{-}2 \ \ldots \ 2 \ 1) \tag{6}$$

Lemma 6. *For* $n \geq 3$, $\lceil \frac{n-1}{2} \rceil + 1 \leq d_{\tau_p \tau_s}(\pi_n^6) \leq n - \lfloor \frac{n}{4} \rfloor$.

Proof. First note that $\lceil \frac{n-1}{2} \rceil$ is a valid lower bound because $b_{\tau_p \tau_s}(\pi_n^6) = n - 1$ and $d_{\tau_p \tau_s}(\pi) \geq \lceil \frac{b_{\tau_p \tau_s}(\pi)}{2} \rceil$ for any π [18]. However, it is not tight and the distance is at least $\lceil \frac{n-1}{2} \rceil + 1$. Due to space restrictions this proof is omitted.

The upper bound is true because of the algorithm presented by Dias and Meidanis [8] that sorts η_n using at most $n - \lfloor \frac{n}{4} \rfloor$ prefix transpositions. $\qquad \square$

Algorithm 1. An algorithm to sort π_n^1 with reversals and transpositions

---

Input: $\pi = \pi_n^1$, $n \geq 4$
if $n \mod 2 = 0$ **then**
$\quad \pi \leftarrow \pi \cdot \rho(\pi_{n-2}^{-1}, \pi_{n-1}^{-1})$;
$\quad$ **for** $i \leftarrow n-4$ **down to** 2 **by** -2 **do**
$\quad\quad \lfloor \; \pi \leftarrow \pi \cdot \tau(\pi_i^{-1}, \pi_n^{-1}, \pi_{i-1}^{-1})$;
$\quad \pi \leftarrow \pi \cdot \rho(1, n)$;
else
$\quad \pi \leftarrow \pi \cdot \tau(\pi_{n-3}^{-1}, \pi_n^{-1}, \pi_{n-4}^{-1})$;
$\quad$ **for** $i \leftarrow n-5$ **down to** 2 **by** -2 **do**
$\quad\quad \lfloor \; \pi \leftarrow \pi \cdot \tau(\pi_i^{-1}, \pi_{n-1}^{-1}, \pi_{i-1}^{-1})$;
$\quad \pi \leftarrow \pi \cdot \tau(1, 3, n+1) \cdot \rho(1, n-2)$;

Algorithm 2. An algorithm to sort π_n^2 with prefix reversals and prefix transpositions

---

Input: $\pi = \pi_n^2$, $n \geq 7$
if $n \mod 2 = 0$ **then**
$\quad \pi \leftarrow \pi \cdot \rho_p(2) \cdot \tau_p(5, \pi_{n-3}^{-1} + 1)$;
else
$\quad \lfloor \; \pi \leftarrow \pi \cdot \tau_p(3, \pi_{n-4}^{-1}) \cdot \tau_p(2, \pi_n^{-1})$;
while $\pi_1 \neq 2$ **do**
$\quad \lfloor \; \pi \leftarrow \pi \cdot \tau_p(2, \pi_{\pi_1 - 1}^{-1})$;
$\pi \leftarrow \pi \cdot \tau_p(\pi_n^{-1} + 1, n+1) \cdot \rho_p(\pi_2^{-1}) \cdot \rho_p(2)$;

Algorithm 3. An algorithm to sort π_n^4 with prefix reversals and suffix reversals

---

Input: $\pi = \pi_n^4$, $n \geq 8$
$\pi \leftarrow \pi \cdot \rho_p(n-1) \cdot \rho_p(n-3) \cdot \rho_s(2) \cdot \rho_s(\pi_{\pi_n+1}^{-1} + 1) \cdot \rho_p(\pi_{\pi_1-1}^{-1} - 1) \cdot \rho_s(\pi_n^{-1})$;
while $\pi_1 \neq 1$ **do**
$\quad \lfloor \; \pi \leftarrow \pi \cdot \rho_p(\pi_{\pi_1+1}^{-1} - 1)$;

Algorithm 4. An algorithm to sort π_n^5 with signed prefix reversals and signed suffix reversals

---

Input: $\pi = \pi_n^5$, $n \geq 5$
if $n \mod 2 = 0$ **then**
$\quad \lfloor \ \pi \leftarrow \pi \cdot \bar{\rho}_p(1);$
for $i \leftarrow 3 - (n \mod 2)$ **to** $n - 1$ **by** 2 **do**
$\quad \lfloor \ \pi \leftarrow \pi \cdot \bar{\rho}_p(i) \cdot \bar{\rho}_p(1);$
$\pi \leftarrow \pi \cdot \bar{\rho}_p(n);$
for $i \leftarrow n - 1$ **down to** 2 **by** -2 **do**
$\quad \lfloor \ \pi \leftarrow \pi \cdot \bar{\rho}_s(i);$

$$\pi_n^7 = \begin{cases} (n-1\ n-3\ n-5\ \ldots\ 5\ 3\ 6\ 8\ 10\ \ldots\ n\ 2\ 4\ 1) & \text{if } n \text{ is even} \\ (n\ n-2\ n-4\ \ldots\ 5\ 3\ 6\ 8\ 10\ \ldots\ n-1\ 2\ 4\ 1) & \text{if } n \text{ is odd} \end{cases} \tag{7}$$

Lemma 7. For $n \geq 6$, $\lceil \frac{n}{2} \rceil \leq d_{\bar{\rho}_p \tau_p \bar{\rho}_s \tau_s}(\pi_n^7) \leq \lceil \frac{n}{2} \rceil + 1$ if n is even and $d_{\bar{\rho}_p \tau_p \bar{\rho}_s \tau_s}(\pi_n^7) = \lceil \frac{n}{2} \rceil + 1$ if n is odd.

Proof. First note that $d_{\bar{\rho}_p \tau_p \bar{\rho}_s \tau_s}(\pi_n^7) \leq \lceil \frac{n}{2} \rceil + 1$ because of Alg. 5. Also, we have that $d_{\bar{\rho}_p \tau_p \bar{\rho}_s \tau_s}(\pi_n^7) \geq \lceil \frac{n-1}{2} \rceil$ because $b_{\bar{\rho}_p \bar{\rho}_s}(\pi_n^7) = n - 1$ and $d_{\bar{\rho}_p \tau_p \bar{\rho}_s \tau_s}(\pi) \geq \lceil \frac{b_{\bar{\rho}_p \bar{\rho}_s}(\pi)}{2} \rceil$ for any π [18]. When n is odd, for the lower bound to be tight, the sorting should only have operations that remove two breakpoints. There is only one possibility of doing this in the first step. After that, there is also always one possibility, which is placing the even number i which is in the first position between the odd numbers $i + 1$ and $i - 1$. This takes $\lceil \frac{n}{2} \rceil - 3 + 1$ moves, leaves the permutation in the form $(n\ n - 1\ n - 2\ \ldots\ 7\ 6\ 5\ 3\ 2\ 4\ 1)$, and three more operations are needed to sort it. $\quad\square$

$$\pi_n^8 = (-1\ +2\ -3\ +4\ \ldots\ -(n-1)\ +n) \tag{8}$$

Lemma 8. For $n \geq 6$ and n even, $\frac{n}{2} \leq d_{\bar{\rho}_p \tau_p \bar{\rho}_s \tau_s}(\pi_n^8) \leq n$.

Proof. The lower bound is true because $b_{\bar{\rho}_p \bar{\rho}_s}(\pi_n^8) = n - 1$ and $d_{\bar{\rho}_p \tau_p \bar{\rho}_s \tau_s}(\pi) \geq \lceil \frac{b_{\bar{\rho}_p \bar{\rho}_s}(\pi)}{2} \rceil$ for any π [17]. The upper bound is true because of Alg. 6. $\quad\square$

Lemma 9. For $n \geq 5$ and n odd, $\frac{n-1}{2} \leq d_{\bar{\rho}_p \tau_p \bar{\rho}_s \tau_s}(\pi_n^5) \leq n$.

Proof. The lower bound is true because $b_{\bar{\rho}_p \bar{\rho}_s}(\pi_n^5) = n - 1$ and $d_{\bar{\rho}_p \tau_p \bar{\rho}_s \tau_s}(\pi) \geq \lceil \frac{b_{\bar{\rho}_p \bar{\rho}_s}(\pi)}{2} \rceil$ for any π [17]. The upper bound is true because of Alg. 7. $\quad\square$

Algorithm 5. An algorithm to sort π_n^7 with prefix reversals, prefix transpositions, suffix reversals, and suffix transpositions

Input: $\pi = \pi_n^7$, $n \geq 6$
if $n \mod 2 = 0$ **then**
    while $\pi_1 \neq 5$ **do**
        $\pi \leftarrow \pi \cdot \tau_p(2, \pi_{\pi_1+1}^{-1})$;
    $\pi \leftarrow \pi \cdot \tau_p(n-1, n) \cdot \tau_p(3, 4) \cdot \tau_p(n-1, n+1) \cdot \rho_p(2)$;
else
    $\pi \leftarrow \pi \cdot \tau_p(\pi_3^{-1}, \pi_4^{-1}) \cdot \tau_s(\pi_2^{-1}, \pi_{n-2}^{-1}) \cdot \tau_p(2, \pi_1^{-1})$;
    while $\pi_1 \neq n-1$ **do**
        $\pi \leftarrow \pi \cdot \tau_p(2, \pi_{\pi_1-1}^{-1})$;
    $\pi \leftarrow \pi \cdot \rho_p(n-1) \cdot \rho_p(2)$;

Algorithm 6. An algorithm to sort π_n^8 with signed prefix reversals, prefix transpositions, signed suffix reversals, and suffix transpositions when n is even

Input: $\pi = \pi_n^8$, $n \geq 8$
for $i \leftarrow 1$ **to** $n-1$ **by** 2 **do**
    $\pi \leftarrow \pi \cdot \bar{\rho}_p(1) \cdot \tau_p(3, n+1)$;

Algorithm 7. An algorithm to sort π_n^5 with signed prefix reversals, prefix transpositions, signed suffix reversals, and suffix transpositions when n is odd

Input: $\pi = \pi_n^5$, $n \geq 7$
$\pi \leftarrow \pi \cdot \tau_p(2, n+1) \cdot \bar{\rho}_p(1) \cdot \tau_p(\pi_{n-5}^{-1}, n+1)$;
for $i \leftarrow n-5$ **down to** 4 **do**
    $\pi \leftarrow \pi \cdot \bar{\rho}_p(1) \cdot \tau_p(3, n+1)$;
$\pi \leftarrow \pi \cdot \bar{\rho}_p(1)$;
$\pi \leftarrow \pi \cdot \tau_p(\pi_{n-3}^{-1} + 1, n+1) \cdot \bar{\rho}_p(n-1) \cdot \tau_p(\pi_1^{-1}, n+1)$;

4   Bounds

This section presents the bounds we found for the diameter of the problems we are interested in this paper. Table 1 shows the known values for the diameter of these problems and it was filled with the results given by the Rearrangement Distance Database[1] of Galvão and Dias [12].

For the problem of Sorting by Signed Reversals and Transpositions there were previous results concerning the diameter. The values presented in Table 1 show that the conjecture of Meidanis et $al.$ [19] that $D_{\bar{\rho}\tau}(n) = \lfloor \frac{n}{2} \rfloor + 2$ is not true. Besides that, $D_{\bar{\rho}\tau}(7) = 6$ but $d_{\bar{\rho}\tau}(\bar{\iota}_7) = 5$, $D_{\bar{\rho}\tau}(9) = 7$ but $d_{\bar{\rho}\tau}(\bar{\iota}_9) = 6$, and $D_{\bar{\rho}\tau}(10) = 8$ but $d_{\bar{\rho}\tau}(\bar{\iota}_{10}) = 7$. Therefore, the other part of their conjecture, which stated that $D_{\bar{\rho}\tau}(n) = d_{\bar{\rho}\tau}(\bar{\iota}_n)$, is also not true.

Table 1. Diameter values for each n [12]. A '-' means the value is still unknown.

n	$D_{\rho\tau}(n)$	$D_{\bar{\rho}\tau}(n)$	$D_{\rho_p\tau_p}(n)$	$D_{\bar{\rho}_p\tau_p}(n)$	$D_{\tau_p\tau_s}(n)$	$D_{\rho_p\rho_s}(n)$	$D_{\bar{\rho}_p\bar{\rho}_s}(n)$	$D_{\rho_p\tau_p\rho_s\tau_s}(n)$	$D_{\bar{\rho}_p\tau_p\bar{\rho}_s\tau_s}(n)$
1	0	1	0	1	0	0	1	0	1
2	1	2	1	3	1	1	3	1	2
3	1	3	2	4	2	2	4	1	3
4	2	4	2	5	3	3	6	2	4
5	3	4	3	6	3	4	7	3	5
6	3	5	4	7	4	5	8	4	6
7	4	6	5	8	5	7	10	5	7
8	4	6	5	8	6	8	11	5	7
9	5	7	6	9	6	9	13	6	8
10	5	8	7	10	7	10	14	6	9
11	6	-	7	-	8	11	-	7	-
12	6	-	8	-	8	12	-	7	-
13	7	-	9	-	9	13	-	8	-

Lemma 10. *For $n \geq 4$, $D_{\rho\tau}(n) \geq \lceil \frac{n}{2} \rceil$ and for $n \geq 9$, $D_{\rho\tau}(n) \leq \lfloor \frac{2n-2}{3} \rfloor$.*

Proof. The lower bound is true because of family π_n^1, as Lemma 1 shows. The upper bound is true because $D_{\rho\tau}(n) \leq \min\{D_\rho(n), D_\tau(n)\}$, since $d_{\rho\tau}(\pi) \leq \min\{d_\rho(\pi), d_\tau(\pi)\}$ for all π. This is true because any sorting sequence for Sorting by Reversals or for Sorting by Transpositions is also valid for Sorting by Reversals and Transpositions, but it is not necessarily optimum.   □

Lemma 11. *For $n \geq 7$, $D_{\rho_p\tau_p}(n) \geq \lceil \frac{n}{2} \rceil$ and for $n \geq 1$, $D_{\rho_p\tau_p}(n) \leq n - \log_{\frac{9}{2}} n$.*

Proof. The lower bound is true because of family π_n^2, as Lemma 2 shows. The upper bound is true because $D_{\rho_p\tau_p}(n) \leq \min\{D_{\rho_p}(n), D_{\tau_p}(n)\}$, since $d_{\rho_p\tau_p}(\pi) \leq \min\{d_{\rho_p}(\pi), d_{\tau_p}(\pi)\}$ for any π.   □

Lemma 12. *For $n \geq 5$, $D_{\bar{\rho}_p\tau_p}(n) \geq \lceil \frac{n}{2} \rceil + 1$ and for $n \geq 16$, $D_{\bar{\rho}_p\tau_p}(n) \leq \frac{18n}{11} + O(1)$.*

---

[1] Available at http://mirza.ic.unicamp.br:8080/bioinfo/index.jsf

Proof. The lower bound is true because of family π_n^3, as Lemma 3 shows. The upper bound is true because $D_{\bar{\rho}_p \tau_p}(n) \leq D_{\bar{\rho}_p}(n)$, since $d_{\bar{\rho}_p \tau_p}(\pi) \leq d_{\bar{\rho}_p}(\pi)$ for any π. Note that a sorting sequence for Sorting by Prefix Transpositions is not valid for this problem, since it does not handle signs. $\qquad\square$

Lemma 13. *For* $n \geq 8$, $D_{\rho_p \rho_s}(n) \geq n-1$ *and for* $n \geq 1$, $D_{\rho_p \rho_s}(n) \leq \frac{18n}{11}+O(1)$.

Proof. The lower bound is true because of family π_n^4, as Lemma 4 shows. The upper bound is true because $D_{\rho_p \rho_s}(n) \leq D_{\rho_p}(n)$, since $d_{\rho_p \rho_s}(\pi) \leq d_{\rho_p}(\pi)$ for any π. This is true because any sorting sequence for Sorting by Prefix Reversals is valid for Sorting by Prefix Reversals and Suffix Reversals. $\qquad\square$

Lemma 14. *For* $n \geq 5$, $D_{\bar{\rho}_p \bar{\rho}_s}(n) \geq n$ *and for* $n \geq 16$, $D_{\bar{\rho}_p \bar{\rho}_s}(n) \leq 2n - 6$.

Proof. The lower bound is true because of family π_n^5, as Lemma 5 shows. The upper bound is true because $D_{\bar{\rho}_p \bar{\rho}_s}(n) \leq D_{\bar{\rho}_p}(n)$, since $d_{\bar{\rho}_p \bar{\rho}_s}(\pi) \leq d_{\bar{\rho}_p}(\pi)$ for any π. This is true because any sorting sequence for Sorting by Signed Prefix Reversals is valid for Sorting by Signed Prefix Reversals and Suffix Reversals. $\quad\square$

Lemma 15. *For* $n \geq 3$, $\lceil \frac{n-1}{2} \rceil +1 \leq D_{\tau_p \tau_s}(n) \leq n - \log_{\frac{9}{2}} n$.

Proof. The lower bound is true because of family π_n^6, as Lemma 6 shows. The upper bound is true because $D_{\tau_p \tau_s}(n) \leq D_{\tau_p}(n)$, since $d_{\tau_p \tau_s}(\pi) \leq d_{\tau_p}(\pi)$ for any π. $\qquad\square$

Lemma 16. *For* $n \geq 6$, $D_{\rho_p \tau_p \rho_s \tau_s}(n) \geq \lceil \frac{n}{2} \rceil$ *and for* $n \geq 1$, $D_{\rho_p \tau_p \rho_s \tau_s}(n) \leq n - \log_{\frac{9}{2}} n$.

Proof. The lower bound is true because of family π_n^7, as Lemma 7 shows. The upper bound is true because $D_{\rho_p \tau_p \rho_s \tau_s}(n) \leq \min\{D_{\rho_p \rho_s}(n), D_{\tau_p \tau_s}(n)\}$ $\leq \min\{D_{\rho_p}(n), D_{\tau_p}(n)\}$, since $d_{\rho_p \tau_p \rho_s \tau_s}(\pi) \leq \min\{d_{\rho_p \rho_s}(\pi), d_{\tau_p \tau_s}(\pi)\} \leq \min \{d_{\rho_p}(\pi), d_{\tau_p}(\pi)\}$ for any π. $\qquad\square$

Lemma 17. *For* $n \geq 7$, $D_{\bar{\rho}_p \tau_p \bar{\rho}_s \tau_s}(n) \geq \lceil \frac{n-1}{2} \rceil$ *and for* $n \geq 1$, $D_{\bar{\rho}_p \tau_p \bar{\rho}_s \tau_s}(n) \leq n+1$.

Proof. The lower bound is true because of families π_n^8 and π_n^5, as Lemmas 8 and 9 show. The upper bound is true because $D_{\bar{\rho}_p \tau_p \bar{\rho}_s \tau_s}(n) \leq D_{\bar{\rho}_p \bar{\rho}_s}(n) \leq D_{\bar{\rho}_p}(n)$, since $d_{\bar{\rho}_p \tau_p \bar{\rho}_s \tau_s}(\pi) \leq d_{\bar{\rho}_p \bar{\rho}_s}(\pi) \leq d_{\bar{\rho}_p}(\pi)$ for any π. $\qquad\square$

By Table 1, we can see that

1. $D_{\rho \tau}(n) = \lceil \frac{n}{2} \rceil$ for $4 \leq n \leq 13$;
2. $D_{\rho_p \rho_s}(n) = n$ for $7 \leq n \leq 13$;
3. $D_{\bar{\rho}_p \bar{\rho}_s}(n) = n + \lfloor \frac{n-1}{2} \rfloor$ for $5 \leq n \leq 10$; and
4. $D_{\rho_p \tau_p \rho_s \tau_s}(n) = \lceil \frac{n}{2} \rceil + 1$ for $6 \leq n \leq 13$.

It is also possible to validate that

1. $d_{\rho_p \rho_s}(\pi_n^4) = n$ for $8 \leq n \leq 15$;
2. $d_{\bar{\rho}_p \bar{\rho}_s}(\pi_n^5) = n + \lfloor \frac{n-1}{2} \rfloor$ for $5 \leq n \leq 12$; and
3. $d_{\rho_p \tau_p \rho_s \tau_s}(\pi_n^7) = \lceil \frac{n}{2} \rceil + 1$ for $6 \leq n \leq 15$.

The next conjectures are directly based on the results shown above.

Conjecture 18. For $n \geq 4$, $D_{\rho \tau}(n) = d_{\rho \tau}(\pi_n^1) = \lceil \frac{n}{2} \rceil$.

Conjecture 19. For $n \geq 8$, $D_{\rho_p \rho_s}(n) = d_{\rho_p \rho_s}(\pi_n^4) = n$.

It is worth noticing that when $n = 7$, the only two permutations for which $d_{\rho_p \rho_s}(\pi) = D_{\rho_p \rho_s}(7) = 7$ are $\pi = (7\ 3\ 5\ 2\ 6\ 4\ 1)$ and $\pi = (7\ 4\ 2\ 6\ 3\ 5\ 1)$.

Conjecture 20. For $n \geq 5$, $D_{\bar{\rho}_p \bar{\rho}_s}(n) = d_{\bar{\rho}_p \bar{\rho}_s}(\pi_n^5) = n + \lfloor \frac{n-1}{2} \rfloor$.

Conjecture 21. For $n \geq 6$, $D_{\rho_p \tau_p \rho_s \tau_s}(n) = d_{\rho_p \tau_p \rho_s \tau_s}(\pi_n^7) = \lceil \frac{n}{2} \rceil + 1$.

For some problems, we found families whose distance match the known diameters but we did not find algorithms to sort them with the required distances. However, we also believe in the next conjectures due to the following:

1. $D_{\bar{\rho}_p \tau_p}(n) = d_{\bar{\rho}_p \tau_p}(\pi_n^3)$ for $2 \leq n \leq 10$, and when $n = 7$ the only two permutations for which $d_{\bar{\rho}_p \tau_p}(\pi) = D_{\bar{\rho}_p \tau_p}(7) = 8$ are $\pi = \pi_7^3$ and $\pi = (+3\ +2\ +1\ -4\ -5\ -6\ -7)$;
2. $D_{\tau_p \tau_s}(n) = d_{\tau_p \tau_s}(\pi_n^6 = \eta_n)$ for $1 \leq n \leq 12$;
3. $D_{\bar{\rho}_p \tau_p \bar{\rho}_s \tau_s}(n) = d_{\bar{\rho}_p \tau_p \bar{\rho}_s \tau_s}(\pi_n^8)$ for $n \in \{8, 10\}$ and $D_{\bar{\rho}_p \tau_p \bar{\rho}_s \tau_s}(n) = d_{\bar{\rho}_p \tau_p \bar{\rho}_s \tau_s}(\pi_n^5)$ for $n \in \{7, 9\}$.

Conjecture 22. For $n \geq 2$, $D_{\bar{\rho}_p \tau_p}(n) = d_{\bar{\rho}_p \tau_p}(\pi_n^3)$.

Conjecture 23. For $n \geq 1$, $D_{\tau_p \tau_s}(n) = d_{\tau_p \tau_s}(\eta_n) = n - \lfloor \frac{n}{3} \rfloor$.

Conjecture 24. For $n \geq 8$ and n even, $D_{\bar{\rho}_p \tau_p \bar{\rho}_s \tau_s}(n) = d_{\bar{\rho}_p \tau_p \bar{\rho}_s \tau_s}(\pi_n^8)$. For $n \geq 7$ and n odd, $D_{\bar{\rho}_p \tau_p \bar{\rho}_s \tau_s}(n) = d_{\bar{\rho}_p \tau_p \bar{\rho}_s \tau_s}(\pi_n^5)$.

Table 2 summarizes these results. For each sorting problem, represented by the allowed operations, it shows the lower and the upper bounds for the diameter and the family whose distance, represented only by d, we conjectured to be equal to the diameter, represented only by D.

5   Conclusion and Future Work

In this paper we presented results on the diameter of problems for which more than one rearrangement is allowed. In special, we presented bounds for problems where prefix and suffix rearrangements are allowed. For the nine problems we considered, only one had previous results regarding its diameter.

We also gave some conjectures on the exact values of the diameters. We intend to keep studying these problems and trying to prove these conjectures.

Table 2. Summary of the results

Operations	Lower bound	Upper bound	Family and conjectured diameter
$\rho\,\tau$	$\lceil \frac{n}{2} \rceil,\, n \geq 4$	$\lfloor \frac{2n-2}{3} \rfloor,\, n \geq 9$	$D(n) = d(\pi_n^1) = \lceil \frac{n}{2} \rceil,\, n \geq 4$
$\rho_p\,\tau_p$	$\lceil \frac{n}{2} \rceil,\, n \geq 7$	$n - \log_{9/2} n,\, n \geq 1$	$-$
$\bar{\rho}_p\,\tau_p$	$\lceil \frac{n}{2} \rceil + 1,\, n \geq 2$	$\frac{18n}{11} + O(1),\, n \geq 16$	$D(n) = d(\pi_n^3),\, n \geq 2$
$\rho_p\,\rho_s$	$n - 1,\, n \geq 8$	$\frac{18n}{11} + O(1),\, n \geq 16$	$D(n) = d(\pi_n^4) = n,\, n \geq 8$
$\bar{\rho}_p\,\bar{\rho}_s$	$n,\, n \geq 5$	$2n - 6,\, n \geq 16$	$D(n) = d(\pi_n^5) = n + \lfloor \frac{n-1}{2} \rfloor,\, n \geq 5$
$\tau_p\,\tau_s$	$\lceil \frac{n-1}{2} \rceil + 1,\, n \geq 3$	$n - \log_{9/2} n,\, n \geq 1$	$D(n) = d(\pi_n^6),\, n \geq 6$
$\rho_p\,\tau_p\,\rho_s\,\tau_s$	$\lceil \frac{n}{2} \rceil,\, n \geq 6$	$n - \log_{9/2} n,\, n \geq 1$	$D(n) = d(\pi_n^7) = \lceil \frac{n}{2} \rceil + 1,\, n \geq 6$
$\bar{\rho}_p\,\tau_p\,\bar{\rho}_s\,\tau_s$	$\lceil \frac{n-1}{2} \rceil,\, n \geq 7$	$n + 1,\, n \geq 1$	$D(n) = \begin{cases} d(\pi_n^8) & \text{if } n \geq 8,\, n \text{ even} \\ d(\pi_n^5) & \text{if } n \geq 7,\, n \text{ odd} \end{cases}$

Acknowledgements. This work was partially supported by São Paulo Research Foundation - FAPESP (grants 2013/01172-0 and 2013/08293-7) and National Counsel of Technological and Scientific Development - CNPq (grants 477692/2012-5 and 483370/2013-4).

References

1. Bafna, V., Pevzner, P.A.: Genome Rearrangements and Sorting by Reversals. In: Proceedings of the 34th Annual Symposium on Foundations of Computer Science (FOCS 1993), pp. 148–157 (1993)
2. Bulteau, L., Fertin, G., Rusu, I.: Pancake Flipping is Hard. In: Rovan, B., Sassone, V., Widmayer, P. (eds.) MFCS 2012. LNCS, vol. 7464, pp. 247–258. Springer, Heidelberg (2012)
3. Bulteau, L., Fertin, G., Rusu, I.: Sorting by Transpositions is Difficult. SIAM Journal on Computing 26(3), 1148–1180 (2012)
4. Caprara, A.: Sorting Permutations by Reversals and Eulerian Cycle Decompositions. SIAM Journal on Discrete Mathematics 12(1), 91–110 (1999)
5. Chitturi, B., Fahle, W., Meng, Z., Morales, L., Shields, C.O., Sudborough, I.H., Voit, W.: An $(18/11)n$ Upper Bound for Sorting by Prefix Reversals. Theoretical Computer Science 410(36), 3372–3390 (2009)
6. Chitturi, B., Sudborough, I.H.: Bounding Prefix Transposition Distance for Strings and Permutations. Theoretical Computer Science 421, 15–24 (2012)
7. Cibulka, J.: On Average and Highest Number of Flips in Pancake Sorting. Theoretical Computer Science 412(8-10), 822–834 (2011)
8. Dias, Z., Meidanis, J.: Sorting by Prefix Transpositions. In: Laender, A.H.F., Oliveira, A.L. (eds.) SPIRE 2002. LNCS, vol. 2476, pp. 65–76. Springer, Heidelberg (2002)

9. Elias, I., Hartman, T.: A 1.375-Approximation Algorithm for Sorting by Transpositions. 375-Approximation Algorithm for Sorting by Transpositions 3(4), 369–379 (2006)
10. Eriksson, H., Eriksson, K., Karlander, J., Svensson, L., Wastlund, J.: Sorting a Bridge Hand. Discrete Mathematics 241(1-3), 289–300 (2001)
11. Fertin, G., Labarre, A., Rusu, I., Tannier, É., Vialette, S.: Combinatorics of Genome Rearrangements. In: Computational Molecular Biology. MIT Press (2009)
12. Galvão, G.R., Dias, Z.: Computing Rearrangement Distance of Every Permutation in the Symmetric Group. In: Chu, W.C., Wong, W.E., Palakal, M.J., Hung, C.C. (eds.) Proceedings of the 26th ACM Symposium on Applied Computing (SAC 22011), pp. 106–107. ACM (2011)
13. Gates, W.H., Papadimitriou, C.H.: Bounds for Sorting by Prefix Reversal. Discrete Mathematics 27(1), 47–57 (1979)
14. Hannenhalli, S., Pevzner, P.A.: Transforming Cabbage into Turnip: Polynomial Algorithm for Sorting Signed Permutations by Reversals. Journal of the ACM 46(1), 1–27 (1999)
15. Heydari, M.H., Sudborough, I.H.: On the Diameter of the Pancake Network. Journal of Algorithms 25(1), 67–94 (1997)
16. Labarre, A.: Edit Distances and Factorisations of Even Permutations. In: Halperin, D., Mehlhorn, K. (eds.) ESA 2008. LNCS, vol. 5193, pp. 635–646. Springer, Heidelberg (2008)
17. Lintzmayer, C.N., Dias, Z.: On Sorting of Signed Permutations by Prefix and Suffix Reversals and Transpositions. In: Dediu, A.H., Martín-Vide, C., Truthe, B. (eds.) Proceedings of the 1st International Conference on Algorithms for Computational Biology (AlCoB 2014), Tarragona, Spain, pp. 1–12. Springer (2014)
18. Lintzmayer, C.N., Dias, Z.: Sorting Permutations by Prefix and Suffix Versions of Reversals and Transpositions. In: Pardo, A., Viola, A. (eds.) LATIN 2014. LNCS, vol. 8392, pp. 671–682. Springer, Heidelberg (2014)
19. Meidanis, J., Walter, M.M.T., Dias, Z.: A Lower Bound on the Reversal and Transposition Diameter. Journal of Computational Biology 9(5), 743–745 (2002)
20. Sharmin, M., Yeasmin, R., Hasan, M., Rahman, A., Rahman, M.S.: Pancake Flipping with Two Spatulas. In: International Symposium on Combinatorial Optimization (ISCO 2010). Electronic Notes in Discrete Mathematics, vol. 36, pp. 231–238 (2010)
21. Walter, M.E.M.T., Dias, Z., Meidanis, J.: Reversal and Transposition Distance of Linear Chromosomes. In: Proceedings of the 5th International Symposium on String Processing and Information Retrieval (SPIRE 1998), pp. 96–102. IEEE Computer Society, Santa Cruz (1998)

Efficiently Enumerating All Connected Induced Subgraphs of a Large Molecular Network

Sean Maxwell[1], Mark R. Chance[1], and Mehmet Koyutürk[1,2]

[1] Center for Proteomics and Bioinformatics
[2] Department of Electrical Engineering and Computer Science
Case Western Reserve University, Cleveland, Ohio, USA
{sean.maxwell,mark.chance,mxk331}@case.edu

Abstract. In systems biology, the solution space for a broad range of problems is composed of sets of functionally associated biomolecules. Since connectivity in molecular interaction networks is an indicator of functional association, such sets can be identified from connected induced subgraphs of molecular interaction networks. Applications typically quantify the relevance (e.g., modularity, conservation, disease association) of connected subnetworks using an objective function and use a search algorithm to identify sets of subnetworks that maximize this objective function. Efficient enumeration of connected subgraphs of a large graph is therefore useful for these applications, and many existing search algorithms can be used for this purpose. However, there is a lack of non-heuristic algorithms that minimize the total number of subgraphs evaluated during the search for subgraphs that maximize the objective function. Here, we propose and evaluate an algorithm that reduces the computations necessary to enumerate subgraphs that maximize an objective function given a monotonically decreasing bounding function.

Keywords: connected subgraph enumeration, protein interaction networks, branch-and-bound algorithms.

1 Introduction

For many applications in systems biology, the connected induced subgraphs of molecular interaction networks are of particular interest since they represent sets of functionally associated biomolecules. For example, in the context of the systems biology of complex diseases, medical scientists are interested in identifying "dysregulated protein subnetworks", i.e., sets of proteins connected to each other via protein-protein interactions that exhibit collective differential expression between different phenotypes [3,4,5,6]. Similarly, gene set enrichment analysis aims to evaluate the statistical significance of the aggregate disease association of sets of genes that are defined *a priori*, and the connected subgraphs of molecular networks provide excellent candidate gene sets since they are functionally related through physical and functional interactions [15]. At the evolutionary scale, sets of orthologous proteins that induce connected subgraphs on networks of different species are shown to be useful in gaining insights into the conservation and modularity of biological processes across diverse taxa [7,11,16].

A.-H. Dediu, C. Martín-Vide, and B. Truthe (Eds.): AlCoB 2014, LNBI 8542, pp. 171–182, 2014.
© Springer International Publishing Switzerland 2014

In all of these applications, an objective function is defined to score any given subnetwork in terms of its relevance to what is being sought by the application. For example, in the identification of disease-associated subnetworks, connected subnetworks that contain a large number of disease-associated gene products are of interest. This scoring function may be computed based on the network topology alone, or may also incorporate other data, such as gene expression [5], genome-wide association [10], or sequence homology [11]. Then the problem is abstracted as one of finding high-scoring (e.g., globally optimal, locally optimal, or above a certain threshold) subnetworks according to this scoring function.

Due to computational considerations, most methods designed to tackle these problems implement heuristic algorithms to search the space of connected induced subgraphs of a network. However, it was shown that exhaustive search may lead to the identification of more biologically relevant patterns as compared to those identified by simple heuristics [4,16,18]. Furthermore, it is often desirable to identify many high-scoring subnetworks as candidates to be further evaluated for statistical significance, as opposed to identifying a single subnetwork with maximum score.

The objective of this work is to develop efficient algorithms for enumerating all sets of vertices that induce a connected subgraph in a large network. Our main motivation is to facilitate effective exploration of the subnetwork space of molecular interaction networks by enabling pruning of the search space in large chunks. For this purpose, we focus on the case where the scoring function satisfies a *hereditary property*. A hereditary property in a graph $G = (V, E)$ is a property such that if a set $S \subseteq V$ of nodes satisfies the property, then all subsets $S' \subseteq S$ also satisfy the property [2]. For example, being a clique is a hereditary property because any induced subgraph of a clique is also a clique. The bounding functions used by branch-and-bound algorithms also exploit hereditary properties. For the purpose of finding all maximal cliques or finding all maximal vertex sets $S \subseteq V$ that "score" greater than a specified threshold, the hereditary property is useful for pruning out the search space. This is because, if the property does not hold for a vertex set S, then no superset of S needs to be evaluated.

Many well established algorithms exist to enumerate connected induced subgraphs, such as ReverseSearch [1] and Algorithm447 [8] which are both variations of depth first search, and new algorithms such as ConSubG[12] are an area of active research. For the purpose of exploiting a hereditary property to prune out the search space, conventional depth first enumeration algorithms exhibit an inherent drawback: These algorithms do not enumerate vertex sets in an order that will allow evaluation of a vertex set after all of its subsets are evaluated. In other words, if a connected induced subgraph $S \subseteq V$ does not satisfy the hereditary property, depth first enumeration methods are likely to needlessly enumerate many $S' \supset S$ either before or after S is evaluated and rejected. We refer to such redundant computations as "*unnecessary rejections*".

Here, we propose an enumeration algorithm that introduces two novel techniques to reduce the number of unnecessary rejections while searching for connected induced subgraphs satisfying a hereditary property: 1) We use *anchor vertices* to seed the search, with a view to enabling easy tracking of the connectedness of the set of vertices being enumerated. 2) We use a *breadth-first discovery, depth-first extension* approach to enumerate sets of vertices, with a view to enabling evaluation of most vertex sets before their supersets are enumerated.

We systematically evaluate the ability of the proposed algorithm in reducing the number of unnecessary rejections and the resulting earnings in terms of runtime. Our results show that the proposed method significantly reduces the number of unnecessary rejections without introducing additional overhead into the enumeration itself.

2   Methods

2.1   Problem Definition and Observations

Let $G = (V, E)$ be an undirected graph. A set $V' \subseteq V$ is said to be a *connected vertex set* if the subgraph induced by V' is connected, i.e., if for every pair of vertices $\{u, v\} \in V'$, there is a path in G from u to v that goes only through nodes in V'. Throughout this work we refer to connected node sets as S where it is implied that $S \subseteq V$ and S induces a connected subgraph of G.

Let $f : 2^V \to \mathbb{R}$ be a function used to score vertex sets. For example, if we are interested in identifying maximal cliques, then we can define $f(S) = |S|$ if S induces a clique, and 0 otherwise. If we are interested in identifying disease-associated subnetworks such that the disease association of vertex $v \in V$ is quantified as $\sigma(v)$, then we can define $f(S) = \sum_{v \in S} \sigma(v)/\sqrt{|S|}$ [9].

We consider a problem setup where we are given a threshold t, and we are interested in enumerating all connected node sets $S \subseteq V$ such that $f(S) \geq t$. We assume that we are given a bounding function $f_b : 2^V \to \mathbb{R}$ such that, for any $S' \supseteq S$, $f(S') \leq f_b(S)$. We say that node set S is rejected if $f_b(S) < t$. The bounding function is useful for pruning out the search space using a bottom-up enumeration algorithm, since $f_b(S) < t$ implies $f(S') < t$ for all $S' \supseteq S$, i.e., once S is evaluated and rejected, there is no need to generate and evaluate any superset of S. In general terms, this problem can be viewed as one of generating all maximal connected vertex sets that satisfy a given hereditary property.

In order to efficiently generate all maximal vertex sets that satisfy a hereditary property, we need an algorithm to enumerate the solution space correctly and efficiently. Any algorithm that solves this problem has to satisfy the following criteria in order to be correct and optimal:

- *Completeness:* All connected vertex sets S in G for which $f(S) \geq t$ should be generated and all generated vertex sets should be connected.
- *No redundant subgraph generation:* Each connected node set in G should be generated exactly once.
- *Optimal order of enumeration:* If S' and S are connected node sets and $S' \subset S$, then S' should be generated before S so that if $f_b(S') < t$ we try to avoid generating S.

The "completeness" criterion relates to the correctness of the algorithm while the "no redundant subgraph generation" and "optimal order of enumeration" criteria relate to efficiency. The "no redundant subgraph generation" criterion asserts that each candidate solution in the solution space should be considered exactly once since additional considerations will lead to redundant computation. The "optimal order of enumeration" criterion, on the other hand, facilitates optimal pruning of the search space by ensuring that all subsets of a connected node set are considered before the node set itself is considered.

While depth-first enumeration approaches satisfy the first two criteria, they lead to many unnecessary rejections because the depth-first order of enumeration does not satisfy criterion three. Avoiding all unnecessary rejections likely requires a breadth-first enumeration, but memory can be a limiting factor for breadth-first approach. Here, we propose a more balanced approach that keeps the size of the problem manageable while reducing the number of redundant rejections.

2.2   Anchor Vertices

We first observe that the "completeness" and "no redundant subgraph genera-tion" criteria can be satisfied by selecting a single $v \in V$ as an *anchor vertex* and enumerating all subgraphs containing v before removing v from G. In this way each $v \in V$ is chosen as a starting point and all subgraphs containing it are enumerated before v is removed from G. When $V \equiv \emptyset$ all subgraphs have been enumerated. An example of enumerating connected induced subgraphs from an anchor vertex is shown in Figure 1.

It is clear that, for some $S' \subset V$, this process generates many $S \supset S'$ before S' itself, and thus it does not satisfy the criterion of "optimal order of enumera-tion". The number of these unnecessary rejections can be reduced using heuris-tic choices for the anchor vertex based on measures of centrality (e.g., degree or betweenness centrality). In this work, we rather focus on reducing unneces-sary rejections within each search anchored at a given vertex. In the following, we first describe our approach for reducing unnecessary rejections in the local search comprised of the anchor vertex and its neighbors, and then generalize our method to all connected induced subgraphs that contain the anchor vertex.

2.3   Efficient Enumeration of Spokes

Observe that, for a given anchor vertex $v \in V$, the neighbors of v can be treated as a set because v and any combination of its neighbors induce a connected subgraph of G. This collection of subgraphs ("spokes") can be represented as a *binomial tree*. A binomial tree is a data structure that can be used to enumerate all subsets of a set [14]. Any node n of a binomial tree has children that are copies of all branches rooted at siblings that precede n in the tree. The binomial tree that enumerates all spokes around the anchor vertex has a root node r labeled by the anchor vertex v and all descendants are labeled by neighbors of v. Since all vertices labeling nodes in T are connected to the vertex labeling the root, the

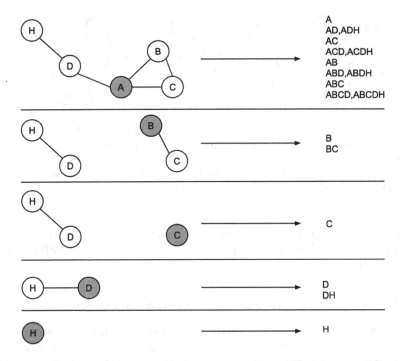

Fig. 1. Example illustrating the enumeration of all connected induced subgraphs of a graph using anchor vertices. On the left is the graph as each vertex becomes the anchor used for enumeration of all connected subgraphs that contain the anchor before it is subsequently removed from G. On the right are the connected subgraphs generated from each anchor vertex.

set of vertices that label each path from the root r of T to a node n represents a connected subgraph of G.

A simple method that *reduces* unnecessary rejections using a binomial tree based approach constructs the tree by adding each neighbor vertex to the root as a new node n, and then adding copies of the branches rooted at each sibling of n as children of n. Copying a branch terminates whenever the set S represented by the path does not satisfy $f_b(S) \geq t$. The resulting local search is similar in spirit to the set enumeration tree (SE-tree) search of Rymon [19]. However, our approach is more closely related to the binomial tree because we construct an explicit tree where the set is defined by the path from the root to a node in the tree. It is important to note that depending on how rejections occur, T may no longer meet the definition of a binomial tree so moving forward we will refer to T as a *local search tree*. An example of constructing a local search tree is shown in Figure 2.

The local search method can be extended to vertices beyond the direct neighbors of the anchor vertex by following a path of T and treating the vertices that label the path as an *anchor set* around which another local search tree is constructed as shown in Figure 3. This leverages the local search for each anchor set

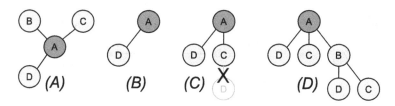

Fig. 2. Creating the local search tree T. **(A)** The input graph G with the anchor vertex A. **(B)** Exploring D with no previous branches yields the D branch. **(C)** Exploring C with the previous D branch evaluates ACD which is rejected resulting in branches C and D. **(D)** Exploring B with previous branches evaluates ABD and ABC (avoiding ABCD which contains the previously rejected ACD).

but the information is not used globally. In order to use the information from previous rejections globally we must modify our procedure as outlined in the following section.

2.4   Efficient Enumeration of All Connected Subgraphs

With slight modification we can combine the local search tree strategy with a conventional depth first approach to enumerate all subgraphs that contain the anchor vertex. Rather than using neighbors to construct the tree, we use the branches generated by depth first search through each neighbor to generate the tree. The top level procedure performs the depth portion of the search and it marks all neighbors as visited so they cannot be reached by continued depth search. The search space of neighbors is explored by the procedure that builds the local search tree from depth branches which we consider the breadth procedure. The only modification required to ensure correctness is that the breadth procedure stops cloning branches at nodes labeled by unvisited vertices adjacent to the vertex labeling the node that is appending the branch. This method represents our solution to the general case and is formalized in the **B**readth-first **D**iscovey, **D**epth-first **E**xploration (BDDE) algorithm. An example of the tree constructed by BDDE is shown in Figure 4 (C). An example of why the stop condition is necessary is show in Figure 4 (B).

2.5   Correctness

The following theorems are based on supporting lemmas in the supplementary materials[1]. Theorem 1 guarantees that our method satisfies the "completeness" and "no redundant subgraph generation" criteria during exhaustive enumeration, i.e, when f_b is satisfied by any S. Theorem 2 guarantees that our method satisfies the "optimal order of enumeration" criterion during exhaustive enumeration. Theorem 3 guarantees that our method satisfies the "completeness" criterion when f_b is selective.

---

[1] http://statler.case.edu/smaxwell/alcob2014/supplement.pdf

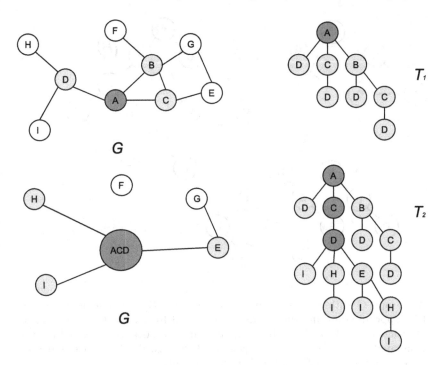

Fig. 3. Example illustrating the key idea of *anchor sets*. Initial tree T_1 generated from anchor vertex A in G (top) is extended by following path ACD in T_1 and treating the vertex set {ACD} as an anchor set. The anchor set can be isolated in G (bottom) the same way as an anchor vertex and a new local search tree is constructed anchored at ACD and appended to T_1 resulting in T_2. In this way all $S \subseteq V$ that contain the anchor vertex can be enumerated.

Theorem 1. *Given an input graph G, an anchor vertex $v \in V$ and a function f_b s.t. for any S, $f_b(S) \geq t$, BDDE uniquely enumerates all $S \subseteq V$ containing v.*

Proof: By Lemma 4 we know that the set represented by any path $P(n_k)$ in T induces a connected subgraph of G and by Lemma 5 we know that every path in T represents a unique set. By Lemma 7 we know that all $S \subseteq V$ containing v are represented by a path $P(n_k)$ in T. Therefore, we can conclude that because BDDE enumerates all paths of T, BDDE uniquely enumerates all $S \subseteq V$ containing v. $\square$

Theorem 2. *Given an input graph G, an anchor vertex $v \in V$ and a function f_b where $f_b(S) \geq t$ for any $S \subseteq V$, BDDE enumerates all connected induced subgraphs of G containing v in an order such that all $S' \subset S$ containing v are enumerated before S.*

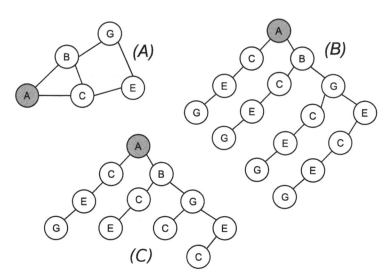

Fig. 4. (A) Graph G with anchor vertex A highlighted. (B) The enumeration tree generated by appending an un-pruned branch generated from S =AC to the branch generated through S =AB which exhibits several redundant instances of G and E. (C) The tree generated by pruning the branch generated through S =AC as it is added to the branch generated through S =AB and subsequent children.

Proof: By Theorem 1 we know that all $S \subseteq V$ containing v that induce a connected subgraph of G are enumerated, and by Lemma 8 we know that any $S' \subset S$ containing v must be generated before S.   □

Theorem 3. *Given an input graph G, an anchor vertex v and a function f_b, if $f_b(S') < t$ all $S \not\supset S'$ are still enumerated by BDDE.*

Proof: We know by Theorem 1 that all $S \subseteq V$ containing v are enumerated by BDDE when no rejections occur, and by Lemma 9 we know that when a rejection of S' occurs it only eliminates $S \supset S'$. Therefore, we conclude that if an S' is rejected all $S \not\supset S'$ are still enumerated.   □

3   Experimental Results

In order to systematically evaluate the performance of BDDE, we define a problem with a simple objective function that allows investigation of the effect of various parameters on performance. For this purpose, we use real-world networks to ensure that the network topology is practically relevant. We assign positive weights to all vertices of each network from a Gaussian distribution with mean m and standard deviation ρ. We define the objective function $f(S) = 1/\sum_{v \in S} w(v)$, where $w(v)$ denotes the weight of vertex v. Then, for a given threshold t and a maximum subgraph size k, we search for all connected subgraphs $S \subseteq V$ with $|S| \leq k$ and $f(S) \geq 1/t$ (note $1/t$ is used because we want to enumerate all subgraphs with a total weight less than t).

Datasets. In the experiments reported in this section, we utilize two real-world networks: 1) A citation network generated by Leskovec *et al.* [17] from the on-line arXiv journal, which consists of 5,241 vertices and 28,958 edges. 2) The protein-protein interaction (PPI) network obtained from the Human Protein Reference Database (HPRD) [13], which consists of 9,455 vertices and 37,080 edges.

Results. For each network, we set $m = 10$ and generate ten instances each for different values of ρ ranging from 1 to 9. On each instance, we perform enumerative search using the proposed algorithm and a standard DFS-based algorithm for t ranging from 5 to 50 for arXiv and 5 to 40 for HPRD. For each combination of ρ and t, we report the average of the performance measures across the ten randomized instances. The results for the HPRD network are shown in Figure 5. The results for the arXiv network follow similar trends and are available in the online supplemental materials.

Figure 5 shows the *rejection rate* for each algorithm. Here, rejection rate is defined as the fraction of subgraphs S with $f(S) < 1/t$ among all subgraphs that are enumerated. The rejection rate for BDDE is consistently lower than that of DFS though the relationship is more obvious at higher values of ρ. This relationship between ρ and rejection rate is indeed expected, since pruning is less effective when the weight distribution is more uniform across vertices. At the extreme case, when $\rho=0$ (all vertices have equal weight), only the leaf nodes of the enumeration tree are rejected and neither algorithm can take advantage of the knowledge on smaller subgraphs to prune out larger subgraphs. But as ρ grows, the enumeration tree becomes more imbalanced, and the benefit of the order of generation implemented by BDDE becomes more apparent.

Our computational tests also show that the two algorithms have similar runtimes for smaller values of the threshold t (when larger subgraphs are less likely to satisfy the objective criterion) and smaller values of ρ (when the vertex weights are more uniform). However, for larger values of t and ρ, BDDE consistently outperforms the DFS-based method. While runtime comparisons are largely implementation dependent we find this is a positive outcome and include additional figures and analysis in the supplementary material.

4   Conclusion

We have investigated the problem of reducing the number of subgraphs evaluated while enumerating all connected induced subgraphs $S \subseteq V$ that satisfy a hereditary property. Our proposed method displays a significant decrease in total number of subgraphs evaluated during enumeration compared to a classical depth first branch and bound approach. In addition, Theorems 1, 2 and 3 provide proof of correctness that all connected induced subgraphs S that satisfy $f_b(S) \geq t$ are enumerated. However, due to the potential for our method to use space exponential to the maximum size of S being enumerated, our method is best suited to enumerating all $|S| \leq k$ from G where k is chosen appropriate to the problem and the available memory.

Algorithm 1. The BDDE algorithm. Enumerates all S that contain anchor vertex v and satisfy $f_b(S) \geq t$. Returns the root node of the enumeration tree T. Entry point is DEPTH($\emptyset,v,[\,]$).

```
1: procedure BREADTH(S, n, U)
2:     if v_n ∈ U then                              ▷ Prune branch by stop condition
3:         return null
4:     end if
5:
6:     S' ← S ∪ v_n                                 ▷ Prune branch by bounding function
7:     if f_b(S') < t then
8:         return null
9:     end if
10:
11:    n' ← Υ(v_n)                                  ▷ Create a new tree node labeled by v_n
12:    for all {n* : nn* ∈ B} do                    ▷ Recursively copy child branches
13:        n'' ← BREADTH(S', n*, U)
14:        if n'' ≠ null then
15:            B ← B ∪ n'n''                        ▷ Append child branch
16:        end if
17:    end for
18:    return n'
19: end procedure

20: procedure DEPTH(S, v, β)
21:    S' ← S ∪ v
22:    if f_b(S') < t then
23:        return null
24:    end if
25:    n ← Υ(v)
26:    β' ← [ ]
27:    for i = 1 to |β| do
28:        n' ← BREADTH(S', β[i], χ_n)
29:        if n' ≠ null then
30:            B ← B ∪ nn'
31:            push(β', n')
32:        end if
33:    end for
34:    for all v ∈ χ_n do                           ▷ Note: Derive χ_n from S and v
35:        n' ← DEPTH(S', v, β')
36:        if n' ≠ null then
37:            B ← B ∪ nn'
38:            push(β', n')
39:        end if
40:    end for
41:    return n
42: end procedure
```

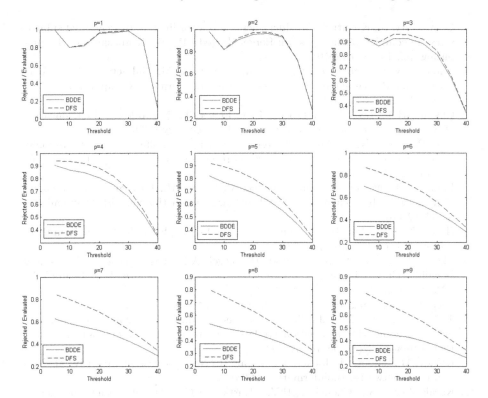

Fig. 5. Rejection rate analysis for enumerating all subgraphs up to size 4 satisfying $f(S) \geq 1/t$ in the HPRD network. Each pane plots the average rejection rate (fraction of subgraphs S with $f(S) < 1/t$ among all subgraphs that are enumerated) for each algorithm versus threshold $5 \leq t \leq 40$ where the node scores were sampled from a Gaussian distribution with mean $m=10$ and standard deviation $1 \leq \rho \leq 9$.

Acknowledgements. We would like to thank Harold Connamacher for his invaluable feedback on our method and related work. This work is supported in part by the Case Comprehensive Cancer Center (Grant P30-CA-043703) and US National Science Foundation (Grant CCF-0953195). This publication was also made possible by the Clinical and Translational Science Collaborative of Cleveland, UL1TR000439 from the National Center for Advancing Translational Sciences (NCATS) component of the National Institutes of Health and NIH roadmap for Medical Research. Its contents are solely the responsibility of the authors and do not necessarily represent the official views of the NIH.

References

1. Avis, D., Fukuda, K.: Reverse search for enumeration. Discrete Applied Mathematics (1993)
2. Bollobas, B.: Hereditary properties of graphs asymptotic enumeration global structure and colouring. Documenta Mathematica, 333–342 (1998)

3. Chowdhury, S., Koyuturk, M.: Identification of coordinately dysregulated subnetworks in complex phenotypes. In: Berger, B. (ed.) Pacific Symposium on Biocomputing, pp. 133–144 (2010)
4. Chowdhury, S., Nibbe, R., Chance, M., Koyuturk, M.: Subnetwork state functions define dysregulated subnetworks in cancer. Journal of Computational Biology 18(3), 263–281 (2011)
5. Chuang, H.Y., Lee, E., Yu-Tsueng, L.D., Ideker, T.: Network-based classification of breast cancer metastasis. Molecular Systems Biology (2007)
6. Dao, P., Wang, K., Collins, C., Ester, M., Lapuk, A., Sahinalp1, S.C.: Optimally discriminative subnetwork markers predict response to chemotherapy. Bioinformatics (July 2011)
7. Flannick, J., Novak, A., Srinivasan, B., McAdams, H., Batzoglou, S.: Graemlin: General and robust alignment of multiple large interaction networks. Genome Research (2006)
8. Hopcroft, J., Tarjan, R.: Efficient algorithms for graph manipulation. Communications of the ACM 16(6) (1973)
9. Ideker, T., Ozier, O., Schwikowski, B., Siegel, A.F.: Discovering regulatory and signalling circuits in molecular interaction networks. Bioinformatics 18(suppl. 1), S233–S240 (2002),
   http://dx.doi.org/10.1093/bioinformatics/18.suppl_1.s233
10. Jia, P., Zheng, S., Long, J., Zheng, W., Zhao, Z.: dmGWAS: dense module searching for genome-wide association studies in protein-protein interaction networks. Bioinformatics 27(1), 95–102 (2011)
11. Kalaev, M., Smoot, M., Ideker, T., Sharan, R.: Networkblast: comparative analysis of protein networks. Bioinformatics (2008)
12. Karakashian, S., Choueiry, B.Y., Hartke, S.G.: An algorithm for generating all connected subgraphs with k vertices of a graph (May 2013),
   http://www.math.unl.edu/~shartke2/math/papers/k-subgraphs.pdf
13. Kesheva, P., et al.: Human protein reference database: 2009 update. Nucleic Acids Research 37, 767–772 (2009)
14. Knuth, D.: The Art of Computer Programming, Combinatorial Algorithms Part 1, vol. 4. Addison-Wesley (2012)
15. Konga, B., Yanga, T., Chenb, L., Qin Kuanga, Y., Wen Gua, J., Xiaa, X., Chenga, L., Hai Zhang, J.: Proteinprotein interaction network analysis and gene set enrichment analysis in epilepsy patients with brain cancer. Journal of Clinical Neuroscience (2013)
16. Koyutürk, M., Kim, Y., Subramaniam, S., Szpankowski, W., Grama, A.: Detecting conserved interaction patterns in biological networks. Journal of Computational Biology (2006)
17. Leskovec, J., Kleinberg, J., Faloutsos, C.: Graph evolution: Densification and shrinking diameters. ACM Transactions on Knowledge Discovery from Data (2007)
18. Patel, V., Gokulrangan, G., Chowdhury, S., Chen, Y., Sloan, A., Koyutrk, M., Barnholtz-Sloan, J., Chance, M.: Network signatures of survival in glioblastoma multiforme. PLOS Computational Biology 9 (2013)
19. Rymon, R.: Search through systematic set enumeration. Tech. rep., University of Pennsylvania (August 1992)

On Algorithmic Complexity
of Biomolecular Sequence Assembly Problem

Giuseppe Narzisi[1], Bud Mishra[1,2], and Michael C. Schatz[1]

[1] Simons Center for Quantitative Biology
One Bungtown Road, Cold Spring Harbor Laboratory, NY, 11724, USA
[2] Courant Institute of Mathematical Sciences, New York University
New York, NY, 10012, USA
{gnarzisi,mschatz}@cshl.edu, mishra@nyu.edu

Abstract. Because of its connection to the well-known $\mathcal{NP}$-complete shortest superstring combinatorial optimization problem, the Sequence Assembly Problem (SAP) has been formulated in simple and sometimes unrealistic string and graph-theoretic frameworks. This paper revisits this problem by re-examining the relationship between the most common formulations of the SAP and their computational tractability under different theoretical frameworks. For each formulation we show examples of logically-consistent candidate solutions which are nevertheless unfeasible in the context of the underlying biological problem. This material is hoped to be valuable to theoreticians as they develop new formulations of SAP as well as of guidance to developers of new pipelines and algorithms for sequence assembly and variant detection.

Keywords: Genome Assembly, Sequence Assembly Problem, Optimality, $\mathcal{NP}$-complete Problem.

1 Introduction

The ability to sequence a genome and reconstruct its DNA sequence is changing human genetics research [14]. Recent advances in DNA sequencing technology have driven the cost of sequencing a complete human genome to below $1000 US[1], and the potential applications to biology and medicine have rekindled enormous interest in several classical algorithmic problems at the core of genomics and computational biology, especially the DNA sequence assembly problem (SAP). Two decades back, in the context of the Human Genome Project, the problem had received unprecedented scientific prominence: its computational complexity and intractability were thought to have been well understood; various competitive heuristics, thoroughly explored and the necessary software, properly implemented and validated. However, recent studies on the experimental validation of de novo assemblers, have highlighted several limitations [19,4,2].

The process of reducing/relating the problem of reconstructing the genome sequence into a well-defined computer science problem is not straightforward: for

---

[1] http://dx.doi.org/doi:10.1038/nature.2014.14530

A.-H. Dediu, C. Martín-Vide, and B. Truthe (Eds.): AlCoB 2014, LNBI 8542, pp. 183–195, 2014.

instance, limited or incomplete knowledge of the original biological problem, can lead to erroneous formulations. Consequently, a perfectly well-defined "optimal solution" in the computational setting may turn out to be irrelevant, infeasible or incorrect, when translated back to the original biological setting. The sequence assembly problem is in fact a *wicked*[2] problem: incomplete, contradictory, changing requirements (e.g., genome structure) lead to incomplete and biologically incorrect formulations.

This paper carefully examines the most popular formulations for SAP over the last 20 years. Each formulation is rigorously defined. Similarity and differences among paradigms are explained, demonstrating a strong connection between the different formalisms. More importantly, we present examples of logically consistent solutions in each of this formulations which are intractable or unfeasible in the context of biology.

2   The Dovetail-Path Framework

The *dovetail-path framework* was first introduced by Myers in [16]. The output of a sequencing project consists of a set of reads $F = \{r_1, r_2, \ldots, r_N\}$, where each read r_i is a string over the alphabet $\Sigma = \{A, C, G, T\}$. Each read is associated a pair of integers (s_i, e_i), $i \in [1, |F|]$ where s_i and e_i are respectively the starting and ending points of the read r_i in the reconstructed string R (to be computed by the assembler), such that $1 \leq s_i, e_i \leq |R|$. The order of s_i and e_i encodes the orientation of the read (whether r_i was sampled from Watson or Crick strand of the DNA molecule).

The overlaps between pairs of reads capture where the suffix of the first matches the prefix of the second within some maximum error rate, and may be computed using the Smith-Waterman algorithm [24] with match, mismatch and gap penalty scores dependent on the error model of the sequencing technology. Thanks to the high throughput of next-generation sequencing technology, overlaps computed using exact-match are now adequately informative for short reads, although emerging third-generation long read sequencing requires in-exact matching algorithms [11,23]. The complete description of an overlap π is given by specifying:

1. the substrings $\pi.A[\pi.s_A, \pi.e_A]$ and $\pi.B[\pi.s_B, \pi.e_B]$ of the two reads that are involved in the overlap;
2. the offsets from the left-most and right-most positions of the reads $\pi.A_{hang}$ and $\pi.B_{hang}$;
3. the relative directions of the two reads: Normal (N), Innie (I);
4. a binary predicate $suffix_\pi(r)$ on a read r such that:

$$suffix_\pi(r) = \begin{cases} true & \textit{iff } \text{suffix of } r \text{ participates in the overlap } \pi \\ false & \textit{iff } \text{prefix of } r \text{ participates in the overlap } \pi \end{cases} \quad (1)$$

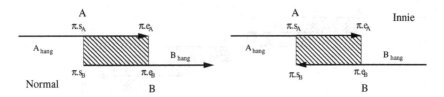

Fig. 1. Two possible overlaps (illustration): left overlap is *normal* (with both reads pointing to the same direction) right overlap is *innie* (with both reads pointing against each other); The suffix predicate for the left (normal) overlap is s.t. $suffix_\pi(A) = true$ and $suffix_\pi(B) = false$.

Figure 1 illustrates two possible overlaps. Because of the double-stranded nature of the DNA molecule, each read can be sampled from either the Watson or Crick strands and they have different orientation.

Definition 1 (Layout). *The layout L associated to a set of reads F is defined as:*

$$L_F = r_1 \overset{\pi_1}{\rightleftharpoons} r_2 \overset{\pi_2}{\rightleftharpoons} r_3 \overset{\pi_3}{\rightleftharpoons} \ldots \overset{\pi_{N-1}}{\rightleftharpoons} r_N \qquad (2)$$

Informally a layout is simply a sequence of reads with each neighboring read pair connected by overlap relations. The previous definition assumes that there are no *containments*[3]; without any loss of correctness or generality, contained reads can be initially removed (in a preprocessing step) and then reintroduced later after the layout has been created. Among all the possible layouts (possibly, exponential in the number of reads), it is imperative to efficiently identify the ones that are consistent according to the following definition:

Definition 2 (Consistency Property). *A layout L is **consistent** if the following property holds for $i = 2, \ldots, N - 1$:*

$$\overset{\pi_{i-1}}{\rightleftharpoons} r_i \overset{\pi_i}{\rightleftharpoons} \text{ iff } suffix_{\pi_{i-1}}(r_i) \neq suffix_{\pi_i}(r_i) \qquad (3)$$

The consistency property imposes a directionality for traversing the sequence of reads in the layout. The directionality of each internal read in the layout must be preserved so that the left and right overlaps have opposite values for the *suffix* predicate. Figure 2 shows an example of layout arising from 7 overlapping reads.

Appealing to parsimony, we are typically interested in a layout whose length is minimal (although we will see that this assumption is biologically incorrect). The following theorem shows the correlation between the length of a layout and the sizes of its overlaps. Let us define the weight of a layout L to be the sum

---

[2] A problem is wicked, if from its original formulation, one is led to a "correct" solution that reveals the incorrectness, incompleteness or inconsistencies in the formulation of the problem [22].

[3] Reads that are proper subsequences of another read.

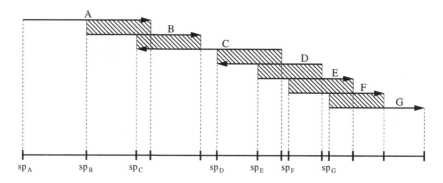

Fig. 2. Example of layout for a set of fragments $F = \{A, B, C, D, E, F, G\}$ with sequence of overlaps $\pi^N_{(A,B)}, \pi^I_{(B,C)}, \pi^N_{(C,D)}, \pi^I_{(D,E)}, \pi^N_{(E,F)}, \pi^N_{(F,G)}$

of the lengths of its overlaps, $weight(L) = \sum_{\pi \in L} length(\pi)$, then the following theorem holds [25,26]:

Theorem 3 (Min-length reconstruction). *A layout of maximum weight results in a reconstruction of minimum length.*

3   Shortest Superstring Problem (*SSP*)

Researchers first approximated the shotgun sequence assembly problem as one of finding the shortest common superstring of a set of sequences. This formulation was encouraged by the results of the previous theorem and the growing body of literature on efficient algorithms to solve the *SSP*.

Definition 4 (Shortest Superstring Problem). *Given a set of strings $S = \{r_1, r_2, \ldots, r_n\}$ find the shortest string R (reconstruction) such that $\forall i$, r_i is a substring of R.*

This formulation led to a simple theoretical abstraction, but by being oblivious to how biological sequences are organized by evolution, it often yielded biologically implausible and incorrect solutions. Its inability to correctly model the assembly problem is owed to a multitude of reasons, but primarily because:

1. the shortest-superstring formulation does not account for possible errors arising during the process of sequencing the fragments,
2. it does not model fragment orientation (the sequence source can be one of the two DNA strands), and
3. most importantly, it fails in the presence of *repeats*, as it encourages repeat-induced compressions.

Elaborating on the last point it is of interest to consider Richard Karp's statement in 2003 [9]: *The shortest superstring problem [is an] an elegant but flawed abstraction: [since it defines assembly problem as finding] a shortest string containing a set of given strings as substrings.* Figure 3 shows an example of the kind of errors that such formulation could lead to. Since strings contained inside a repeat regions cannot be disambiguated, multiple copies of a repeat are compressed into a single one.

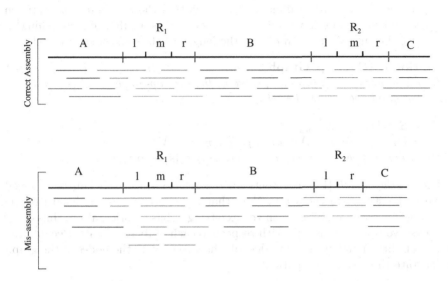

Fig. 3. Example of compression: the two copies of repeat (R_1 and R_2) are compressed into one leading to a shorter but misassembled sequence

Because of the theoretical computational intractability ($\mathcal{NP}$-completeness [5]) of the *SSP*, most of the approaches for genome sequence assembly have resorted to greedy and heuristic methods that, by definition, restrict themselves to near-optimal solutions, where the "nearness" may be guaranteed within a multiplicative competitiveness factor. The best known greedy algorithm for the *SSP* has an approximation factor of $2\frac{2}{3}$ [1].

4   Graph-Theoretic Formulation

Differently from string-based approaches, graph-theoretic formulations convert sequence assembly into solving specific problems for general graphs constructed using the overlap information of the input set of reads. This mapping has the advantage of allowing us to apply the large collection of algorithms and heuristics that have been developed in graph theory for many decades. However, this formulation still fails to completely cure the problems and limitations of the *SSP* model, since it can produce mis-assembly errors (as shown later). In this section we introduce the two most used graphical models for the sequence assembly problem: *string graph* and *De Bruijn graph*. But before formally specifying these graphs, we need to give a few basic definitions.

4.1   Strings, Overlaps and Overlap Graph

Let x and y be two strings over the alphabet Σ. Let us denote the length of x by $|x|$. The ith character of x is denoted by $x[i]$. If $1 \leq i \leq j \leq |x|$, we use $x[i, j]$ to denote the substring of x starting at position i and ending at position j. Given two strings x and y over the alphabet Σ, we say that there is an *overlap* between x and y, and we denote it with $x \rightleftharpoons y$, if there exists a suffix of x matching[4] a prefix of y. Let us denote with $o(x, y)$ the length of the longest such match.

Definition 5 (Overlap Graph). *Given a set of strings $S = \{r_1, r_2, \ldots, r_n\}$ and a minimum overlap threshold value k, the overlap-graph for S is a weighted bidirected graph $OG^k = (V, E)$ where:*

- $V = S = \{r_1, r_2, \ldots, r_n\}$;
- $E = \{(r_i, r_j) : (r_i \rightleftharpoons r_j) \wedge o(r_i, r_j) \geq k, r_i, r_j \in V\}$;
- *the weight of each edge (r_i, r_j) is $w(r_i, r_j) = |s_j| - o(r_i, r_j)$.*

The *overlap graph* [16] represents all the relationships that can be inferred between the strings in the set S. Note that $|r_j| - o(r_i, r_j)$ is the length of the overhang[5] for string r_j, Since each vertex/string r_i has an orientation, thus every edge has two orientations, one with respect to each of its endpoints. Because the graph is bidirectional, we need to describe how to explore the nodes of the graph to generate the set of *valid* paths.

Definition 6 (Path validity). *A path $P = \langle r_1 \overset{e_1}{\rightleftharpoons} r_2 \overset{e_2}{\rightleftharpoons} r_3 \overset{e_3}{\rightleftharpoons} \ldots \overset{e_{m-1}}{\rightleftharpoons} r_m \rangle$ in G is valid if $\forall i, 2 \leq i \leq m - 1$, e_{i-1} and e_i have opposite directions at r_i.*

Note that this definition is equivalent to the *consistency property* for a layout. In order to traverse a node in the graph we need that the entry edge and the exit edge have opposite directions at the node. So we are allowed to enter a node x even if the edge e_i is pointing out of the node as long as we use an edge e_j with opposite direction to e_i when we exit the node (see figure 4 for an example of overlap graph).

Given any path P in the overlap graph, we associate a *path-string* to P that consist of the concatenation of the strings according to the order in the path, where only one copy of the overlap is kept. Clearly the weight of a path P is given by the sum of the weights of its edges:

$$w(P) = \sum_{(r_i, r_j) \in P} w(r_i, r_j) = \sum_{(r_i, r_j) \in P} (|r_j| - o(r_i, r_j)) \tag{4}$$

Note that because of the weight function associated to the edges of the graph, a path of minimum weight defines a path-string of minimum length.

---

[4] The matching does not have to be perfect and it can be approximated allowing up to ϵ percent error on real data.

[5] A relaxation to an overlap, such that some small number of bases at the beginning or end of the read are excluded from the overlap region, typically because of a high error rate.

4.2   String Graph

The size of the overlap graph can be dramatically reduced by a sequence of transformations whose goal is to eliminate edges that can be *transitively* inferred.

Definition 7 (transitively inferable edge). *If $x \stackrel{e_1}{\rightleftharpoons} y \stackrel{e_2}{\rightleftharpoons} z$ and $x \stackrel{e_3}{\rightleftharpoons} z$ are* **mutually consistent** *overlaps among nodes x, y and z then the edge e_3 is said to be transitively inferable from the sequence of edges e_1 and e_2.*

Informally the overlap between strings x and z is implied by the composition of the overlaps between x, y and z. It is important to note the edges must be mutually consistent: entry edge and the exit edge must have opposite directions. The *string graph* is a particular graph where all the contained string and transitively inferable edges are removed [17].

Definition 8 (String Graph). *Given a set of strings $S = \{r_1, r_2, \ldots, r_n\}$ and a minimum overlap threshold value k, the string graph SG^k for S is obtained from the overlap graph OG^k by removing contained strings (strings that are substrings of other strings) and transitively inferable edges [17].*

Such transformation can be computed in polynomial time using the algorithm proposed by Myers in [17]. In order to correctly apply the transitivity reduction step to the graph, it is important to first mark all transitively inferable edges and then remove all marked edges in a distinct phase. This is because this process is not Church-Rosser [3] and any arbitrary strategy would fail to remove some of the transitively inferable edges. Equipped with the notion of string graph, the sequence assembly problem can be formulated as follows:

Definition 9 (Sequence Assembly Problem). *Given a set of fragment or reads $S = \{r_1, r_2, \ldots, r_n\}$ and a minimum overlap threshold k, the Sequence Assembly Problem (SAP) is the problem of finding an Hamiltonian Path in the string graph SG^k for S such that its weight is minimum.*

The problem is clearly a special case of the Traveling Salesman Problem (TSP) with the following two differences: (1) instead a looking for a Hamiltonian cycle we look for an Hamiltonian path; (2) we work with bi-directed graphs instead of undirected or directed graphs. However, for circular genomes (such as plasmids and bacterial genomes), the first difference does not apply anymore as we need to find an Hamiltonian cycle as well.

Note that this formulation differs from the one presented in [18]. Specifically Nagarajan and Pop define the sequence assembly problem as one of finding a generalized Hamiltonian path (every node is visited at least once) of minimum weight in the string graph of the reads. This is in accordance to the solution proposed in [17] where they seek a cyclic tour. In such model each edge has assigned a selection constraint c that says how many times the edge should appear in the target solution: *exact* edge ($c = 1$), *required* edge ($c \geq 1$) and *optional* edge ($c \geq 0$). Note that, even if we allow a read to be potentially used more than once, the appeal to parsimony (min weight) could compromise the correctness of the layout.

Before discussing the complexity of this problem it is important to observe that this graph-theoretical formulation suffers from the same kind of problems of the shortest superstring approach. Figure 4 show an example of string graph where all the possible Hamiltonian paths create mis-assembly error due to the presence of a repeat. The compression error is due to the fact that repeats can induce *false positive transitively inferable edges*. For example consider the reads 3, 7 and 8 in figure 4, we have that $7 \rightleftharpoons 3$, $3 \rightleftharpoons 8$ and $7 \rightleftharpoons 8$, so the edge $7 \rightleftharpoons 8$ is removed with the negative effect to merging together reads that belong to two different copies of the repeat R_2. In particular, after removal of the transitively inferable edges, there is more than one path that traverses all the nodes and it always produces mis-assembled layouts. Note that edge $2 \rightleftharpoons 6$ cannot be removed because, although there are edges $2 \rightleftharpoons 7$ and $7 \rightleftharpoons 6$, the directions at node 7 do no match and so it cannot be traversed (the edges are not *mutually consistent*).

This example also shows another problem associated to this framework. Even if it would be possible to efficiently compute the Hamiltonian path, the string graph might have many different Hamiltonian paths (as in this example) of minimum length and all these paths represent a possible reconstruction of the genome. Additional information, such as mate-pairs, can sometimes be used to help resolve this ambiguity, although since mate-pairs are generally at most 10 to 20kbp long, in general they do not fully resolve the ambiguity except for the smallest genomes lacking any large repeats. The problem of finding a minimum weight Hamiltonian path in a directed or undirected graph is known to be $\mathcal{NP}$-complete. Since directed graphs are special types of bidirected graphs, we have:

Theorem 10. *The Sequence Assembly Problem is $\mathcal{NP}$-complete.*

4.3   De Bruijn Graph

In a de Bruijn graph the notions of nodes and edges are somehow inverted compared to the overlap graph. A de Bruijn graph is formally defined as follows.

Definition 11 (De Bruijn Graph). *Given a set of strings $S = \{r_1, r_2, \ldots, r_n\}$ and a minimum overlap threshold value k, the de Bruijn graph for S is a directed graph $BG^k = (V, E)$ where:*

- $V = \{d \in \Sigma^k \mid \exists i \text{ s.t. } d \text{ is a substring of } r_i \in S\}$;
- $E = \{(d_i, d_j) : \text{if the prefix of length } k - 1 \text{ of } d_i \text{ is a suffix of } d_j\}$;

Informally the set of vertices of BG^k is the set of k-mers for the set of input strings S (the *spectrum L*), and the edges correspond to their perfect $k - 1$ overlap. Clearly every read $r_i \in S$ is translated into a path composed of $(|r_i| - k)$ nodes. Let us call such a path a *walk* and define it $w(r_i)$. Also note that there is no weight associated to the edges (the overlap weight is $k - 1$ for all the edges and it can be omitted). Specifically, we create one node for each k-mer in the set L and a directed edge from node x_1 to node x_2 if the $k - 1$ suffix of x_1 is a prefix

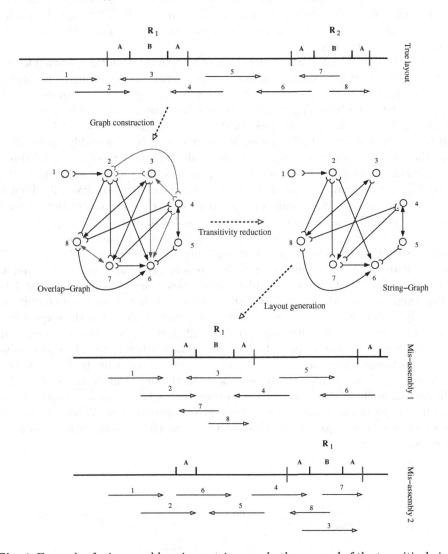

Fig. 4. Example of mis-assembly using a string graph: the removal of the transitively inferable edges (in red) produces a string graph where every (Hamiltonian) paths through all nodes creates mis-assemblies. The layouts for two of these paths are shown at the bottom: the first one with compression and the second one with both compression and inversion.

of x_2 and we label the edge with the remaining rightmost string in x_2. Hence, in this graph each edge corresponds to one of the k-mers and so the general problem consists of finding a path that visits all the edges exactly once, namely, an *Eulerian path*. The string S corresponding to a path in this graph can be reconstructed by concatenating k-mer sequence of the first node, in order, with all the labels of the edges in the path.

The de Bruijn Graph framework is currently the most popular approach for assembling the shorter reads coming from next-generation sequencing technologies such as Illumina [6,12]. Moreover, it is now becoming more and more important to model the haplotyipic structure of DNA, specifically in the context of detecting DNA mutations such as short insertions and deletions of bases (IN-DELs). Recent works [21,13,8] demonstrate how sequence assembly approaches are the most promising methods for this task. However, repetitive structures, in particular near-perfect repeats, within genomes can produce artifacts in the assembly graphs that mislead such methods to make false-positive calls. Figure 5 shows an example of a near-perfect repeat that can be misinterpreted as a large deletion. The key observation is that the beginning of this sequence is a nearly perfect 69bp repeat. There is just 1bp difference between the two copies that are 15bp apart. The sequence is segmented as 19-C-49-A-14-19-T-49-G-21 where 19 and 49 are 19bp and 49bp perfect repeats, separated by a 15bp unique sequence (A, C, T, G are the regular bases). Since the longest exact repeat is 49bp long, one would expect that using k-mer=55 should be large enough to correctly assemble reads sampled from this sequence. However, if the sequencing data also contains reads with sequence 19-C-49-G, it can be wrongly interpreted as a long 84bp deletion of the A-14-19-T-49 segment when instead it is just a single base change. Since the de Bruijn graph is constructed using perfect matches of length $k - 1 = 54$ (no mismatches allowed), the only way to connect all the 55-mers from these two sequences is to construct a false bubble jumping form the first copy of the near-perfect repeat to the second copy. When aligned to the reference, the sequence associated to the branch will show a false-positive deletion.

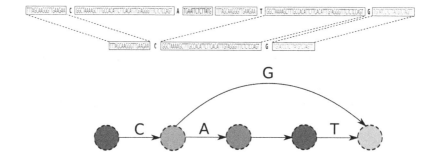

Fig. 5. Example of false bubble in a De Bruijn graph

Finally, it is important to note that in the de Bruijn framework, similarly to the String Graph framework, the graph can have more then one Eulerian path and choosing the correct one is not trivial. Indeed the number of valid paths may be extremely large, and bounded only by the product of the factorial of the degrees of the nodes times the number of potential spanning trees of the graph [10]. Although an Eulerian path can be computed in polynomial time (using the Hierholzer's algorithm [7]), it might not represent a correct assembly of the input reads (the path may not be read-coherent [17]). However, as mentioned before, each read correspond to a particular walk in the de Bruijn graph, and any walk that contains all the reads as subwalks (a *superwalk*) represents a possible assembly of the reads. In this framework a parsimonious solution corresponds to a superwalk of minimum length:

Definition 12 (Superwalk Problem). *Given a set of reads $S = \{r_1, \ldots, r_n\}$ find a minimum length superwalk in the De Bruijn graph BG^k of S.*

It can be shown that this problem is also $\mathcal{NP}$-complete by reduction from the Shortest Superstring Problem [15]:

Theorem 13. *The Superwalk Problem is $\mathcal{NP}$-complete.*

5   Discussion

The process of abstracting a problem from its biological interpretation is a powerful tool to better investigate a biological problem. However, as demonstrated in this paper for the sequence assembly problem, it is very important to develop (biologically) correct formulations. The shortest superstring formulation was an elegant theoretical abstraction, but it was clearly oblivious to what biology needs to make a correct interpretation of genomic data. The subsequent graph-theoretical formulations, although more powerful than the simpler SSP model, still suffer from similar problems when dealing with repeat structures. We have presented examples of many popularly accepted formulations that can lead to miss-assembly errors. Although all the SAP formations presented in this paper lead to computationally intractable problems ($\mathcal{NP}$-complete), approximated solutions can be efficiently computed using graph search methods (BFS vs DFS) often in combination with branch-and-bound method [20,21]. A better understanding and modeling of the sequence composition (e.g., repeats) contained within genomes has been one of the key factors to improve accuracy in computational genomics, but much work needs to be done to achieve the goal of an *error-free* reconstruction. Finally there is now the urgency to model the haplotypic structure of the human genome which introduces another level of complexity for example in the algorithms seeking to discover genetic mutations.

References

1. Armen, C., Stein, C.: A 2 2/3-approximation algorithm for the shortest superstring problem. In: Hirschberg, D.S., Meyers, G. (eds.) CPM 1996. LNCS, vol. 1075, pp. 87–101. Springer, Heidelberg (1996)

2. Bradnam, K., et al.: Assemblathon 2: evaluating de novo methods of genome assembly in three vertebrate species. GigaScience 2(1), 10 (2013)
3. Church, A., Rosser, J.B.: Some properties of conversion. Transactions of the American Mathematical Society 39(3), 472–482 (1936)
4. Earl, D.A., et al.: Assemblathon 1: A competitive assessment of de novo short read assembly methods. Genome Research (2011)
5. Gallant, J., Maier, D., Astorer, J.: On finding minimal length superstrings. Journal of Computer and System Sciences 20(1), 50–58 (1980)
6. Gnerre, S., et al.: High-quality draft assemblies of mammalian genomes from massively parallel sequence data. Proceedings of the National Academy of Sciences 108(4), 1513–1518 (2011)
7. Hierholzer, C., Wiener, C.: Ueber die mglichkeit, einen linienzug ohne wiederholung und ohne unterbrechung zu umfahren. Mathematische Annalen 6(1), 30–32 (1873)
8. Iqbal, Z., Caccamo, M., Turner, I., Flicek, P., McVean, G.: De novo assembly and genotyping of variants using colored de bruijn graphs. Nature Genetics 44(2), 226–232 (2012)
9. Karp, R.M.: The role of algorithmic research in computational genomics. In: Computational Systems Bioinformatics Conf, p. 10. IEEE Computer Society (2003)
10. Kingsford, C., Schatz, M., Pop, M.: Assembly complexity of prokaryotic genomes using short reads. BMC Bioinformatics 11(1), 21 (2010)
11. Koren, S., et al.: Hybrid error correction and de novo assembly of single-molecule sequencing reads. Nature Biotechnology 30(7), 693–700 (2012)
12. Li, R., et al.: De novo assembly of human genomes with massively parallel short read sequencing. Genome Research 20(2), 265–272 (2010)
13. Li, S., Li, R., Li, H., Lu, J., Li, Y., Bolund, L., Schierup, M., Wang, J.: Soapindel: Efficient identification of indels from short paired reads. Genome Research (2012)
14. Mardis, E.R.: The impact of next-generation sequencing technology on genetics. Trends in Genetics 24(3), 133–141 (2008)
15. Medvedev, P., Georgiou, K., Myers, G., Brudno, M.: Computability of models for sequence assembly. In: Giancarlo, R., Hannenhalli, S. (eds.) WABI 2007. LNCS (LNBI), vol. 4645, pp. 289–301. Springer, Heidelberg (2007)
16. Myers, E.W.: Toward simplifying and accurately formulating fragment assembly. Journal of Computational Biology 2, 275–290 (1995)
17. Myers, E.W.: The fragment assembly string graph. Bioinformatics 21(suppl. 2), ii79–ii85 (2005)
18. Nagarajan, N., Pop, M.: Parametric complexity of sequence assembly: theory and applications to next generation sequencing. Journal of Computational Biology 16(7), 897–908 (2009)
19. Narzisi, G., Mishra, B.: Comparing de novo genome assembly: The long and short of it. PLoS ONE 6(4), e19175 (2011)
20. Narzisi, G., Mishra, B.: Scoring-and-unfolding trimmed tree assembler: concepts, constructs and comparisons. Bioinformatics 27(2), 153–160 (2011)
21. Narzisi, G., O'Rawe, J.A., Iossifov, I.: ha Lee, Y., Wang, Z., Wu, Y., Lyon, G.J., Wigler, M., Schatz, M.C.: Accurate detection of de novo and transmitted indels within exome-capture data using micro-assembly. bioRxiv (2013)
22. Rittel, H.W.J., Webber, M.M.: Dilemmas in a general theory of planning. Policy Sciences 4, 155–169 (1973)
23. Roberts, R., Carneiro, M., Schatz, M.: The advantages of smrt sequencing. Genome Biology 14(7), 405 (2013)
24. Smith, T.F., Waterman, M.S.: Identification of common molecular subsequences. Journal of Molecular Biology 147(1), 195–197 (1981)

25. Tarhio, J., Ukkonen, E.: A greedy approximation algorithm for constructing short-est common superstrings. Theor. Comput. Sci. 57(1), 131–145 (1988)
26. Turner, J.S.: Approximation algorithms for the shortest common superstring prob-lem. Inf. Comput. 83(1), 1–20 (1989)

A Closed-Form Solution for Transcription Factor Activity Estimation Using Network Component Analysis

Amina Noor[1], Aitzaz Ahmad[2], Bilal Wajid[1], Erchin Serpedin[1],
Mohamed Nounou[3], and Hazem Nounou[4]

[1] Department of Electrical and Computer Engineering, Texas A&M University
College Station, TX, 77840 USA
[2] Corporate Research & Development
Qualcomm Technologies Inc., San Diego, CA 92121, USA
[3] Department of Chemical Engineering, Texas A&M University at Qatar
Doha, Qatar
[4] Department of Electrical Engineering, Texas A&M University at Qatar
Doha, Qatar

Abstract. Non-iterative network component analysis (NINCA), proposed by Jacklin *at.al*, employs convex optimization methods to estimate the transcription factor control strengths and transcription factor activities. While NINCA provides good estimation accuracy and higher consistency, the costly optimization routine used therein renders a high computational complexity. This correspondence presents a closed form solution to estimate the connectivity matrix which is tens of times faster, and provides similar accuracy and consistency, thus making the closed form NINCA (CFNINCA) algorithm useful for large data sets encountered in practice. The proposed solution is assessed for accuracy and consistency using synthetic and yeast cell cycle data sets by comparing with the existing state-of-the-art algorithms. The robustness of the algorithm to the possible inaccuracies in prior information is also analyzed and it is observed that CFNINCA and NINCA are much more robust to erroneous prior information as compared to FastNCA.

Keywords: Gene Regulatory Network, transcription factor activity, convex optimization.

1 Introduction

Transcription regulation is an important biological process which governs the transcription and translation of genes using transcription factors (TFs). The binding of TFs to genes causes them to express themselves, and these expression levels are measured using the DNA microarray technology or RNA-Seq. These well-established methods quantify the expression levels in the form of gene expression data, which are widely used in the inference of gene regulatory networks [1,2,3,4]. TF activities (TFAs), defined as the concentration of the subpopulation with DNA binding ability, are hard to measure experimentally, owing to the

A.-H. Dediu, C. Martín-Vide, and B. Truthe (Eds.): AlCoB 2014, LNBI 8542, pp. 196–207, 2014.
© Springer International Publishing Switzerland 2014

change in correlation between TFs and TFAs at the post-transcriptional stage [5,6]. This necessitates the use of computational methods for their estimation.

The gene regulatory network which captures the interactions between the genes and TFs is mathematically modeled as [7]

$$X = AS + \Gamma ,\tag{1}$$

where X ($N \times K$) denotes the gene expression data matrix, A ($N \times M$) stands for the control strength matrix between genes and TFs, S ($M \times K$) indicates the TFAs and Γ, representing the modeling errors, is assumed Gaussian noise. Genes are known to interact nonlinearly and in a dynamic fashion. However, the variation of the TFAs is much slower than that of gene expression. The log-linear model in 1, therefore, provides a good approximation [8].

The estimation of A and S from the gene expression data matrix X has been performed using principal component analysis (PCA) [9] and independent component analysis (ICA) [10]. These algorithms assume properties of orthogonality and independence, respectively, which do not conform to the biological signals. To model gene networks more accurately, prior information about the connectivity matrix available from the ChIP-chip data should be incorporated in the system model [11,7]. Towards this end, network component analysis (NCA) was proposed by [7] which can be formulated as:

$$\min_{\mathbf{A,S}} \ ||\mathbf{X} - \mathbf{AS}||_F^2 \quad \text{s.t.} \ \mathbf{A}(I) = 0 ,\tag{2}$$

where $||.||_F$ denotes the Frobenius norm and I stands for the set of indices corresponding to the entries of A that are known to be zero a-priori. This optimization problem yields a unique solution up to a scaling ambiguity provided that certain conditions, referred to as *NCA criteria*, are met. These conditions are as follows: (i) matrix A has full column rank, (ii) matrix S has full row rank, and (iii) removing a node from the regulatory layer and the corresponding entries of matrices A and X should still result into a full column-rank matrix A.

NCA problem was first solved in [7] by performing alternating least squares (ALS). Since a high dimensional matrix is optimized at each step, this method entails prohibitive computational complexity. ROBNCA was proposed in [12] which significantly reduces the complexity and provides an additional advantage of being robust to the outliers in datasets. NCA has been successfully applied to gene network inference problems in various scenarios [13,14,15]. FastNCA, which was proposed for the first time in [8], reduces the complexity and it yields a very efficient solution to the gene regulatory network reconstruction problem. However, shows poor consistency in the estimation of TFAs [16]. To counter this problem, a convex optimization based non-iterative method: NINCA was proposed in [16]. NINCA estimates the signals with higher consistency even in the presence of high correlation. However, this algorithm estimates the connectivity matrix A by resorting to a costly optimization routine, and the resulting high computational complexity may limit its usefulness for large data sets encountered in practice. In order to alleviate the computational load, this correspondence presents a closed-form solution to the optimization problem, herein correspondence referred to as Closed-Form NINCA (CFNICA), exhibiting a significantly reduced complexity. Simulations are performed over synthetic as well as real data to test the performance of the proposed CFNICA solution. It is observed that the CFNICA solution for the estimation of connectivity matrix A presents the same superior estimation performance as that offered by NINCA and leads to a significant reduction in computational complexity.

2 CFNINCA: NINCA with Closed Form Solutions

CFNINCA is a two step algorithm which first estimates the matrix A and once it is available, the problem of estimating S is reduced to a simple least-squares algorithm. The following subsections explain the estimation of the two matrices.

2.1 Estimating Connectivity Matrix A

First, the estimation of A is performed by making use of the gene expression data X and the available prior connectivity information [16]. This step is accomplished by separating the signal and noise subspaces from X. The kth column of the system model in (1) is expressed as

$$x_k = As_k + \gamma_k \quad k = 1, 2, ..., K .\tag{3}$$

Defining $R_s = \mathbb{E}\{s_k s_k^T\}$, the autocovariance of gene expression data vector is expressed as

$$R_x = \mathbb{E}\{x_k x_k^T\} = AR_s A^T + \sigma_\gamma^2 I .\tag{4}$$

Since the matrix $AR_s A^T$ is positive semi-definite and symmetric, it can be factored using the eigenvalue decomposition in terms of a unitary matrix U, whose columns represent the eigenvectors of A, and a diagonal matrix Λ as follows:

$$R_x = U(\Lambda + \sigma_\gamma^2)U^T .\tag{5}$$

The matrix U can be further partitioned into U_s and U_0, which consist of the first M dominant eigenvectors corresponding to the M largest eigenvalues and spanning the signal subspace and the remaining $N - M$ eigenvectors spanning the noise subspace, respectively. Since the matrices A and R_s are full rank, by virtue of NCA criterion, we obtain

$$U_0^T AR_s A^T = 0 \Rightarrow U_0^T A = 0.\tag{6}$$

The subspace separation does not result into a unique solution. However, it was shown in [16] given the a-priori information about the connectivity matrix, the constrained subspace solution for

$$U_0^T A = 0, \quad A(I) = 0\tag{7}$$

is unique up to a scaling ambiguity. Since the rows of X and A can always be reordered, the mth column of A can be rewritten as

$$a_m = \begin{bmatrix} \bar{a}_m \\ \mathbf{0}_{L_m \times 1} \end{bmatrix}, \tag{8}$$

where L_m is the number of zeros in a_m.

Lemma 1. *Under NCA conditions (1) and (2), a solution to $U_0^T a_m = 0$, subject to a_m defined in (8), is unique up to a scale ambiguity [16].*

The lemma also implies that a solution to the constrained optimization problem in (7) can be obtained column-wise instead of estimating the entire matrix A. The trivial solution of $a_m = \mathbf{0}$ can be avoided by introducing a normalization constraint.

However, this subspace based approach requires ensemble average R_x, and since the data available to us is of finite length, only approximations are used. Towards this end, the left subspace of A is estimated using the singular value decomposition (SVD). In the standard SVD notation, the matrix X can be factorized as $X = U \Sigma V^T$. As previously, the matrix U can be factored as $U = \begin{bmatrix} \hat{U}_s & \hat{U}_0 \end{bmatrix}$, and the estimate $\hat{U}_0$ of U_0 will be used henceforth for estimation purposes. The constrained optimization problem can therefore be stated as:

$$\min_{a_m} \ \|U_0^T a_m\|_p \quad \text{s.t. } a_m = \begin{bmatrix} \bar{a}_m \\ \mathbf{0}_{L_m \times 1} \end{bmatrix}, \quad \mathbf{1}^T . a_m = 1, \tag{9}$$

Remark 2. The optimization problem in (9) was solved using convex optimization algorithms for $p = 1, 2$ in [16]. However, for real data sets, the vector a_m is usually large and its optimization entails significant computational complexity. Hence, a closed form solution is desired to improve the complexity and efficiency of the subspace based approach.

In this correspondence, we derive a closed form solution for $p = 2$ using convex optimization techniques. Define an $L_m \times N$ matrix C_m such that

$$C_m = \begin{bmatrix} \mathbf{0}_{L_m \times (N - L_m)} & I_{L_m} \end{bmatrix}. \tag{10}$$

Using the above definition, the optimization problem (9) can be equivalently written as

$$\hat{a}_m = \arg\min_{a_m} \ \|U_0^T a_m\|_2^2$$
$$\text{such that} \quad C_m a_m = \mathbf{0}, \quad \mathbf{1}^T a_m = 1 \tag{11}$$

Define now the substitute vector $\bar{a}_m$ via the following equation:

$$a_m = D_m \bar{a}_m, \tag{12}$$

where the $N \times L_m$ matrix D_m is constructed such that it lies in the null space of the matrix C_m, i.e., $C_m D_m = 0$. The matrix D_m is, therefore, given by

$$D_m = \begin{bmatrix} I_{(N-L_m)} \\ 0_{L_m \times (N-L_m)} \end{bmatrix}. \tag{13}$$

Upon substituting a_m from (12) in (11), we note that the first constraint is always satisfied by virtue of the construction of matrix D_m. The resulting optimization problem can be rewritten as

$$\hat{\bar{a}}_m = \arg\min_{\bar{a}_m} \; \frac{1}{2} \bar{a}_m^T D_m^T Q D_m \bar{a}_m$$
$$\text{such that} \quad 1^T \bar{a}_m = 1, \tag{14}$$

where $Q = U_0 U_0^T$. The Lagrangian function can be expressed as

$$\mathcal{L} = \frac{1}{2} \bar{a}_m^T D_m^T Q D_m \bar{a}_m - \mu \left(1^T \bar{a} - 1 \right). \tag{15}$$

The Karush-Kuhn-Tucker (KKT) conditions can be written as

$$D_m^T Q D_m \bar{a}_m - \mu 1 = 0$$
$$1^T \bar{a}_m = 1. \tag{16}$$

It can be shown that the KKT conditions are necessary and sufficient [17]. It follows from the first condition that

$$\bar{a}_m = \mu \left(D_m^T Q D_m \right)^{-1} 1 \tag{17}$$

where the matrix $D_m^T Q D_m$ is indeed invertible since D_m has full column rank and Q is a product of unitary matrices. Substituting (17) into (16), the Lagrange multiplier can be expressed as

$$\mu = \frac{1}{1^T \left(D_m^T Q D_m \right)^{-1} 1}. \tag{18}$$

The symmetric invertible matrix Q is partitioned as follows

$$Q = \begin{bmatrix} Q_{11} & Q_{12} \\ Q_{21} & Q_{22} \end{bmatrix},$$

where the invertible matrix Q_{11} stands for the upper left-corner $(N - L_m) \times (N - L_m)$ submatrix of Q. From the structure of D_m, the matrix $D_n^T Q D_m$ can be reduced to

$$D_m^T Q D_m$$
$$= \begin{bmatrix} I_{(N-L_m)} & 0_{(N-L_m) \times L_m} \end{bmatrix} \begin{bmatrix} Q_{11} & Q_{12} \\ Q_{21} & Q_{22} \end{bmatrix} \begin{bmatrix} I_{(N-L_m)} \\ 0_{L_m \times (N-L_m)} \end{bmatrix}$$
$$= Q_{11}. \tag{19}$$

The constrained solution for $\bar{a}_m$ is therefore given by

$$\bar{a}_m = \frac{Q_{11}^{-1} 1}{1^T Q_{11}^{-1} 1} \tag{20}$$

Remark 3. The closed form solution (20) only requires the inversion of the $(N - L_m) \times (N - L_m)$ submatrix Q_{11}, which is typically a much smaller matrix, since there are a few non-zero entries in a_m. The matrix inversion requires $O\left((N - L_m)\right)^3$ operations. The numerator in (20) requires $O\left((N - L_m)\right)^2$ operations. The denominator requires $O\left((N - L_m)\right)^2 + O\left((N - L_m)\right)$ operations. Hence, the complexity of the closed form solution is approximately $O\left((N - L_m)\right)^3$ for large $(N - L_m)$.

2.2   Estimating the TFA Matrix S

Once an estimate $\hat{A}$ is available, S can now be estimated using a least squares criterion. The optimization problem can be expressed as

$$S = \arg \min_{S} \| X - \hat{A} S \|_F^2 . \tag{21}$$

By setting the derivative of (21) equal to zero and solving for S, the estimate is obtained as

$$S = \left(\hat{A}^T \hat{A}\right)^{-1} \hat{A}^T X . \tag{22}$$

Since closed form solutions are available for the estimates of both A and S, CFNINCA exhibits much lower computational complexity than NINCA.

3   Simulation Results

In this section, the performance of the proposed algorithm is evaluated in comparison with the existing state-of-the-art algorithms ALS [7], FastNCA [8] and NINCA [16] for synthetic as well as real yeast cell cycle data set.

3.1   Synthetic and Hemoglobin Test Data

The algorithm is first investigated for the Hemoglobin test data set used by the original NCA paper [7] and modified in [16], which assumed spectroscopy data obtained by mixing Hemoglobin solutions. This data set is used because the underlying network structure follows the gene network very closely. Moreover, the knowledge of the original source solutions aids in the performance evaluation of the algorithms. The data set consists of $M = 3$ source solutions which result into $N = 7$ mixtures where the spectra are measured for $K = 321$ data points. The presence of a source solution in the mixture solutions indicates the presence of the respective connection in the network connectivity matrix A.

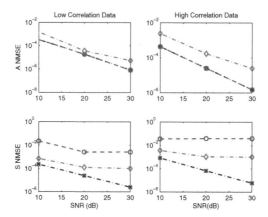

Fig. 1. Normalized mean square error for the estimation of *A* and *S* using CFNINCA (green), NINCA (blue), FastNCA (red), and ALS (magenta)

This data set is used to evaluate the estimation performance of the algorithms with mean square error (MSE) as the fidelity criterion for *A* and *S* matrices. Experiments are performed for low and high correlated data over varying signal-to-noise ratio (SNR) and the Normalized MSEs are depicted in Fig. 1. The noise is assumed to be additive white Gaussian (AWGN). For the estimation of *A*, CFNINCA, NINCA and FastNCA perform comparably and provide lower NMSE than the ALS algorithm. CFNINCA and NINCA yield the lowest NMSE for the estimation of *S* as well, however, the performance of FastNCA and ALS deteriorates significantly. For the estimation of *A*, the MSE decreases with the increase in SNR for all the algorithms, however, FastNCA and ALS exhibit an error floor for estimation of *S*. The better performance of these algorithms in estimating *A* can be attributed to the availability of prior information.

3.2   Results S.Cerevisiae Cell Cycle Data

This section compares the previously mentioned algorithms in the presence of Yeast cell cycle data from [18] aand [19].

This data consists of measurements from three synchronization experiments performed on a large number of genes and TFs. TFA estimation is evaluated for 11 TFs of interest that are considered in [8] and [16]. These TFs are Ace2, Fkh1, Fkh2, Mbp1, Mcm1, Ndd1, Skn7, Stb1, Swi4, Swi5, and Swi6. The first experiment performs synchronization by elutriation and consists of one cell cycle from 0 to 390. The second synchronization experiment with two cell cycles from 0 to 119 is by α−factor arrest. The third experiment carried out under cdc15 temperature sensitive mutant consists of three cycles from 0 to 300 mins. The data in the three experiments contain 14, 18, and 15 samples, respectively, and are stacked together to form one large data set.

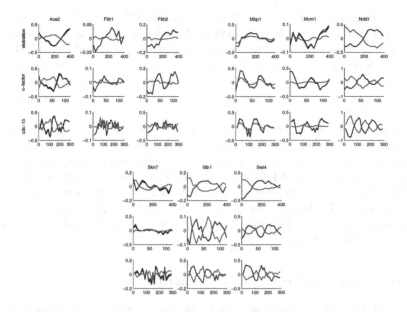

Fig. 2. TFAs Reconstruction: Estimation of 11 TFAs (9 shown) of cell-cycle regulated yeast TFs. Average values of the TFs are shown for the four subnetworks. The results offered by CFNINCA (black), FastNCA (red) and NINCA (blue) are displayed.

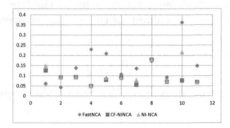

Fig. 3. Consistency Evaluation for S. Cerevisiae Data: Average disagreement from the subsets for TFA estimation.

Subnetwork analysis is performed here to assess the consistency of the algorithms, where the data set is divided into four overlapping subsets similar to [5]. Each subset consists of 40 TFs, while the 11 TFs under consideration are present in all of them. The number of genes is set to be between 921 to 1247. TFAs are estimated using the four subsets and the difference in their estimation indicates higher degree of inconsistency. This enables us to analyze the robustness of the algorithm to minor modifications in the TFs and genes under consideration [5]. The disagreement is measured using the same metric as in [16] which is

$$\text{disagreement}(i) = \frac{1}{K} \sum_i \left[\max_n s_{n,i}(k) - \min_n s_{n,i}(k) \right] \tag{23}$$

where $s_{n,i}$ denote the TFA in the nth subset and i is TF index. The average of the TFAs estimated using the four subsets is plotted in Fig. 2. The rows depict the results of the three synchronization experiments. It is observed that CFNINCA and NINCA result in estimating the same TFA profiles and recovering one, two and three cycles for the three cycles, respectively. FastNCA yields estimates that are either opposite to the other algorithms for most TFAs or it does not reveal their periodicity.

In order to further corroborate the results, a consistency comparison study is performed and the disagreement between the subset estimates is shown in Fig. 3. It is observed that FastNCA yields much larger disagreement, and therefore, it is less consistent than CFNINCA and NINCA. Therefore, it can be stated that CFNINCA is able to estimate the TFAs with a higher degree of accuracy and consistency. It should also be mentioned that the large size of data set in this experiment prohibits the use of ALS for comparison due to its high computational complexity.

3.3   Robustness to Errors in Prior Information

The prior information about connectivity matrix A helps in obtaining a unique solution. However, it is important to study the reliability of the results in case of inaccuracies present in prior knowledge which is a possible scenario [20]. In this analysis, we consider the missed connections only. Suppose that the prior for connectivity matrix erroneously misses some of the true connections and is denoted by A^*. Then, the mth column of this matrix is given by

$$a_m^* = \begin{bmatrix} \bar{a}_m^* \\ 0_{L_m^* \times 1} \end{bmatrix} . \tag{24}$$

where L_m^* is the number of zeros in a_m^*. The constrained optimization solution can now be stated as

$$\min_{a_m^*} \ \|U_0^T a_m^*\|_p \quad \text{s.t. } a_m^* = \begin{bmatrix} \bar{a}_m^* \\ 0_{L_m^* \times 1} \end{bmatrix} , \quad 1^T.a_m^* = 1, \tag{25}$$

Following the same steps as in Section 2.1, the solution for this problem is obtained as

$$\bar{a}_m^* = \frac{Q_{11}^{-1*} 1}{1^T Q_{11}^{-1*} 1} . \tag{26}$$

where Q_{11}^* is $(N - L_m^*) \times (N - L_m^*)$ Let the error in m^{th} column be $e_m = a_m - a_m^*$. Then the error in estimation of A and S is calculated as

$$E_A = \sum_{m=1}^{M} \|e_m\|_2^2 , \tag{27}$$

and

$$E_S = \|S - S^*\|_F^2 . \tag{28}$$

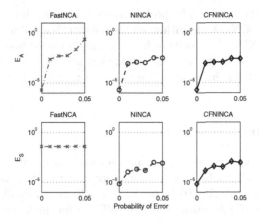

Fig. 4. Robustness to Imperfect Prior: Error in the estimation of A and S matrices with missed connections in prior

respectively. The SNR for this experiment is kept at 30dB. The estimation performance for FastNCA, NINCA and CFNINCA is evaluated using the same Hemoglobin data used in the previous subsection. It is noted in Fig. 4, that as the probability of error in the prior increases, the MSE for the estimation of all the algorithms increases for A. However, NINCA and CFNINCA give much lower MSE than FastNCA for estimation of TFA matrix S. NINCA and CFN-INCA, therefore ,show more robustness to imperfect knowledge of prior. However, CFNINCA offers these advantages at a much lower computational cost.

3.4   Run Time Comparison

Table 1. Average Computational Time in Seconds for *S. Cerevisiae.*

Subset	1	2	3	4
FastNCA	0.2	0.2	0.24	0.2
CFNINCA	6	3	3	6
NINCA	71	30	125	97
ALS	Exceeds memory limit			

Gene regulatory networks require working with large data sets and therefore a lower computational time for the algorithms is a very appealing feature. We compare the average run time for the algorithms discussed previously for the four subsets of the real data set. These simulations were carried out using Matlab 7.10.0 on a Windows 7 system with a 1.90 GHz Intel Core i7 processor. It is observed that CFNINCA is tens of times faster than NINCA. FastNCA has advantage in terms of lower complexity. However, as noted in the previous simulations, FastNCA suffers from poor estimation accuracy and consistency. The complexity of ALS is known to be prohibitive and is added here only for comparison. Hence, the CFNINCA algorithm avoids the drawback of high computational complexity of the NINCA algorithm by providing a closed form solution to estimate A, while maintaining the same estimation accuracy and consistency. This makes CFNINCA well suited for TFA estimation for large data sets encountered in practice.

4   Conclusions

This paper presented a closed form solution to a non-iterative network component analysis algorithm which uses convex optimization techniques to estimate the control strength matrix [16]. The NINCA algorithm exhibits superior consistency in terms of TFA estimation but suffers from high computational complexity. The proposed closed form CFNINCA solution considerably speeds up the algorithm while offering comparable estimation accuracy and consistency to NINCA. The performance of CFNINCA is compared to NINCA, FastNCA, and ALS over synthetic data and yeast cell cycle data. The conducted simulations confirm CFNINCA's advantages in terms of lower run time, robustness to imperfect prior and comparable or better estimation accuracy with respect to the existing state-of-the-art algorithms.

Acknowledgments. This work was supported by NSF Award No. 1318338.

References

1. Cai, X., Wang, X.: Stochastic modeling and simulation of gene networks. IEEE Signal Process. Mag. 24(1), 27–36 (2007)
2. Shmulevich, I., Saarinen, A., Yli-Harja, O., Astola, J.: Inference of genetic regulatory networks via best-fit extensions. In: Computational and Statistical Approaches to Genomics, pp. 197–210 (2003)
3. Lähdesmäki, H., Shmulevich, I., Yli-Harja, O.: On learning gene regulatory networks under the boolean network model. Machine Learning 52(1), 147–167 (2003)
4. Noor, A., Serpedin, E., Nounou, M.N., Nounou, H.N.: Inferring gene regulatory networks via nonlinear state-space models and exploiting sparsity. IEEE/ACM Trans. Comput. Biology Bioinform. 9(4), 1203–1211 (2012)
5. Yang, Y.L., Suen, J., Brynildsen, M.P., Galbraith, S.J., Liao, J.C.: Inferring yeast cell cycle regulators and interactions using transcription factor activities. BMC Genomics 6(1), 90 (2005)
6. Meng, J., Zhang, J.M., Chen, Y., Huang, Y.: Bayesian non-negative factor analysis for reconstructing transcription factor mediated regulatory networks. Proteome Science 9, 1–14 (2011)
7. Liao, J., Boscolo, R., Yang, Y., Tran, L., Sabatti, C., Roychowdhury, V.: Network component analysis: Reconstruction of regulatory signals in biological systems. Proceedings of the National Academy of Sciences 100(26), 15522–15527 (2003)
8. Chang, C., Ding, Z., Hung, Y., Fung, P.: Fast network component analysis (FastNCA) for gene regulatory network reconstruction from microarray data. Bioinformatics 24(11), 1349–1358 (2008)

9. Jolliffe, I.T.: Principal component analysis, vol. 487. Springer, New York (1986)
10. Comon, P.: Independent component analysis. Higher-Order Statistics, 29–38 (1992)
11. Tan, M., Alshalalfa, M., Alhajj, R., Polat, F.: Influence of prior knowledge in constraint-based learning of gene regulatory networks. IEEE/ACM Transactions on Computational Biology and Bioinformatics 8(1), 130–142 (2011)
12. Noor, A., Ahmad, A., Serpedin, E., Nounou, M., Nounou, H.: ROBNCA: robust network component analysis for recovering transcription factor activities. Bioinformatics 29(19), 2410–2418 (2013)
13. Tran, L.M., Brynildsen, M.P., Kao, K.C., Suen, J.K., Liao, J.C.: gNCA: A framework for determining transcription factor activity based on transcriptome: Identifiability and numerical implementation. Metabolic Engineering 7(2), 128–141 (2005)
14. Tran, L., Hyduke, D., Liao, J.: Trimming of mammalian transcriptional networks using network component analysis. BMC Bioinformatics 11(1), 511 (2010)
15. Galbraith, S.J., Tran, L.M., Liao, J.C.: Transcriptome network component analysis with limited microarray data. Bioinformatics 22(15), 1886–1894 (2006)
16. Jacklin, N., Ding, Z., Chen, W., Chang, C.: Noniterative convex optimization methods for network component analysis. IEEE/ACM Transactions on Computational Biology and Bioinformatics 9(5), 1472–1481 (2012)
17. Boyd, S., Vandenberghe, L.: Convex Optimization. Cambridge University Press (2004)
18. Lee, T.I., Rinaldi, N.J., Robert, F., Odom, D.T., Bar-Joseph, Z., Gerber, G.K., Hannett, N.M., Harbison, C.T., Thompson, C.M., Simon, I., et al.: Transcriptional regulatory networks in saccharomyces cerevisiae. Science Signalling 298(5594), 799 (2002)
19. Spellman, P.T., Sherlock, G., Zhang, M.Q., Iyer, V.R., Anders, K., Eisen, M.B., Brown, P.O., Botstein, D., Futcher, B.: Comprehensive identification of cell cycle–regulated genes of the yeast saccharomyces cerevisiae by microarray hybridization. Molecular Biology of the Cell 9(12), 3273–3297 (1998)
20. Wang, C., Xuan, J., Shih, I.M., Clarke, R., Wang, Y.: Regulatory component analysis: A semi-blind extraction approach to infer gene regulatory networks with imperfect biological knowledge. Signal Processing 92(8), 1902–1915 (2012)

SVEM: A Structural Variant Estimation Method Using Multi-mapped Reads on Breakpoints

Tomohiko Ohtsuki[1], Naoki Nariai[2], Kaname Kojima[2], Takahiro Mimori[2], Yukuto Sato[2], Yosuke Kawai[2], Yumi Yamaguchi-Kabata[2], Testuo Shibuya[1], and Masao Nagasaki[2],[⋆]

[1] Human Genome Center, Institute of Medical Science, University of Tokyo
4-6-1 Shirokanedai, Minato-ku, Tokyo, 108-8639, Japan
[2] Department of Integrative Genomics
Tohoku Medical Megabank Organization, Tohoku University
2-1 Seiryo-machi, Aoba-ku, Sendai, Miyagi, 980-8573, Japan
nagasaki@megabank.tohoku.ac.jp

Abstract. Recent development of next generation sequencing (NGS) technologies has led to the identification of structural variants (SVs) of genomic DNA existing in the human population. Several SV detection methods utilizing NGS data have been proposed. However, there are several difficulties in analysis of NGS data, particularly with regard to handling reads from duplicated loci or low-complexity sequences of the human genome. In this paper, we propose SVEM, a novel statistical method to detect SVs with a single nucleotide resolution that can utilize multi-mapped reads on breakpoints. SVEM estimates the amount of reads on breakpoints as parameters and mapping states as latent variables using the expectation maximization algorithm. This framework enables us to handle ambiguous mapping of reads without discarding information for SV detection. SVEM is applied to simulation data and real data, and it achieves better performance than existing methods in terms of precision and recall.

1 Introduction

Structural variants (SVs) are common genomic differences among individual genomes and include various types, such as insertions, deletions, inversions, tandem duplications, and translocations [1]. It has been reported that SVs affect not only phenotypes but also the occurrence of diseases, such as schizophrenia [2] and cancer [3]. When classifying SV types according to their size, larger SVs are called microscopic variants, because they can be detected by using microscopes [4]. Chemical and biological methods, such as karyotyping, chromosome painting, and FISH-based techniques, are used with the microscope observation for SV detection [5]. The microscopic approach can identify SVs that are larger than several Mbp. To find smaller SVs, which are referred to as submicroscopic

---

[⋆] Corresponding author.

A.-H. Dediu, C. Martín-Vide, and B. Truthe (Eds.): AlCoB 2014, LNBI 8542, pp. 208–219, 2014.

SVs, DNA microarray analyses, such as arrayCGH [6] and single nucleotide polymorphism (SNP) arrays [7] are commonly used [8]. Although they are useful for identifying medium-sized SVs with copy number changes, such as tandem duplications and deletions, they are not suitable for detecting small SVs and SVs without copy number changes [9]. Genome-wide discovery of novel SVs is also difficult with array based methods. To find such SV events, sequencing-based methods are preferable. The first sequencing-based method was performed with capillary sequencing, and short insertion and deletion polymorphisms were discovered [10].

Recently, because of its high-throughput capability, next generation sequencing (NGS) has been used for discovery of SVs and SNPs with high coverage and resolution. Several methods for detecting SVs from NGS data have been proposed. These methods use one or more of the following four types of approaches; read depth approach (CNVnator [11]), read pair approach (DELLY [12] and BreakDancer [13]), split read approach (DELLY, Pindel [14], and ClipCrop [15]) and *de novo* assemble approach (SOAPdenovo2 [16]). Although the read depth approach identifies deletions and copy number variants based on the depth of coverage of mapped reads on the reference genome, it cannot identify SV events without copy number changes, such as inversions, insertions and translocations. It is also difficult to find precise locations of SVs, called breakpoints, on the reference genome. The read pair approach identifies SVs by detecting discordance of read pairs around breakpoints in terms of their distance and orientations when they are aligned to the reference genome. This approach can detect various types of SVs, such as deletions, insertions, inversions, translocations, and partial duplications. However, precise positions of breakpoints cannot be located, and hence detectable SV sizes are limited to several kbp. The split read approach detects SVs using partially aligned reads on the reference genome. A clipped part of the read (portion of the read that is not aligned to the reference sequence) is re-aligned to the reference genome, and breakpoints of SVs within the reference genome are identified. Although this approach has a high resolution for detecting positions of breakpoints, it requires extensive computational resources and is not appropriate for detecting larger insertions and deletions. Moreover, because there are many regions whose sequences are similar to each other, such as duplicated loci or low-complexity sequences, some reads can be aligned to more than one region of the reference genome. This multi-mappability complicates the usage of the split read approach. The *de novo* assemble approach is very unique from the other approaches in that it determines actual genome sequences directly without mapping to the reference genome. However, *de novo* assembly requires extensive computational resources, in terms of both memory usage and CPU time [5].

In this paper, we propose a new method based on the split read approach, called SVEM, which utilizes multi-mapped reads on breakpoints under a statistical framework. In SVEM, the most likely breakpoints where the multi-mapped reads are generated are estimated by the expectation maximization (EM) algorithm. To evaluate the performance of SVEM, we create simulated read data with artificial SVs and sequencing errors. We also apply SVEM to real data obtained from a human sample in the 1000 Genomes Project [17] and evaluate its performance.

2 Methods

2.1 Preprocessing

In the preprocessing step, short reads are aligned to the reference genome with alignment tools such as BWA [18]. At starting or ending positions of SVs in the reference genome, which are hereafter called "breakpoints", there exist reads that are partially aligned and contain unmapped fragments. Such unmapped fragments are called "clipped" fragments, and reads containing clipped fragments are called clipped reads. In addition to the positions where the clipped reads aligned, we obtain candidate breakpoints by the realignment of clipped fragments of reads to the reference genome. Because the clipped fragments and original reads can be aligned to multiple positions in many cases, naïve approaches have difficulties in uniquely determining the breakpoints for each SV. In the following, we describe the estimation of the starting and ending points of SVs with SVEM which considers the read generation and abundance of reads around the candidate breakpoints in a probabilistic manner.

2.2 Modeling for Paired-End Reads

The generative model for sequence reads used for SV detection is shown in Figure 1. The read abundance parameter, selection of the breakpoint, read start position, fragment length, and the first and second sequences of read n are represented by θ, B_n, S_n, F_n, R_n^1 and R_n^2, respectively. Each read is associated with two latent variables B_n and S_n.

The complete likelihood function of our model is decomposed as

$$P(B_n, S_n, F_n, R_n^1, R_n^2 \mid \theta) = P(B_n \mid \theta)P(S_n \mid B_n)P(F_n \mid B_n)$$
$$\cdot\ P(R_n^1 \mid B_n, S_n, F_n)P(R_n^2 \mid B_n, S_n, F_n).$$

$P(B_n \mid \theta)$ represents the probability that read n is generated from the breakpoint B_n, and it is calculated as

$$P(B_n = b \mid \theta) = \theta_b, \quad \text{where} \sum_b \theta_b = 1.$$

$P(S_n \mid B_n)$ represents the probability of the start position of read n which spans breakpoint B_n. Let the length of each read be l. Because there are $l - 1$ possible start positions for reads spanning breakpoint B_n, under the assumption of equal occurrence of these $l - 1$ start positions, $P(S_n \mid B_n)$ is given as

$$P(S_n \mid B_n) = \begin{cases} 1/(l-1) & \text{if the distance between } S_n \text{ and } B_n < l \\ 0 & \text{otherwise} \end{cases}.$$

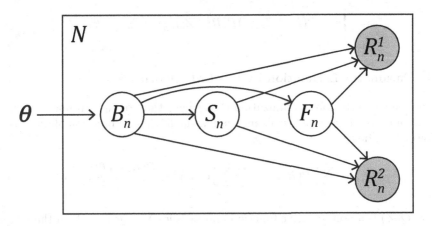

Fig. 1. Generative model of paired-end DNA sequence data around breakpoints

$P(F_n \mid B_n)$ represents the probability of a fragment length from breakpoint B_n. We assume that F_n is normally distributed and calculate $P(F_n \mid B_n)$ as

$$P(F_n = f_n \mid B_n) = \frac{\exp(-\frac{(f_n-\mu)^2}{2\sigma^2})}{\sum_{x=2l}^{f_{max}} \exp(-\frac{(x-\mu)^2}{2\sigma^2})},$$

where f_n is the fragment length of read n and f_{max} is the maximum fragment length, whose value is dependent on the conditions around the corresponding breakpoint. μ and σ^2 are the mean and variance of the fragment length, respectively, and are given by pre-set values according to experimental protocols, or estimated from uniquely aligned reads.

$P(R_n^1 \mid B_n, S_n, F_n)$ represents the probability of the first sequence of read n given B_n, S_n, and F_n. Here, we use the Phred quality score [19] Q_i at position i which indicates the reliability of base calling at each base. We calculate

$$P(R_n^1 \mid B_n, S_n, F_n) = \prod_{i=1}^{l} r_{ni}^1,$$

where

$$r_{ni}^1 = \begin{cases} 1 - 10^{-Q_i/10} & \text{if } i\text{th nucleotide of read } n \text{ is the same as the reference} \\ 10^{-Q_i/10} & \text{otherwise} \end{cases}.$$

Similarly, $P(R_n^2 \mid B_n, S_n, F_n)$ is calculated for the second sequence of read n.

Let Z_{nbsf} be a binary variable that takes the value of one if $(B_n, S_n, F_n) = (b, s, f)$ holds for the read n and zero otherwise. By using Z_{nbsf}, the complete likelihood of the model can be given as:

$$\prod_{n,b,s,f} P(R_n^1, R_n^2, Z_{nbsf} \mid \theta)$$

$$= \prod_{n,b,s,f} \{P(R_n^1 \mid Z_{nbsf})P(R_n^2 \mid Z_{nbsf})P(Z_{nbsf} \mid \theta)\}^{Z_{nbsf}}.$$

2.3 Parameter Estimation by the EM Algorithm

We use the expectation maximization (EM) algorithm for the parameter estimation of our model. From Jensen's inequality, the likelihood function of our model is lower bounded by

$$\log \sum_Z P(R, Z \mid \theta) \geq \sum_Z Q(Z) \log \frac{P(R, Z \mid \theta)}{Q(Z)}, \tag{1}$$

where $Q(Z)$ is a non-negative function and satisfies $\sum_Z Q(Z) = 1$. In the EM algorithm, the lower bound of the maximum likelihood is increased by alternately applying E-Step and M-Step until a convergence criterion is satisfied. In E-Step, the lower bound in equation (1) is maximized with respect to $Q(Z)$. $Q(Z)$ maximizing the lower bound is given by $P(Z \mid R, \theta)$ from the equality condition in Jensen's inequality. In M-Step, the lower bound in equation (1) is maximized with respect to θ. Let $\theta_{(old)}$ be θ at the previous step. Because the lower bound is given by

$$\sum_Z P(Z \mid R, \theta_{(old)}) \log P(R, Z \mid \theta) - \sum_Z P(Z \mid R, \theta_{(old)}) \log P(Z \mid R, \theta_{(old)}),$$

θ maximizing the following formula is obtained by

$$\underset{\theta}{\operatorname{argmax}} \sum_Z P(Z \mid R, \theta_{(old)}) \log P(R, Z \mid \theta). \tag{2}$$

In the following, more details of the EM algorithm for our model are described.

E-Step for our Model. Let ρ_{nbsf} be the probability of generating read n given $(B_n, S_n, F_n) = (b, s, f)$:

$$\rho_{nbsf} = P(B_n \mid \theta_{(old)})P(S_n \mid B_n)P(F_n \mid B_n)P(R_n^1 \mid Z_{nbsf})P(R_n^2 \mid Z_{nbsf}).$$

By using ρ_{nbsf}, $P(Z_{nbsf} \mid R_n^1, R_n^2, \theta_{(old)})$ is given by

$$P(Z_{nbsf} \mid R_n^1, R_n^2, \theta_{(old)}) = \prod_{n,b,s,f} r_{nbsf}^{Z_{nbsf}}, \tag{3}$$

where

$$r_{nbsf} = \begin{cases} \dfrac{\rho_{nbsf}}{\sum_{(b',s',f') \in \pi_n} \rho_{nb's'f'}} & \text{if } (b, s, f) \in \pi_n \\ 0 & \text{otherwise} \end{cases}.$$

Here, π_n is a set of (b, s, f) for all possible alignments of read n.

M-Step for our Model. The parameter to be estimated in our model is θ. Equation (2) in our model is written as follows:

$$\sum_Z P(Z \mid R, \theta_{(old)}) \log P(R, Z \mid \theta) = \sum_{n,b,s,f} E[Z_{nbsf}] \log P(R_n^1, R_n^2, Z_{nbsf} = 1 \mid \theta),$$

where $E[Z_{nbsf}]$ is the expectation of Z_{nbsf} on $P(Z \mid R_n^1, R_n^2, \theta_{(old)})$. Because $E[Z_{nbsf}]$ is given by r_{nbsf} from equation (3), θ_b for each breakpoint b is updated by

$$\theta_b = \frac{\sum_{n,s,f} E[Z_{nbsf}]}{\sum_{n,b',s,f} E[Z_{nb'sf}]}.$$

Initialization and Convergence Criterion. At the initialization step, $E[Z_{nbsf}]$ is set to $1/$(the size of π_n). We use the difference between the current and previous lower bounds after the M-Step as the convergence criterion. Let $\theta_{(i)}$ be θ estimated at the ith M-Step. The lower bound of the likelihood after the ith M-Step $l(\theta_{(i)})$ is given by

$$l(\theta_{(i)}) = \sum_{n,b,s,f} E[Z_{nbsf}] \log P(R_n^1, R_n^2, Z_{nbsf} = 1 | \theta_{(i)})$$
$$- \sum_{n,b,s,f} E[Z_{nbsf}] \log E[Z_{nbsf}].$$

The lower bound is updated until the following criterion is satisfied for a small value ϵ:

$$l(\theta_{(i)}) - l(\theta_{(i-1)}) \leq \epsilon.$$

We set $\epsilon = 10^{-4}$ for the convergence criterion, and after the algorithm converged, we use the threshold for $E[Z_{nbsf}]$ to determine whether or not the breakpoint exists as three in the following experiments.

Identification of SVs from the Estimated Breakpoints. After breakpoints were estimated, deletions, inversions, and insertions are identified according to the initial alignments of reads and the second alignments of the clipped reads to the reference genome. A schematic diagram is shown in Figure 2.

3 Results

3.1 Simulation Data Analysis

We evaluate the performance of SVEM by using synthetically generated NGS data. We prepared an artificial DNA sequence of chromosome 21 from the human reference genome (GRCh37) with one SNP per 1,000 bp, which is based on the average base diversity in human genomes [20].

Paired-end sequencing data of 100 bp were generated, under the assumption that the fragment size was normally distributed with a mean value of 350 bp

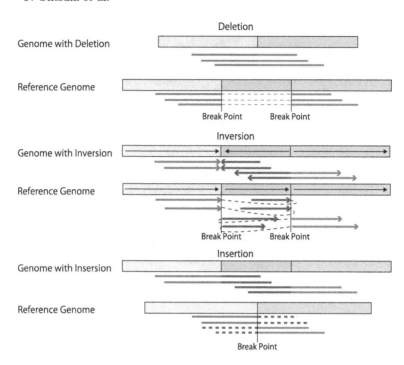

Fig. 2. Identification of SVs from breakpoints and re-alignments of clipped reads

and standard deviation of 50 bp. Base substitution errors of 0.1% were randomly added to each position of reads to simulate sequencing errors. The number of generated reads was based on "read coverage", which represents the average number of overlapping reads at a base position. We prepared a data set whose read coverage was 50x. The generated reads were aligned to the reference genome using the Burrows-Wheeler Aligner [18], and the produced bam file containing the alignment information was used as the input for each SV caller considered in the following analysis.

Among the SV candidates predicted by each method, if SVs whose start and end points were sufficiently close (within 10 bp) to those of true SVs, they were classified as true positives (TPs), and the other SVs were classified as false positives (FPs). However, some SVs can be expressed in multiple ways, and such ambiguous cases are difficult to classify as TP or FP. For example, when the sequence "GAG" is deleted from "CTGAGAGTC", the remaining sequence is uniquely represented as "CTAGTC", but the start and end points of the deletion can be expressed in two ways: the deletion from the third "G" to the fifth "G", or that from the fifth "G" to the seventh "G". Because such cases are possible with more complex and large SVs, we defined true positive SVs in a more relaxed manner: SVs with start and end points among such possible positions were classified as true positives; otherwise, they were classified as false positives.

Under this relaxed definition of true positives and false positives, we calculated the recall rate and precision rate as follows:

$$\text{Recall} = \frac{\#\ \text{TP}}{\#\ \text{true SV}}, \quad \text{Precision} = \frac{\#\ \text{TP}}{\#\ \text{detected SV}}.$$

Because a trade-off exists between recall and precision, we used F-measure to evaluate overall performance, which was given by a harmonic mean of precision and recall as follows:

$$\text{F-measure} = \frac{2 \times \text{Recall} \times \text{Precision}}{\text{Recall} + \text{Precision}}.$$

Performance Evaluation for Medium-Size SVs. We evaluate the performance of SV detection for a data set containing both homozygous and heterozygous SVs. 350 true SVs, whose sizes were set to 500 bp, were randomly assigned as homozygous or heterozygous SVs as follows:

- Among 250 deletions, 88 were homozygous, and the other 162 were heterozygous.
- Among 50 inversions, 13 were homozygous, and the other 37 were heterozygous.
- Among 50 insertions, 25 were homozygous, and the other 25 were heterozygous.

SVEM detected 331 SVs in total, among which 300 of them were true positives and the other 31 were false positives, and the precision, recall, and F-measure were 0.91, 0.86, and 0.88, respectively. The same statistics for BreakDancer, DELLY, and Pindel are summarized in Table 1 (top). These results show that SVEM performed better in both precision and recall than the other methods.

Performance Evaluation for Larger SVs. We additionally generated simulation read data containing SVs with a larger size (1,000 bp) to evaluate the influence of SV sizes on precision and recall. Here, we considered 100 deletions, 20 insertions, and 20 inversions for true SVs, i.e., the total number of true SVs was 140. Note that to maintain the total length of SVs, we reduced the number of SVs compared with the other experiments above. These 140 SVs were randomly assigned as homozygous or heterozygous as follows:

- Among 100 deletions, 29 were homozygous, and the other 71 were heterozygous.
- Among 20 inversions, six were homozygous, and the other 14 were heterozygous.
- Among 20 insertions, eight were homozygous, and the other 12 were heterozygous.

SVEM predicted 123 SVs in total, among which 121 were true positives and the other two are false positives, and the precision, recall, and F-measure were 0.98, 0.86, and 0.92, respectively. The same statistics for BreakDancer, DELLY, and Pindel are summarized in Table 1 (bottom). The precision of SVEM and Pindel was better for larger SVs than in a data set with medium-size SVs. Moreover, DELLY and BreakDancer showed improved precision compared to the results for the data set with medium-size SVs, which implies that the read pair based approach used in these methods works well for larger SVs.

3.2   Real Data Analysis

We applied SVEM, BreakDancer, DELLY, and Pindel to the real sequencing data of a HapMap sample and evaluated their recall rates. Paired-end sequencing data of NA12878, a sample in the 1000 Genomes Project [17], sequenced with the Illumina HiSeq 2000 system, were used in our analysis. The depth of coverage, read length, and fragment size of the data were 45x, 101 bp, and 315 bp, respectively. In this analysis, 617 experimentally validated SVs from [21] were used for calculation of the recall rates. Because the validated SV set did not cover all of the existing SVs in the sample, we focused on performance in recall assuming that the validated SVs were the gold standard. Note that the estimated precision was also provided for the reference.

SVEM predicted 7,082 SVs in total, among which 214 were validated, and 6,868 were not. As a result, the recall and precision were 0.35 and 0.030, respectively. Note again that the precision calculated here would not reflect the true precision for each method. The same statistics for BreakDancer, DELLY, and Pindel are summarized in Table 2. Therefore, the recall with SVEM was higher than that with the other methods. Moreover, fewer non-validated SVs were detected by SVEM than those by the other methods, which implies that SVEM detected SVs with higher precision than the other methods.

3.3   Computational Resources

All of the experiments were performed on a computer with an Intel Xeon CPU E5-2670 processor (2.60GHz) with the Red Hat Enterprise Linux Server release 6.2 operating system. SVEM is implemented in Java 1.7.0_17 and executed on a single core environment. The SV calling tools used in these experiments were as follows: Pindel 0.2.4, BreakDancer 1.1, and DELLY 0.2.1. All mapping processes were performed by BWA 0.7.5 [18]. In the experiments for the simulated data sets containing both homozygous and heterozygous SVs, the execution time for SVEM was 75 seconds in the first experiment, whereas Pindel, BreakDancer, and DELLY required 10 minutes, 90 seconds, and 90 minutes, respectively. Although SVEM required some preprocessing, which took about one minute, it required two minutes and 15 seconds in total and was still much faster than Pindel and DELLY. Regarding memory consumption, 48 GBytes were required for running SVEM, one GBytes for Pindel, 5.6 GBytes for BreakDancer, and 47.7 GBytes for DELLY. Thus, both Pindel and BreakDancer had better memory efficiency than

Table 1. Results with simulated data of medium-size (top) and larger (bottom) SVs

Method	Predicted SVs	TP	FP	Precision	Recall	F-measure
SVEM	331	300	31	0.91	0.86	0.88
BreakDancer	506	3	503	0.0059	0.0086	0.0070
DELLY	894	50	844	0.056	0.14	0.080
Pindel	375	298	77	0.79	0.85	0.82

Method	Predicted SVs	TP	FP	Precision	Recall	F-measure
SVEM	123	121	2	0.98	0.86	0.92
BreakDancer	119	10	109	0.084	0.071	0.076
DELLY	314	105	209	0.33	0.75	0.46
Pindel	123	119	4	0.97	0.85	0.91

Table 2. Results with NA12878 data obtained from the 1000 genomes project

Method	Predicted SVs	TP	Recall	Estimated TP ratio
SVEM	7081	214	0.35	0.030
BreakDancer	3213	21	0.034	0.0065
DELLY	206968	43	0.070	0.00021
Pindel	288783	205	0.33	0.0007

SVEM, and the memory requirements for DELLY and SVEM were almost identical. SVEM achieved high precision and recall with a relatively short execution time and practical memory requirements.

4   Discussion

We proposed SVEM, a new statistical structural variant caller that considers split reads and their read mapping uncertainty at breakpoints. In an analysis of simulation data, SVEM was evaluated using Pindel, BreakDancer, and DELLY considering simulated sequencing errors, the zygosity, and the size of SVs. Through a careful comparison, we showed that SVEM outperformed existing methods with regard to both precision and recall under various simulation conditions. SVEM and the existing methods were also applied to the human sequencing data of a HapMap sample, NA12878, and predictions were compared with SVs that were validated by biological experiments. The higher recall rate with SVEM may be explained by successful incorporation of multi-mapped reads under the statistical model. Although utilizing multi-mapped reads has been shown to be effective for quantifying gene expression levels from RNA-Seq data [22,23], we showed in this paper that this strategy was also effective to improve the performance of SV detection from DNA sequencing data. Because BreakDancer is based on the paired-end read approach, it is suitable for detecting larger SVs, but the start and end positions of SVs are not precisely determined. Although the split read approach can detect precise breakpoints of SVs with a high resolution, it requires extensive computational resources and is not efficient at identifying novel insertion sequences. DELLY, based on both the read pair approach and split read approach, performs better than BreakDancer, although it requires the memory and CPU time most intensively. Although SVEM is faster than Pindel and DELLY, some of the drawbacks listed above still exist. Combination of other approaches with the split read approach may be worth considering [24]. Finally, additional information such as the pedigree information of individuals can be useful for improving the prediction of SVs and SNPs [25]. The above topics will be investigated in our future study.

Acknowledgements. The super-computing resource was provided by Human Genome Center, Institute of Medical Science, University of Tokyo. This work was supported (in part) by MEXT Tohoku Medical Megabank Project.

References

1. Feuk, L., Carson, A.R., Scherer, S.W.: Structural variation in the human genome. Nat. Rev. Genet. 7(2), 85–97 (2006)
2. Xu, B., Roos, J.L., Levy, S., Van Rensburg, E.J., Gogos, J.A., Karayiorgou, M.: Strong association of de novo copy number mutations with sporadic schizophrenia. Nat. Genet. 40(7), 880–885 (2008)
3. Futreal, P.A., Coin, L., Marshall, M., Down, T., Hubbard, T., Wooster, R., et al.: A census of human cancer genes. Nat. Rev. Cancer 4(3), 177–183 (2004)
4. Reich, D.E., Schaffner, S.F., Daly, M.J., McVean, G., Mullikin, J.C., Higgins, J.M., et al.: Human genome sequence variation and the influence of gene history, mutation and recombination. Nat. Genet. 32(1), 135–142 (2002)
5. Hoogendoorn, E.: Computational methods for the detection of structural variation in the human genome (2012)
6. Pinkel, D., Segraves, R., Sudar, D., Clark, S., Poole, I., Kowbel, D., et al.: High resolution analysis of DNA copy number variation using comparative genomic hybridization to microarrays. Nat. Genet. 20(2), 207–211 (1998)
7. Hehir-Kwa, J.Y., Egmont-Petersen, M., Janssen, I.M., Smeets, D., Van Kessel, A.G., Veltman, J.A.: Genome-wide copy number profiling on high-density bacterial artificial chromosomes, single-nucleotide polymorphisms, and oligonucleotide microarrays: a platform comparison based on statistical power analysis. DNA Res. 14(1), 1–11 (2007)
8. Miller, D.T., Adam, M.P., Aradhya, S., Biesecker, L.G., Brothman, A.R., Carter, N.P., et al.: Consensus statement: chromosomal microarray is a first-tier clinical diagnostic test for individuals with developmental disabilities or congenital anomalies. Am. J. Hum. Genet. 86(5), 749–764 (2010)
9. Alkan, C., Coe, B.P., Eichler, E.E.: Genome structural variation discovery and genotyping. Nat. Rev. Genet. 12(5), 363–376 (2011)
10. Tuzun, E., Sharp, A.J., Bailey, J.A., Kaul, R., Morrison, V.A., Pertz, L.M., et al.: Fine-scale structural variation of the human genome. Nat. Genet. 37(7), 727–732 (2005)

11. Abyzov, A., Urban, A.E., Snyder, M., Gerstein, M.: CNVnator: an approach to discover, genotype, and characterize typical and atypical CNVs from family and population genome sequencing. Genome. Res. 21(6), 974–984 (2011)
12. Rausch, T., Zichner, T., Schlattl, A., Stütz, A.M., Benes, V., Korbel, J.O.: DELLY: structural variant discovery by integrated paired-end and split-read analysis. Bioinformatics 28(18), i333–i339 (2012)
13. Chen, K., Wallis, J.W., McLellan, M.D., Larson, D.E., Kalicki, J.M., Pohl, C.S., et al.: BreakDancer: an algorithm for high-resolution mapping of genomic structural variation. Nat. Methods 6(9), 677–681 (2009)
14. Ye, K., Schulz, M.H., Long, Q., Apweiler, R., Ning, Z.: Pindel: a pattern growth approach to detect break points of large deletions and medium sized insertions from paired-end short reads. Bioinformatics 25(21), 2865–2871 (2009)
15. Suzuki, S., Yasuda, T., Shiraishi, Y., Miyano, S., Nagasaki, M.: ClipCrop: a tool for detecting structural variations with single-base resolution using soft-clipping information. BMC Bioinformatics 12(Suppl. 14), 7 (2011)
16. Luo, R., Liu, B., Xie, Y., Li, Z., Huang, W., Yuan, J., et al.: SOAPdenovo2: an empirically improved memory-efficient short-read de novo assembler. Giga Science 1(1), 18 (2012)
17. Abecasis, G.R., Auton, A., Brooks, L.D., DePristo, M.A., Durbin, R.M., Handsaker, R.E., Kang, H.M., Marth, G.T., McVean, G.A.: An integrated map of genetic variation from 1,092 human genomes. Nature 491(7422), 56–65 (2012) (1000 Genomes Project Consortium)
18. Li, H., Durbin, R.: Fast and accurate short read alignment with Burrows-Wheeler transform. Bioinformatics 25(14), 1754–1760 (2009)
19. Ewing, B., Hillier, L., Wendl, M.C., Green, P.: Base-calling of automated sequencer traces using Phred. I. Accuracy assessment. Genome Res. 8(3), 175–185 (1998)
20. Sachidanandam, R., Weissman, D., Schmidt, S.C., Kakol, J.M., Stein, L.D., Marth, G., et al.: A map of human genome sequence variation containing 1.42 million single nucleotide polymorphisms. Nature 409(6822), 928–933 (2001)
21. Mills, R.E., Walter, K., Stewart, C., Handsaker, R.E., Chen, K., Alkan, C., et al.: Mapping copy number variation by population-scale genome sequencing. Nature 470(7332), 59–65 (2011)
22. Li, B., Ruotti, V., Stewart, R.M., Thomson, J.A., Dewey, C.N.: RNA-Seq gene expression estimation with read mapping uncertainty. Bioinformatics 26(4), 493–500 (2010)
23. Nariai, N., Hirose, O., Kojima, K., Nagasaki, M.: TIGAR: transcript isoform abundance estimation method with gapped alignment of RNA-Seq data by variational Bayesian inference. Bioinformatics 29(18), 2292–2299 (2013)
24. Mimori, T., Nariai, N., Kojima, K., Takahashi, M., Ono, A., Sato, Y., Yamaguchi-Kabata, Y., Nagasaki, M.: iSVP: an integrated structural variant calling pipeline from high-throughput sequencing data. BMC Systems Biology 7(6), 1–8 (2013)
25. Kojima, K., Nariai, N., Mimori, T., Takahashi, M., Yamaguchi-Kabata, Y., Sato, Y., Nagasaki, M.: A statistical variant calling approach from pedigree information and local haplotyping with phase informative reads. Bioinformatics 29(22), 2835–2843 (2013)

Analysis and Classification of Constrained DNA Elements with N-gram Graphs and Genomic Signatures

Dimitris Polychronopoulos[1,4], Anastasia Krithara[2], Christoforos Nikolaou[3], Giorgos Paliouras[2], Yannis Almirantis[1], and George Giannakopoulos[2,*]

[1] Institute of Biosciences and Applications, NCSR Demokritos, 15310 Athens, Greece
[2] Institute of Informatics and Telecommunications
NCSR Demokritos, 15310 Athens, Greece
[3] Department of Biology, University of Crete, 71409 Heraklion, Greece
[4] Department of Biochemistry and Molecular Biology, Faculty of Biology,
National and Kapodistrian University of Athens 15701 Athens, Greece

Abstract. Most common methods for inquiring genomic sequence composition, are based on the bag-of-words approach and thus largely ignore the original sequence structure or the relative positioning of its constituent oligonucleotides. We here present a novel methodology that takes into account both word representation and relative positioning at various lengths scales in the form of n-gram graphs (NGG). We implemented the NGG approach on short vertebrate and invertebrate constrained genomic sequences of various origins and predicted functionalities and were able to efficiently distinguish DNA sequences belonging to the same species (intra-species classification). As an alternative method, we also applied the Genomic Signatures (GS) approach to the same sequences. To our knowledge, this is the first time that GS are applied on short sequences, rather than whole genomes. Together, the presented results suggest that NGG is an efficient method for classifying sequences, originating from a given genome, according to their function.

Keywords: genomic sequence representation, n-gram graphs, conserved non-coding elements, CNEs, UCEs, ultraconserved elements, classification, genomic signatures.

1 Introduction

1.1 Constrained Elements in Eukaryote Genomes

High throughput sequencing at a massive scale combined with comparative genome analysis has led to the discovery of a variety of constrained genomic elements. In fact, there are many more selectively constrained noncoding than protein-coding sequences in mammalian genomes. One of the most interesting discoveries that have arisen from comparative genomics among mammalian

---

* Corresponding author: ggianna@iit.demokritos.gr

A.-H. Dediu, C. Martín-Vide, and B. Truthe (Eds.): AlCoB 2014, LNBI 8542, pp. 220–234, 2014.
© Springer International Publishing Switzerland 2014

genomes are the hundreds of such noncoding elements of more than 200bp in length that show absolute conservation among mammalian orders [1]. These only represent the tip of the iceberg of a much larger class of conserved noncoding elements (CNEs), a general class of sequence elements that are significantly more conserved than protein-coding genes and non-coding RNAs (ncRNAs)[10]. In the following analysis, we implement the term "constrained elements" for both protein-coding and conserved non-coding sequence stretches, while we refer to noncoding as strictly not protein-coding.

Conserved non-coding elements can be found in the literature under various definitions, depending on the percentage of identity between two or more organisms and the minimum length. Increasing evidence suggests that CNEs are selectively constrained and not mutational cold-spots [5] and there is a plethora of studies indicating possible functions of those elements [10]. Among the various reported functional roles of CNEs, enhancers appear to be the most plausible. Nevertheless, the relative abundance, genomic distribution and variable length of CNEs is indicative of various alternatives. CNEs have been shown to bear resemblance to CTCF insulator sites [30], matrix attachment regions [8], while a recent, concise study across 29 mammals revealed constrained elements of smaller sizes that may be directly related to transcriptional regulation as well as to the encoding of functional RNA molecules [19]. CNE existence may be extended even further back in evolutionary time as suggested by recent works including both bony and cartilaginous fish species [18]. While the majority of the analyses have been conducted in mammals, there is growing evidence that CNEs are not a vertebrate innovation and can also be found in invertebrates and plants. Despite the fact that vertebrate and invertebrate CNEs bear no sequence identity, they share common sequence characteristics, indicating a parallel evolution of those sequences in order to perform the same, possibly essential, functions [27].

Among the various attributes of these genomic sequences, DNA composition has been greatly ignored. When compared to non-CNEs and near-promoter sequences, CNEs possess an excess of AT-rich motifs, often containing runs of identical nucleotides. In a recent paper, Walter *et al* have analysed the base composition of human and Fugu CNEs at single nucleotide level [29]. They have found that those elements are A+T rich, much more so than the region they reside in, in contrast to their flanking region just outside their boundaries, which exhibits a marked drop in A+T content that forms a unique pattern. Such compositional extremes are strong indications of functionality as has been shown for gene-dense regions and CpG islands [26,31]. It is therefore of great interest to further investigate the compositional preference of constrained regions in greater detail. To this end, conventional approaches addressing composition through histograms, or bag-of-words approaches tend to overlook the positional information, while probabilistic sequential methods like HMMs are likely to undermine the effect of local sequence boundaries. In the following we describe a novel methodology that is able to address both such aspects with the additional advantage of allowing for similarity measurements between any two sequences.

1.2   Analyzing Sequence Composition through a Combination of Word-Content and Relative Positioning Information

A sequence can easily be considered equivalent to a natural language text, under the assumption that the vocabulary is very limited. Traditionally, natural language processing methods based on n-grams (n-nucleotides correspondingly) have been applied on biological sequences, aiming to support sequence matching [15], indexing [16], analysis of protein sequences [6] and coding and non-coding DNA sequences [20]. The n-gram models of sequences indicate how short subsequences of length n appear in the whole analyzed sequence. Other alternatives of analysis, like Hidden Markov Models or Conditional Random Fields [3], model probabilistically the possible combinations of elements in a sequence.

In this work, we propose the application of the n-gram graph (NGG) representation methodology [7], which manages to capture both local and global characteristics of the analysed sequences. The main idea behind the NGGs is that the neighborhood between sub-sequences in a sequence contains a crucial part of the sequence information. The NGG, as derived from a single sequence, is essentially a histogram of the co-occurrences of symbols. The symbols are considered to co-occur when found within a maximum distance (window) of each other. The size of the window, which is a parameter of the NGG, allows for fuzziness in the representation of co-occurrences within a sequence. The fact that NGGs take into account co-occurrences offers the local descriptiveness, while the fact that they act as a histogram of such co-occurrences provides their global representation potential. We note that at the limit of window size (infinite window size) the NGGs lose the local descriptiveness trait.

As opposed to probabilistic models (e.g. HMM), the NGGs are deterministic. Furthermore, NGGs do not rely on numerous examples to infer model parameters. Third, they treat equally under-represented phenomena which removes the bias of probabilistic approaches towards frequent patterns. As opposed to n-gram models, NGGs offer more information, based on the representation of co-occurrences. Overall, they provide a trade-off between expressiveness and generalization.

The n-gram graph framework, also offers a set of important operators. These operators allow combining individual graphs into a model graph (the update operator), and comparing pairs of graphs providing graded similarity measurements (similarity operators). In the sequence composition setting, the representation and set of operators provide one more means of analysis and comparison, one that is lacking from widely-implemented probabilistic models such as HMMs.

Finally, the NGGs can be combined with vector representation of sequences to allow the application of machine learning methods for the classification of sequences. Within this study, both conserved non-coding and protein coding segments are analyzed through a NGG-based approach and an approach based on the method of genomic signatures [11]. Additionally, datasets of CNEs and protein coding segments (coding exons) are studied along with suitably chosen (see Methods) surrogate sequence sets.

2   Methods

2.1   Datasets Retrieval

We consider various published datasets of constrained sequences. Human, worm and insect denote sequences taken from *H.sapiens*, *C.elegans* and *D.melanogaster* genomes. Given that those datasets are heterogeneous in numbers, we randomly select 1000 elements from each set for our subsequent analysis. Only worm CNEs are studied in their entirety, as this dataset is relatively small. The exonic sequences of human, worm and insect genomes were obtained from the UCSC genome repository based on the *RefSeq* annotation referring to the latest genome assemblies (hg19, ce10, dm3) [21]. The datasets described below along with their surrogates are used in 26, in total, pairwise classification experiments, denoted throughout the text as #1, #2, ...; see Supplementary Spreadsheet[1] where further information is provided. Apart from exonic sequences, the following classes of constrained non-exonic sequences are used:

- UltraConserved Noncoding Elements (UCNEs): These are sequences of at least 200bp in length mapped on the human genome (hg19) that display sequence similarity which is greater than or equal to 95% between human and chicken whole genome alignments [4]
- EU100 nonexonic CNEs (EU100nx CNEs): These are sequences mapped on the human genome (hg18) that are identical over 100bp in at least 3 out of 5 placental mammals (human, mouse, rat, dog and cow) [25]. The whole set is named EU100+ and since we remove elements overlapping exons it will be referred to as EU100nx.
- Amniotic and Mammalian CNEs: These are elements identified by Kim and Pritchard[17]. Mammalian CNEs are sequences that are conserved within mammals but not found in chicken or fish, while Amniotic CNEs are conserved in mammals and chicken but not found in fish. LiftOver [14] is used to convert the coordinates to the most recent release of the human genome (from hg17 to hg19).
- Worm and Insect UCNEs: These are elements mapped on ce10 and dm3 genome releases of *C.elegans* and *D.melanogaster* respectively. Worm UCNEs are DNA stretches longer than 60bp that exhibit sequence similarity greater than 90% among *C.elegans* and *C.japonica*, while Insect UCNEs are stretches longer than 60bp that display sequence similarity greater than 90% among *D.melanogaster* and *D.virilis* (unpublished results, Philipp Bucher's group, EPFL)

2.2   Treatment of Sequences and Extraction of Surrogate Sequences

A useful suite of tools called BEDTools [22] is used in order to extract FASTA sequences from BED files and to calculate overlapping elements. Additionally, we

---

[1] You can download the spreadsheet from
http://users.iit.demokritos.gr/~ggianna/Publications/alcob2014/

make use of the EMBOSS suite to calculate fractional GC content of sequences [24]. It is known from the literature that vertebrate and invertebrate CNEs are of significantly different lengths (the ones belonging to the latter category are considerably smaller) [23]. We make sure that we take segments of equal lengths as follows: for classification experiments involving CNE sets of vertebrate and invertebrate origin, we truncate the vertebrate ones around their middle point to the length of the shortest invertebrate CNE included in the experiment. For each element of each collection under study (CNE or exon), an analogue of it is extracted from the corresponding genome. Every resulting DNA segment (surrogate sequence) is ensured that is of the same length and GC content (within a 1% deviation limit) with its corresponding element in the collection under study. Statistics of the datasets (mean length and GC content of sequences) are available in the Supplementary Spreadsheet. All custom shell scripts are available upon request to interested readers. All the files (coordinates in BED and sequences in FASTA format) are also available upon request.

2.3   From Sequences to the N-Gram Graph Similarity Vector Space

We have followed a set of steps to represent our sequences using NGGs. The idea is that from known (labeled) sequences we form representatives of each class of sequences. Then, we describe all sequences based on their similarity to the representatives of each class. Thus, there are essentially three steps in our application of the NGGs:

- Representation of individual sequences using NGGs.
- Calculation of representative (model) NGGs for each training class used.
- Calculation of similarity between training instances and model graphs.
- Representation of instances using only their similarities, i.e., in a similarity vector space.

In the following paragraphs we elaborate on these steps.

Representation of Sequences. The *n-gram graph (NGG)* is a graph $G =< V^G, E^G, L, W >$, where V^G is the set of vertices, E^G is the set of edges, L is a function assigning a label to each vertex *and to each edge* and W is a function assigning a weight to every edge. The graph has n-grams labeling its vertices $v^G \in V^G$. The edges $e^G \in E^G$ (the superscript G will be omitted where easily assumed) connecting the n-grams indicate proximity of the corresponding vertex n-grams. The weight of the edges can indicate a variety of traits. In our implementation we apply as weight the number of times the two connected n-grams were found to co-occur. It is important to note that in NGGs *each vertex is unique*. To ensure that no duplicate vertices exist, we also require that the labelling function is an one-to-one function. Two vertices are considered equal if and only if their labels are equal.

We repeat that the edges E are assigned weights of $c_{i,j}$ where $c_{i,j}$ is the number of times a given pair S_i, S_j of n-grams happen to be neighbors in a string within some distance D_{win} of each other. The distance d of two n-grams $S^{(i,i+n)}, S^{(j,j+n)}$ is $d = |i - j|$. To create the NGG from a given sequence, a fixed-width window D_{win} of characters (or words) around a given n-gram $N_0 \equiv S^r, r \in \mathbb{N}^*$ is used. All character n-grams within the window are considered to be neighbors of N_0. These neighbors are represented as connected vertices in the text graph. We use a symmetric approach, where a window of length $2 \times D_{\text{win}} + 1$ runs over the text, centered at the beginning of N_0. If the n-gram we are interested in is located at position p_0, then the window will span from $p_0 - D_{\text{win}}$ to $p_0 + D_{\text{win}}$, taking into account *both preceding and succeeding* characters or words. Each edge $e = < a, b >$ is weighted based on the number of co-occurrences of the neighbors within a window in the text: $w(e) = |\{S^{(i,i+n)} = L(a), S^{(j,j+n)} = L(b) : abs(i - j) \leq D_{\text{win}}, i \neq j\}|$

Creation of Representative Graphs. The NGG representation specification indicates how to represent a text using an NGG. However, in sequence analysis it is often required to represent a whole sequence set, i.e. a sequence class. In our applications we have used the *update function U* to represent sets. The update function $U(G_1, G_2, l)$ takes two graphs as input. One graph is considered to be the pre-existing graph G_1 and one that is considered to be the new graph G_2.

The U function takes an additional argument, the *learning factor* $l \in [0, 1]$, which determines the sensitivity of G_1 to the change G_2 brings. The higher the value of learning factor, the higher the impact of the new graph to the existing graph. The definition of the weighting performed in the graph resulting from $U(G_1, G_2, l)$ is: $w(e) = w_1(e) + (w_2(e) - w_1(e)) \times l$ for every edge $e \in E_1 \cup E_2$.

The U function allows creating a representative graph for a set of documents, in analogy to the centroid of a set of vectors. The creation of a set-representative graph G_s, for a given set of graphs $\mathbb{G} = \{G_1, G_2, G_3, ..., G_n\}$ is as follows. We initialize the class graph with the first document of the class $G_s = G_1$. Then, iteratively we update G_s by: $G_s = U(G_s, G_i, \frac{1}{i+1})$ for $i \in 2, n$. After all iterations the weight of every edge $e \in E_s$ will have a weight of: $w_e = \frac{\sum\limits_{i:e \in E_i} w_i}{|\{i:e \in E_i\}|}$

This weight is the average of the weights of edge e over all the graphs G_i, where it appears. This means that graphs where the edge does not appear, *are not taken into account* in this calculation. This essentially means that having a graph with an edge e, where $w(e) = 0$ is different than not having the edge at all.

Measuring Similarity and Vector Space Representation. In the n-gram graph framework there are different ways to measure similarity. We choose the Value Similarity function [7]. This measure quantifies the ratio of common edges between two graphs, taking into account the ratio of weights of common edges. If $|G| = |\{e \in G\}|$ the edge count of a graph G, then in the Value Similarity measure each matching edge e having weight w_e^i in graph G^i contributes $\frac{\text{VR}(e)}{\max(|G^i|, |G^j|)}$ to VS, while not matching edges do not contribute (consider that for an edge $e \notin G^i$ we define $w_e^i = 0$).

The *ValueRatio (VR)* scaling factor is defined as: $\text{VR}(e) = \frac{\min(w_e^i, w_e^j)}{\max(w_e^i, w_e^j)}$. The equation indicates that the *ValueRatio* takes values in $[0, 1]$, and is symmetric. Thus, the full equation for *VS* is: $\text{VS}(G_i, G_j) = \frac{\sum_{e \in G_i} \frac{\min(w_i^e, w_j^e)}{\max(w_i^e, w_j^e)}}{\max(|G_i|, |G_j|)}$

Given the similarity function $\text{VS}(G_i, G_j)$, a sequence instance S with a corresponding graph G_S and class-representative graphs G_i for classes C_i, we can describe S in a similarity vector space. In this space, each dimension reflects the similarity of the instance sequence to a corresponding class. Thus, the vector $\overline{V}(S)$ is of the form $\overline{V}(S) = <\text{VS}(G_S, G_1), \text{VS}(G_S, G_2), ..., \text{VS}(G_S, G_n)>$ where n is the number of classes we have. Thus, the above process allows us to map each sequence in a dataset to a vector space of similarities.

In the following section we discuss another representation of sequences we use in our experiments: genomic signatures.

2.4   The "Genomic Signature" Method

This is a standard methodology for classifying and distinguishing genomes based on the quantification of neighbor preferences in a DNA sequence of an entire genome by computing the vector of the odds ratios for dinucleotides [13]. The odds ratio of each dinucleotide is the quantity: $r_{ij} = f_{ij}/(f_i f_j)$, where f_{ij} and f_i, f_j stand for the frequencies of occurrence in the studied sequence of a dinucleotide and its constituent nucleotides respectively. Therefore, the subscripts i, j represent any pair of A, G, C and T. This is the ratio of the "observed" dinucleotide frequency over the "expected" one under no neighbor effects, thus it expresses the actual neighbor preferences of the given pair of nucleotides. Before computing the odds ratios for a given sequence, this is concatenated to its reverse complement. Consequently, the relevant ratios are only ten, i.e. four for the self-complementary dinucleotides and six for the mutually complementary couples. Karlin and co-workers first proposed that these quantities differentiate between different genomes, according, approximately, to their evolutionary distance. Thus they have assigned to the vector of these ten "first neighbour preferences" the name of genomic signature (GS). It has to be noted that GS filter out mononucleotide composition retaining only the first neighbor preferences. Among the various implementations of GS, dinucleotides are used more widely due to the obvious reason of statistical limitations imposed by sequence size. In our case, we used the same approach for two reasons: a) The examined sequences were of short size. Both exons and CNE have a mean length of no more than 150 nucleotides and addressing them with higher-order k-mers is bound to be compromised by the finite size effects. b) Exons, one of the main functional categories used in this work, are known to be biased towards specific trinucleotides (and multiples thereof such as 6- and 9-mers) due to the inherent structure of the genetic code and specific protein coding preferences. This property, (widely known as codon bias) is much more pronounced in 3-mers than in 2-mers, thus we chose the latter as the basis of our analysis with GS.

For a direct comparison with the n-gram graphs (NGG) approach described in the previous section, we use classification based on genomic signatures (GS) and apply that to the same pairs of genomic sequences. When applying classification, in both the NGG and the GS approaches we use the output of the analysis (similarity vectors, or ratio correspondingly) as input vectors to train a well-known, rule-based classifier: the JRIP implementation [9] of RIPPER [2]. We note that the results provided by RIPPER are comparable to those of other state-of-the-art classifiers we tried (e.g., Support Vector Machines, Random Forest). In the following section we report our findings based on experiments using both analysis methods (NGG and GS).

3   Results and Discussion

In this section we describe a systematic analysis of short genomic segments, which display different functionalities and stem from human, worm and insect genomes. For every classification task, we adopt the technique of NGG and GS as explained previously in order to estimate how different modalities could be separated based on sequence composition.

We note that the NGG representation embeds three parameters. The first is the value of m, which defines the minimum length of the n-grams into which a sequence is split. The second is M, which defines the maximum length of the n-grams into which a sequence is split. The third is D which represents the maximum distance within which we consider the n-grams (n-bases) to be neighbors. Keeping $m = M = D$ simplifies the analysis step reducing the required time, while not significantly altering the results in most cases [7].

An estimator described in the the original NGGs' work [7] was used in order to define these parameters. As it was devised for a different, linguistic task (summarization system evaluation) not taking into account the prediction potential of n-grams for a classification task, we performed an approximate optimization based on exhaustive experiments (for $m \in [2, 9]$) using 10-fold cross validation. The tuning was performed using 6 different datasets.

We measure performance by means of the *F-measure*. F-measure, widely used in machine learning and classification, is the geometric mean of the precision P (the percentage of sequences assigned to a class that truly belong to the class) and recall R (the percentage of the set of sequences that truly belong to a class that were assigned to this class) of a classifier. Thus, $F1 = \frac{2 \times P \times R}{P + R}$. In Figure 1 we illustrate, via F1, the performance of the NGG based classification. Each column corresponds to a parameter value combination. The strong horizontal line in each column indicates the average performance of this combination over the set of 6 datasets. The remaining lines represent the quantile values of the performance. We deduce that there exists a persistent high plateau of performance when $5 \leq m \leq 8$, which shows that the method does not improve significantly after $m = 5$. Another observation is that 2-2-2 performs considerably better than 3-3-3 and almost equally to 4-4-4. This might be related to the importance of first neighbor preferences as reflected in the NGG methodology.

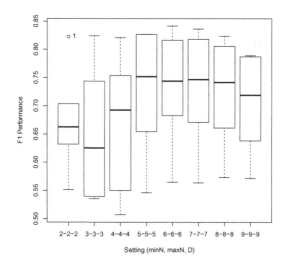

Fig. 1. Boxplots of the performance of n-gram graphs classification over varying parameters

In the tables included throughout the Discussion section, the NGG performance obtained for the optimal combination of parameter values is included; e.g. in Exp #1, Table 1, the value $83, 86$ is given for NGG, corresponding to the parameter choice $(7,7,7)$. We propose $(m, M, D) = (5, 5, 5)$ as the best parameter settings for analyzing short genomic sequences, based on all the 26 performed experiments (see Supplementary Spreadsheet, Overall Statistics tab). Genomic signatures have been used as an alternative to the NGG based classification method. To our knowledge, for such short genomic segments of possible functional importance, not only the NGG method but also the genomic signature approach has not been applied before.

3.1 Inter-Species Comparisons of Background Sequences

For comparisons between human, worm and insect background sequences, we use the surrogate sets described in Methods as representative samples of the different genomes. Comparisons involving *H. sapiens* yield always the best classification rates using both n-gram graphs (NGG) and Genomic Signatures (GS), see Table 1. This may be understood on the grounds of the high difference of neighbor preferences, mainly in CpG and TpA between *H.sapiens* and the invertebrates, while these preferences are found to be quite close between invertebrate species: *D.melanogaster* (insect) and *C.elegans* (worm). GS are exclusively a quantification of first neighbor preferences and as a consequence are not influenced by GC content. On the contrary, NGG are able to conceptually incorporate various components/aspects of sequence composition, such as mononucleotide composition and higher order neighbor preferences (a criterion that is provided as a parameter value in our technique by adjusting D).

In cases where human sequences are included in the comparison, GS perform systematically better than NGG. This difference is also statistically significant with a p-value below 0.10, based on a paired t-test. It is also worth mentioning that the two cases with the highest differences in GC content between the sets involved in the classification experiment are: experiment #22 which is the only one where NGG perform better than GS; and experiment #1 where GS perform better than NGG, but with the lowest difference in their performances. I.e. the cases of the highest relative preponderance of NGG are the ones with the highest differences in GC content, as expected due to the relative sensitivities of the two methods to this composition parameter.

Table 1. Inter-species comparisons of background genomic sequences

Exp	Description	Average length	Average GC	NGG	GS
#1	surrogates for human exons	167.837	0.5155	83.86	85.98
	surrogates for worm exons	169.318	0.4049		
#14	surrogates for human UCNEs	86.094	0.3651	79.38	84.05
	surrogates for insect UCNEs	86.582	0.3949		
#20	surrogates for insect exons	169.318	0.5202	80.48	87.49
	surrogates for human exons	169.816	0.5087		
#22	surrogates for worm exons	213.365	0.4194	73.50	70.35
	surrogates for insect exons	212.858	0.5194		
#23	surrogates for human UCNEs	82.932	0.3648	80.35	83.75
	surrogates for worm UCNEs	82.875	0.4297		
#13	surrogates for worm UCNEs	83.407	0.4265	58.79	64
	surrogates for insect UCNEs	86.582	0.3949		
	Average			**76.06**	**79.27**

3.2 Classification Experiments of Constrained DNA Sequences Versus Their Background Surrogates (Intra-Species Classification)

The following experiments refer to comparisons of constrained sequences against their surrogates (Table 2). Note that surrogates share the same GC% and length with the initial sequences (see Methods). As evidenced from inspection of Table 2, comparisons involving invertebrate constrained sequences are not classified as successfully as their human counterparts using NGG, see experiments #2 and #17. Only insect UCNEs appear to be resistant to this effect, see experiment #12. This finding might be understood on the grounds of several particularities of the warm-blooded animals (often of all vertebrates) especially in their non-functional, non-constrained background genomic fraction. These include a high enrichment in transposable elements, microsatellites, homo-purine/homo-pyrimidine tracts and several other characteristic compositional motifs. Such genomes also present a typical profile of avoided dinucleotides (especially CpG and TpA) that are less avoided in the constrained elements (exons, CNEs), which having functional roles, do not strictly follow the average genomic compositional trends. Note that invertebrate genomes are much less abundant in repeats and less marked by under-representation of specific dinucleotides. In the comparisons by means of GS, the same trend is followed but the differences are minor, due to the sensitivity of these quantities solely upon first neighbor preferences. We know from earlier studies that genomic signatures do not perform well in intra-species comparisons because neighbor preferences remain relatively constant within the same genome. Consequently, in most cases listed, NGG perform better than GS (compare averages, 68,76% versus 61,84%). The difference is statistically significant (p-value below 0.10, based on a paired t-test).

Table 2. Classification of constrained DNA sequences versus background surrogates

Exp	Description	Average length	Average GC	NGG	GS
#2	worm exons	213.365	0.4243	57.7	61.05
	surrogates	213.365	0.4239		
#3	human exons	169.816	0.5190	74.3	63.41
	surrogates	169.816	0.5183		
#17	insect exons	388.822	0.5412	52.36	59.45
	surrogates	381.557	0.5389		
#4	worm UCNEs	82.875	0.4309	56.76	55.62
	surrogates	82.875	0.4308		
#5a	human UCNEs	326.923	0.3676	82.63	72.00
	surrogates	326.923	0.3676		
#5b	human EU100nx CNEs	155.499	0.3783	76.43	63.75
	surrogates	155.499	0.3783		
#5c	amniotic CNEs	289.061	0.3756	78.62	63.00
	surrogates	289.061	0.3756		
#5d	mammalian CNEs	246.488	0.4015	75.85	55.65
	surrogates	246.488	0.4018		
#12	insect UCNEs	86.582	0.3949	64.15	62.65
	surrogates	86.582	0.3949		
	Average			**68.76**	**61.84**

The six last rows of Table 2 denote comparisons of several CNE sequences collections against their surrogates. In general we notice that among constrained sequences, human CNE sequences versus surrogates exhibit relatively higher classification rates, if compared with exonic sequences versus their corresponding surrogates. This might be attributed to the fact that CNEs (and especially human UCNEs that we consider here) are more conserved than exons. In addition, it is known from the literature that CNEs do serve as transcription factor binding sites and bear several motifs [28] that NGG analysis is possibly sensitive enough to detect and thus, increase the obtained CNE-background vs. exon-background classification rates.

3.3 Genomic Signatures Perform Better in Classifying Genomic Segments of Functional Importance and Different Origin

In the experiments included in Table 3 we study the performance of NGG and GS when they are applied on sets of functional sequences (CNEs or exons) of different genomes. We verify that GS perform better in inter-genomic classification and their performance is slightly improved in Table 3, if compared to Table 1. This means that it is not blurred due to the constrained character of the sequences, i.e. the first neighbour preference characterizing different genomes and filtered by GS is clearly retained in exons and CNEs. Once again, we tested the statistical significance in performance between the NGG and GS methods, based on a paired t-test, and there exists statistically significant difference (with a p-value below 0.05).

On the other hand, the performance of NGG drops. This might be associated with the complex set of compositional traits that are characteristic of the NGG method: nucleotide composition of the sequence, first neighbor preferences and also higher neighbor relationships. Thus, in this context, the inter-genomic differences leading to an efficient NGG based classification of background sequences (Table 1, average = 76.06) are blurred by the presence of common compositional constraints related to function as shown by inspection of Table 3 (average = 74.47). Such an example is the known over-representation of adenine and guanine in the composition of protein coding exons (purine loading).

Table 3. Comparisons of functional sequences between genomes

Exp	Description	Average length	Average GC	NGG	GS
#9	worm exons	169.318	0.3886	74.68	82.21
	human exons	169.816	0.5190		
#10	worm UCNEs	83.113	0.4277	77.77	82.29
	human UCNEs	82.640	0.3728		
#15	worm UCNEs	83.086	0.4285	70.89	74.95
	insect UCNEs	86.582	0.3949		
#16	human UCNEs	86.094	0.3704	82.08	86.70
	insect UCNEs	86.582	0.3949		
#19	insect exons	169.318	0.5148	70.03	81.29
	human exons	169.816	0.5089		
#21	worm exons	213.365	0.4196	71.39	72.35
	insect exons	212.858	0.5093		
	Average			**74.47**	**79.97**

4 Conclusions

The problem of assigning functionality to un-annotated sequences is highly relevant in light of the complexity of mammalian genomes. By performing various comparisons between representative sequences of different origin and functionality, we were able to show the potential of the n-gram graphs (NGG) in effectively discriminating between genomic sequence fragments with expected function against the bulk of the genome. NGG were able to quantify the effect of sequence constraint, thus their implementation in the description of other functional sequences of mammalian genomes (such as regions with structural properties, matrix attachment regions and various non-coding RNA species) may provide valuable insight in the interplay between composition and function. When it comes to inter-genomic comparisons between sequences of different origin, NGG are less effective compared to the method of Genomic Signatures (GS). The efficiency of the latter is for the first time demonstrated here at this length scale, as up to now GS had only been applied in more extended genomic regions of 50kb or more [12]. Further work is needed for a better exploitation of the presented results, while combinations of the two methods may result in higher classification rates.

Acknowledgments. D.P. would like to thank Professors Georgios Rodakis and Stavros Hamodrakas (Faculty of Biology, University of Athens) for serving as academic advisors. We would also like to express our gratitude to Dr Slavica Dimitrieva and Dr Philipp Bucher for providing unpublished datasets of worm and insect UCNEs. The research leading to these results has received funding from the European Commission's Seventh Framework Programme (FP7/2007-2013, ICT-2011.4.4(d), Intelligent Information Management, Targeted Competition Framework) under grant agreement n. 318652 (BioASQ challenge).

References

1. Bejerano, G., Pheasant, M., Makunin, I., Stephen, S., Kent, W.J., Mattick, J.S., Haussler, D.: Ultraconserved elements in the human genome. Science 304(5675), 1321–1325 (2004), http://www.ncbi.nlm.nih.gov/pubmed/15131266
2. Cohen, W.W.: Fast effective rule induction. ICML 95, 115–123 (1995)
3. Culotta, A., Kulp, D., McCallum, A.: Gene prediction with conditional random fields, Tech. Rep. UM-CS-2005-028, University of Massachusetts, Amherst (2005)
4. Dimitrieva, S., Bucher, P.: Genomic context analysis reveals dense interaction network between vertebrate ultraconserved non-coding elements. Bioinformatics 28(18), i395–i401 (2012), http://www.ncbi.nlm.nih.gov/pubmed/22962458
5. Drake, J.A., Bird, C., Nemesh, J., Thomas, D.J., Newton-Cheh, C., Reymond, A., Excoffier, L., Attar, H., Antonarakis, S.E., Dermitzakis, E.T., Hirschhorn, J.N.: Conserved noncoding sequences are selectively constrained and not mutation cold spots. Nat. Genet. 38(2), 223–227 (2006), http://www.ncbi.nlm.nih.gov/pubmed/16380714

6. Ganapathiraju, M., Weisser, D., Rosenfeld, R., Carbonell, J., Reddy, R., Klein-Seetharaman, J.: Comparative n-gram analysis of whole-genome protein sequences. In: Proceedings of the Second International Conference on Human Language Technology Research, pp. 76–81. Morgan Kaufmann Publishers Inc. (2002)

7. Giannakopoulos, G., Karkaletsis, V., Vouros, G., Stamatopoulos, P.: Summarization system evaluation revisited: N-gram graphs. ACM Trans. Speech Lang. Process. 5(3), 139 (2008)

8. Glazko, G.V., Koonin, E.V., Rogozin, I.B., Shabalina, S.A.: A significant fraction of conserved noncoding DNA in human and mouse consists of predicted matrix attachment regions. Trends Genet. 19(3), 119–124 (2003), http://www.ncbi.nlm.nih.gov/pubmed/12615002

9. Hall, M., Frank, E., Holmes, G., Pfahringer, B., Reutemann, P., Witten, I.H.: The weka data mining software: an update. ACM SIGKDD Explorations Newsletter 11(1), 10–18 (2009)

10. Harmston, N., Baresic, A., Lenhard, B.: The mystery of extreme non-coding conservation. Philosophical transactions of the Royal Society of London 368(1632), 20130021 (2013), http://www.pubmedcentral.nih.gov/articlerender.fcgi?artid=3826495&tool=pmcentrez&rendertype=abstract

11. Karlin, S., Mrázek, J.: Compositional differences within and between eukaryotic genomes. Proceedings of the National Academy of Sciences of the United States of America 94(19), 10227–10232 (1997), http://www.pubmedcentral.nih.gov/articlerender.fcgi?artid=23344&tool=pmcentrez&rendertype=abstract

12. Karlin, S.: Global dinucleotide signatures and analysis of genomic heterogeneity. Current Opinion in Microbiology 1(5), 598–610 (1998)

13. Karlin, S., Burge, C.: Dinucleotide relative abundance extremes: a genomic signature. Trends in Genetics 11(7), 283–290 (1995)

14. Karolchik, D., Baertsch, R., Diekhans, M., Furey, T.S., Hinrichs, A., Lu, Y., Roskin, K.M., Schwartz, M., Sugnet, C.W., Thomas, D.J., et al.: The ucsc genome browser database. Nucleic Acids Research 31(1), 51–54 (2003)

15. Kim, J.Y., Shawe-Taylor, J.: Fast string matching using an n-gram algorithm. Software: Practice and Experience 24(1), 79–88 (1994)

16. Kim, M.S., Whang, K.Y., Lee, J.G., Lee, M.J.: n-gram/2l: A space and time efficient two-level n-gram inverted index structure. In: Proceedings of the 31st International Conference on Very Large Data Bases, pp. 325–336. VLDB Endowment (2005)

17. Kim, S.Y., Pritchard, J.K.: Adaptive evolution of conserved noncoding elements in mammals. PLoS Genetics 3(9), 1572–1586 (2007), http://www.pubmedcentral.nih.gov/articlerender.fcgi?artid=1971121&tool=pmcentrez&rendertype=abstract

18. Lee, A.P., Kerk, S.Y., Tan, Y.Y., Brenner, S., Venkatesh, B.: Ancient vertebrate conserved noncoding elements have been evolving rapidly in teleost fishes. Mol. Biol. Evol. 28(3), 1205–1215 (2011), http://www.ncbi.nlm.nih.gov/pubmed/21081479

19. Lindblad-Toh, K., et al.: A high-resolution map of human evolutionary constraint using 29 mammals. Nature 478(7370), 476–482 (2011), http://www.ncbi.nlm.nih.gov/pubmed/21993624

20. Mantegna, R., Buldyrev, S., Goldberger, A., Havlin, S., Peng, C.K., Simons, M., Stanley, H.: Systematic analysis of coding and noncoding dna sequences using methods of statistical linguistics. Physical Review E 52(3), 2939 (1995)

21. Pruitt, K.D., Tatusova, T., Maglott, D.R.: Ncbi reference sequences (refseq): a curated non-redundant sequence database of genomes, transcripts and proteins. Nucleic Acids Research 35(suppl. 1), 61–65 (2007)

22. Quinlan, A.R., Hall, I.M.: BEDTools: a flexible suite of utilities for comparing genomic features. Bioinformatics 26(6), 841–842 (2010), http://www.ncbi.nlm.nih.gov/pubmed/20110278
23. Retelska, D., Beaudoing, E., Notredame, C., Jongeneel, C.V., Bucher, P.: Vertebrate conserved non coding DNA regions have a high persistence length and a short persistence time. BMC Genomics 8, 398 (2007), http://www.ncbi.nlm.nih.gov/pubmed/17973996
24. Rice, P., Longden, I., Bleasby, A.: EMBOSS: the European Molecular Biology Open Software Suite. Trends in genetics: TIG 16(6), 276–277 (2000), http://www.ncbi.nlm.nih.gov/pubmed/10827456
25. Stephen, S., Pheasant, M., Makunin, I.V., Mattick, J.S.: Large-scale appearance of ultraconserved elements in tetrapod genomes and slowdown of the molecular clock. Mol. Biol. Evol. 25(2), 402–408 (2008), http://www.ncbi.nlm.nih.gov/pubmed/18056681
26. Touchon, M., Arneodo, A., d'Aubenton Carafa, Y., Thermes, C.: Transcription-coupled and splicing-coupled strand asymmetries in eukaryotic genomes. Nucleic Acids Research 32(17), 4969–4978 (2004)
27. Vavouri, T., Walter, K., Gilks, W.R., Lehner, B., Elgar, G.: Parallel evolution of conserved non-coding elements that target a common set of developmental regulatory genes from worms to humans. Genome Biol. 8(2), R15 (2007), http://www.ncbi.nlm.nih.gov/pubmed/17274809
28. Viturawong, T., Meissner, F., Butter, F., Mann, M.: A DNA-Centric Protein Interaction Map of Ultraconserved Elements Reveals Contribution of Transcription Factor Binding Hubs to Conservation. Cell reports 5(2), 531–545 (2013), http://www.cell.com/cell-reports/fulltext/S2211-1247
29. Walter, K., Abnizova, I., Elgar, G., Gilks, W.R.: Striking nucleotide frequency pattern at the borders of highly conserved vertebrate non-coding sequences. Trends Genet. 21(8), 436–440 (2005), http://www.ncbi.nlm.nih.gov/pubmed/15979195
30. Xie, X., Mikkelsen, T.S., Gnirke, A., Lindblad-Toh, K., Kellis, M., Lander, E.S.: Systematic discovery of regulatory motifs in conserved regions of the human genome, including thousands of CTCF insulator sites. Proc. Natl. Acad. Sci U. S. A. 104(17), 7145–7150 (2007), http://www.ncbi.nlm.nih.gov/pubmed/17442748
31. Zhang, L., Kasif, S., Cantor, C.R., Broude, N.E.: Gc/at-content spikes as genomic punctuation marks. Proceedings of the National Academy of Sciences of the United States of America 101(48), 16855–16860 (2004)

Inference of Boolean Networks from Gene Interaction Graphs Using a SAT Solver

David A. Rosenblueth[1,3], Stalin Muñoz[1],
Miguel Carrillo[1], and Eugenio Azpeitia[2,3]

[1] Instituto de Investigaciones en Matemáticas Aplicadas y en Sistemas,
Universidad Nacional Autónoma de México
Apdo. 20-126, 01000 México, D.F., México
[2] Laboratorio de Genética Molecular, Desarrollo y Evolución de Plantas
Instituto de Ecología, Universidad Nacional Autónoma de México
3er Circuito Universitario Exterior, Junto al Jardín Botánico
Coyoacán, 04510 México D.F., México
[3] Centro de Ciencias de la Complejidad, piso 6, ala norte, Torre de Ingeniería,
Universidad Nacional Autónoma de México
Coyoacán, 04510 México D.F., México
drosenbl@unam.mx, stalinmunoz@fi-b.unam.mx,
{miguel.mcb,emazpeitia}@gmail.com

Abstract. Boolean networks are important models of gene regulatory
networks. Such models are sometimes built from: (1) a gene interaction
graph and (2) a set of biological constraints. A *gene interaction graph*
is a directed graph representing positive and negative gene regulations.
Depending on the biological problem being solved, the *set of biological
constraints* can vary, and may include, for example, a desired set of stationary states. We present a *symbolic*, SAT-based, method for inferring
synchronous Boolean networks from interaction graphs augmented with
constraints. Our method first constructs Boolean formulas in such a way
that each truth assignment satisfying these formulas corresponds to a
Boolean network modeling the given information. Next, we employ a
SAT solver to obtain desired Boolean networks. Through a prototype,
we show results illustrating the use of our method in the analysis of
Boolean gene regulatory networks of the *Arabidopsis thaliana* root stem
cell niche.

Keywords: Boolean network, Gene interaction graph, SAT solver.

1 Introduction

Boolean networks, as simple as they are, have proven to encode meaningful
biological information. Moreover, Boolean networks have emerged as valuable
models of several biological phenomena. With the advent of high-throughput
technologies, the inference of Boolean networks from experimental data has become an increasingly relevant problem. Two main approaches have appeared in

A.-H. Dediu, C. Martín-Vide, and B. Truthe (Eds.): AlCoB 2014, LNBI 8542, pp. 235–246, 2014.

the literature of Boolean-network inference, depending on the input to the algorithm. There are methods, on the one hand, inferring a network from time-series data (or the binarized input-output pairs) [1,5,12,14], and there are methods, on the other hand, employing a gene interaction graph augmented with biological constraints [4,18]. Our objective is to present a *symbolic* approach in the second category, having as input, in addition to an interaction graph, a set of optional constraints, such as a desired set of stationary states (also called fixed points) or other biological restrictions.

A Boolean network can be viewed as a set of Boolean variables representing genes (with values "active" and "inactive", say), together with a "next-state" Boolean function for each gene. In synchronous networks, all such functions are simultaneously applied. We will omit the term "synchronous", as we only treat synchronous Boolean networks here. Usually, the next-state function of a gene depends only on a subset of the genes, called the set of *regulators* of that gene. Intuitively, an interaction graph describes the structure of the network. Such a graph has a node for each gene, and arcs showing: (1) the regulators of each gene, and (2) whether a regulation is "positive" or "negative". An example of an interaction graph appears in Fig. 1 (left) [1]. A gene j positively regulates a gene i if *there exist* values of all genes other than j such that the next value of i is the *same* as the current value of j (for both values of j). Conversely, a gene j negatively regulates a gene i if *there exist* values of all genes other than j such that the next value of i is the *complement* of the current value of j (for both values of j) [19].

Because the positive and negative regulations of a gene only determine some values of the next-state function of such a gene, there may be many Boolean functions satisfying the regulations of that gene. By the same token, there may be many Boolean networks satisfying a given interaction graph. However, from a biology point of view, not all such networks may be meaningful. In practice, therefore, a number of biological constraints are usually imposed.

A minimum requirement for a given Boolean network to be useful is that each stationary state of the network corresponds to a different biological behavior, such as the stable gene configurations observed in the cell types of an organism, as hypothesized by Kauffman [10].

In addition, it may happen that not all experiments necessary to infer an adequate Boolean network have been reported in the literature. Hence, to have a Boolean network with a desired set of stationary states, it may be necessary to add hypothetical regulations [2]. Similarly, we may be willing to admit additional stationary states than those reported in the literature [4].

Several other biological constraints may be considered. For example, we may wish to eliminate Boolean networks in which a gene both positively and negatively regulates the same gene. The reason may be that such a double regulation has rarely been observed in nature. This and other biological constraints are employed, for instance, in [4].

The problem of inferring Boolean networks from interaction graphs (in addition to biological constraints) is important, as a number of Boolean networks have been so inferred in practice [2,7,9,11,13,15]. Such networks, however, have mainly been built manually. Hence, the development of an efficient algorithm for such an inference could have a significant impact.

Two algorithms for inferring Boolean networks from interaction graphs are [4] and [18]. Such algorithms, nevertheless, explicitly represent each individual network. As the search space is vast, these methods traverse a fragment of the space with an algorithm employing a random component. Such methods are only adequate for small networks.

This limitation suggests considering the development of methods based on approaches other than the explicit representation of the Boolean-network search space, such as symbolic techniques, as used in model checking. Moreover, symbolic algorithms have already been utilized in Boolean-network analysis. For instance, [17] uses Binary Decision Diagrams (BDDs) for computing the set of stationary states, while [8] and [20] employ SAT-based algorithms for finding attractors (of any size) and stationary states, respectively.

Central to our method is a symbolic representation of a set of Boolean networks through a set of propositional constraints on "entry" variables. Each assignment of truth values to the entry variables that satisfies the given constraints determines the truth table of a Boolean network. Now, the interaction graph indicates the set of *regulator* genes for each gene (in addition to the sign of the regulation). Normally, only a proper subset of all genes regulate each gene. Hence, when the regulators are incorporated into the entry variables, a reduction in the number of such variables typically occurs. Next, we generate formulas representing the sign of each regulation. The biological constraints, such as the desired set of stationary states, can readily be formalized. The same is true of the fact (which we may or may not wish to enforce) that a gene should not both positively and negatively regulate the same gene. We can also incorporate stationary states of *mutated* versions of the networks of interest. The conjunction of the formulas representing all such constraints is then given to an incremental SAT solver. Such a solver would have to compute not only one solution, but any number of desired solutions.

Many of the queries we present are scalable, as their corresponding formulas are small. The same does not hold for all queries, however. For example, if, given a set of stationary states, we do not wish networks with additional stationary states than those given, we must explicitly prohibit such additional states in the formula. This makes the size of the resulting formula proportional to the number of network states, that is, *exponential* in the number of genes.

This paper is organized as follows. Section 2 gives our method, Section 3 reports some experiments, and Section 4 gives concluding remarks.

2   Boolean Networks via Propositional Logic

In this section, after some introductory definitions, we establish a propositional representation of Boolean networks. Then, we use this representation to develop propositional formulas expressing some properties of Boolean networks.

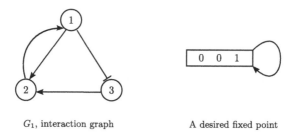

G_1, interaction graph   A desired fixed point

Fig. 1. A possible input to our algorithm (taken from [1]): G_1, an interaction graph (left) and the desired fixed point 001 (right). Of the seven Boolean networks satisfying the interaction graph, only two have 001 as fixed point. Of these, only one has no gene both positively and negatively regulating the same gene.

We use the following notation. $\mathbb{N}^+$ is *the set of positive natural numbers*. Unless differently stated, *we assume that* $n \in \mathbb{N}^+$. $\mathbb{N}_n^+$ is an *initial segment of* $\mathbb{N}^+$, $\mathbb{N}_n^+ = \{x \in \mathbb{N}^+ \mid x \le n\}$. $\mathbb{B} = \{0,1\}$ is a set of *Boolean values*. If $b \in \mathbb{B}$, then b' is the complement of b. If $x \in \mathbb{B}^n$, we say that x is a *state* and x_i denotes the i-th *component* of x. We identify the state x with the string $x_1 \cdots x_n$ denoted by $\ddot{x}$. If $b \in \mathbb{B}$ and $J \subseteq \mathbb{N}_n^+$, $x[b/J]$ is the state resulting from replacing, for all $i \in J$, the i-th component of x by b. We write $x[b/j]$ instead of $x[b/\{j\}]$.

2.1 Boolean Networks and Interaction Graphs

Following Richard et al. [19], we define a *Boolean network with n components* as a function $f: \mathbb{B}^n \to \mathbb{B}^n$. The *$i$-th component of f* is a function $f_i: \mathbb{B}^n \to \mathbb{B}$ such that $f_i(x) = f(x)_i$. We say that x is a *fixed point* (or *stationary state*) of f if $f(x) = x$. The *set of fixed points* of f is defined as $\mathrm{FP}(f) = \{x \in \mathbb{B}^n \mid f(x) = x\}$. Unless differently stated, *we assume that* $f: \mathbb{B}^n \to \mathbb{B}^n$.

Definition 1. *If $j \in \mathbb{N}_n^+$, and $b \in \mathbb{B}$, the mutation of f by $j = b$ is defined as $f^{j=b}: \mathbb{B}^n \to \mathbb{B}^n$, where, for all $x \in \mathbb{B}^n$, $f_j^{j=b}(x) = b$, and $f_i^{j=b}(x) = f_i(x)$ if $i \ne j$.*

To infer Boolean networks, we use interaction graphs defined as follows.

Definition 2. *We say that G is an* interaction graph *with n nodes if $G = \langle \mathbb{N}_n^+, I^+, I^- \rangle$, where: $I^+ \subseteq \mathbb{N}_n^+ \times \mathbb{N}_n^+$ is a set of* positive interactions, *and $I^- \subseteq \mathbb{N}_n^+ \times \mathbb{N}_n^+$ is a set of* negative interactions. *If $x \in I^+ \cap I^-$, we say that x is an* ambiguous interaction.

We use arcs with ordinary arrowheads $\to$ (resp. $\dashv$) to graphically represent positive (resp. negative) interactions. Figure 1 shows the depiction of an interaction graph $G_1 = \langle \mathbb{N}_n^+, I^+, I^- \rangle$. In this example, $n = 3$, $\mathbb{N}_n^+ = \{1, 2, 3\}$, $I^+ = \{(1,2), (2,1), (3,2)\}$, and $I^- = \{(1,3)\}$.

Definition 3 ([19]). *The* interaction graph *of f is $G_f = \langle \mathbb{N}_n^+, I^+, I^- \rangle$ where:*

1. $(j, i) \in I^+$ iff $\exists x \in \mathbb{B}^n$ such that $f_i(x[1/j]) - f_i(x[0/j]) > 0$
2. $(j, i) \in I^-$ iff $\exists x \in \mathbb{B}^n$ such that $f_i(x[1/j]) - f_i(x[0/j]) < 0$

Intuitively, the interaction graph of f, G_f, describes the structure of f. Note that G_f may have both a positive and a negative interaction from j to i.

Definition 4. Let $G = \langle \mathbb{N}_n^+, I^+, I^- \rangle$ be an interaction graph. The set of essential regulators of i in G is defined as $R_G(i) = \{j \in \mathbb{N}_n^+ \mid (j, i) \in I^+ \cup I^-\}$. The set of fictitious regulators of i in G is defined as $R_G^c(i) = \mathbb{N}_n^+ - R_G(i)$. The full set of regulators of i in G is defined as $K_n(i) = \mathbb{N}_n^+$, for all $i \in \mathbb{N}_n^+$.

For the example in Fig. 1, $R_G(1) = \{2\}$, $R_G(2) = \{1, 3\}$ and $R_G(3) = \{1\}$.

2.2 Representation of Boolean Networks

In this subsection, we first give a propositional representation of Boolean networks and then develop propositional formulas expressing constraints.

We use PL for the set of propositional-logic formulas and assume the standard semantics of PL. When necessary, we use PL(V) to emphasize that PL is built from the set of variables V.

Definition 5. A set of variables is any nonempty set. We use a particular set of variables, called entry variables, defined as $\text{Varf}_n = \{f_i \ddot{x} \mid i \in \mathbb{N}_n^+ \text{ and } x \in \mathbb{B}^n\}$.

The variable $f_i \ddot{x}$ represents the value of the f_i at the state x. Hence, Varf_n is a set of variables representing the values of the Boolean network f. For example, given the interaction graph in Fig. 1, the corresponding set of entry variables placed at their represented values can be seen in Fig. 2 (a).

x	$f(x)$
0 0 0	$f_1 000\ f_2 000\ f_3 000$
0 0 1	$f_1 001\ f_2 001\ f_3 001$
0 1 0	$f_1 010\ f_2 010\ f_3 010$
0 1 1	$f_1 011\ f_2 011\ f_3 011$
1 0 0	$f_1 100\ f_2 100\ f_3 100$
1 0 1	$f_1 101\ f_2 101\ f_3 101$
1 1 0	$f_1 110\ f_2 110\ f_3 110$
1 1 1	$f_1 111\ f_2 111\ f_3 111$

x	$f(x)$
0 0 0	0 0 1
0 0 1	0 0 1
0 1 0	1 0 1
0 1 1	1 0 1
1 0 0	0 0 0
1 0 1	0 1 0
1 1 0	1 0 0
1 1 1	1 1 0

x	$f(x)$
0 0 0	$f_1 000\ f_2 000\ f_3 000$
0 0 1	$f_1 000\ f_2 001\ f_3 000$
0 1 0	$f_1 010\ f_2 000\ f_3 000$
0 1 1	$f_1 010\ f_2 001\ f_3 000$
1 0 0	$f_1 000\ f_2 100\ f_3 100$
1 0 1	$f_1 000\ f_2 101\ f_3 100$
1 1 0	$f_1 010\ f_2 100\ f_3 100$
1 1 1	$f_1 010\ f_2 101\ f_3 100$

| (a) | (b) | (c) |

Fig. 2. (a): Entry variables Varf_n of a three-gene example. (b): One of seven Boolean networks satisfying the interaction graph in Fig. 1. (c): Representative variables as determined by the interaction graph in Fig. 1.

Observe in Fig. 1 that gene 1 is only regulated by gene 2. Hence, there is no need to use three different variables for the next value of gene 1: All Boolean networks satisfying the interaction graph will have the same value for both $f_1 000$ and

$f_1 001$, for example. Moreover, by using one variable instead of two in each of the following sets: $\{f_1 000, f_1 001\}$, $\{f_1 010, f_1 011\}$, $\{f_1 100, f_1 101\}$, and $\{f_1 110, f_1 111\}$, we guarantee that the regulation of gene 1 by itself will not be added by the SAT solver.

Since gene 1 only depends on one gene, we only need two variables for f_1. We will arbitrarily select the lexicographically smaller $f_i \ddot{x}$, which we will call the "representative" of $f_i \ddot{x}$. Figure 2 (c) shows the representative variables of Fig. 1.

Definition 6. *Let G an interaction graph and $f_i \ddot{x} \in \mathrm{Varf}_n$ an entry variable. We define the representative variable of $f_i \ddot{x}$ in G as $\mathrm{rep}(G, f_i \ddot{x}) = f_i \ddot{x}[0/R_G^c(i)]$.*

Definition 7. *If v is a propositional variable and $b \in \mathbb{B}$, the literal of (v, b) is defined as $\mathrm{lit}(v, b) = \neg v$ if $b = 0$ and $\mathrm{lit}(v, b) = v$ if $b = 1$.*

Definition 8. *Given a set of variables V, a truth assignment for V is a function $\sigma: V \to \mathbb{B}$. As usual, we identify a truth assignment σ with the set $\{x \in V \mid \sigma(x) = 1\}$. The truth assignment of f, $\sigma_f: \mathrm{Varf}_n \to \mathbb{B}$, is defined as $\sigma_f(f_i \ddot{x}) = f_i(x)$ for all $x \in \mathbb{B}^n$.*

Observe that $f_i \ddot{x}$ is true if $f_i(x) = 1$, and $f_i \ddot{x}$ is false if $f_i(x) = 0$. In the Boolean network in Fig. 2 (b), $f_2 000$ is false and $f_3 000$ is true.

Definition 9. *If $J \subseteq \mathbb{N}_n^+$, the restriction of $\mathbb{B}^n$ to components in J is defined as $\mathbb{B}^n|_J = \{x \in \mathbb{B}^n \mid \forall i \in \mathbb{N}_n^+ - J, \ x_i = 0\}$.*

Hence, $\mathbb{B}^n|_J$ is the set of states having 0 in the positions that are not in J.

We are now in a position to define the formulas expressing desired properties to constrain the inference of Boolean networks.

Definition 10. *Let $R: \mathbb{N}_n^+ \to \mathcal{P}(\mathbb{N}_n^+)$. For each $(j, i) \in \mathbb{N}_n^+ \times \mathbb{N}_n^+$, we define the positive, and negative, interaction formulas of (j, i) w.r.t. R as:*

$$\varphi_R^+(j, i) = \bigvee_{x \in \mathbb{B}^n|_{R(i)}} (f_i \ddot{x}[1/j] \wedge \neg f_i \ddot{x}[0/j])$$

$$\varphi_R^-(j, i) = \bigvee_{x \in \mathbb{B}^n|_{R(i)}} (\neg f_i \ddot{x}[1/j] \wedge f_i \ddot{x}[0/j]).$$

Note that $\varphi_{K_n}^+(j, i)$ is true iff $\exists x \in \mathbb{B}^n$ such that $f_i(x[1/j]) - f_i(x[0/j]) > 0$. A similar remark applies to $\varphi_{K_n}^-(j, i)$. Observe that this definition is analogous to Definition 3 and to the definition of interaction graph in [19].

If R is the regulation of G_1 in Fig. 1, for instance, these formulas will be: $\varphi_{R_G}^+(1, 2) = (f_2 100 \wedge \neg f_2 000) \vee (f_2 101 \wedge \neg f_2 001)$, $\varphi_{R_G}^+(2, 1) = f_1 010 \wedge \neg f_1 000$, $\varphi_{R_G}^+(3, 2) = (f_2 001 \wedge \neg f_2 000) \vee (f_2 101 \wedge \neg f_2 100)$, and $\varphi_{R_G}^-(1, 3) = \neg f_3 100 \wedge f_3 000$.

The following theorem has implications for efficiency in the construction of Boolean networks starting from an interaction graph G. This theorem states that, if f_i is independent of $R_G^c(i)$ for all $i \in \mathbb{N}_n^+$, in accordance with the meaning established by σ_f, the interaction formulas w.r.t. K_n are equivalent to the interaction formulas w.r.t. R_G.

So, to build a Boolean network containing a given interaction graph, we only have to satisfy a disjunction over $\mathbb{B}^n|_{R_G(i)}$ rather than satisfying a disjunction over $\mathbb{B}^n$. Since in practice $\mathbb{B}^n|_{R_G(i)}$ is often much smaller than $\mathbb{B}^n$, the reduction of such a disjunction to $\mathbb{B}^n|_{R_G(i)}$ is significant in terms of efficiency.

Theorem 11. *If* $R: \mathbb{N}_n^+ \to \mathcal{P}(\mathbb{N}_n^+)$ *and* $f: \mathbb{B}^n \to \mathbb{B}^n$ *are such that,* $\forall i \in \mathbb{N}_n^+$, $\forall j \in R^c(i)$, $\forall x \in \mathbb{B}^n$, $f_i(x) = f_i(x[0/j])$ *(i.e.,* $\forall i \in \mathbb{N}_n^+$ f_i *is independent of* $R^c(i)$*), then,* $\forall i, j \in \mathbb{N}_n^+$:

1. $\sigma_f \models \varphi_{K_n}^+(j, i)$ *iff* $\sigma_f \models \varphi_R^+(j, i)$, *and*
2. $\sigma_f \models \varphi_{K_n}^-(j, i)$ *iff* $\sigma_f \models \varphi_R^-(j, i)$.

Definition 12. *Let* G *be an interaction graph, and* $x, y \in \mathbb{B}^n$. *We define the* input-output pair formula *for the input state* x *and the output state* y *as:*

$$\varphi_{\mathrm{IO}}(G, x, y) = \bigwedge_{i \in \mathbb{N}_n^+} \mathrm{lit}(\mathrm{rep}(G, f_i \ddot{x}), y_i).$$

We can use formulas φ_{IO}, for example, to express that x is a fixed point of f

Definition 13. *Let* G *be an interaction graph. If* $x \in \mathbb{B}^n$, *the* fixed-point for-mula *of* x *is defined as:*

$$\varphi_{\mathrm{FP}}(G, x) = \varphi_{\mathrm{IO}}(G, x, x).$$

For example, taking G_1 and the desired fixed point of Fig. 1, we have $\varphi_{\mathrm{FP}}(G_1, x) = \mathrm{lit}(f_1 000, 0) \wedge \mathrm{lit}(f_2 001, 0) \wedge \mathrm{lit}(f_3 000, 1) = \neg f_1 000 \wedge \neg f_2 001 \wedge f_3 000$.

As well as fixed points of the desired networks, fixed points of networks of *mutated organisms* often constitute additional constraints [2]. Instead of using an additional set of entry variables to treat the Boolean network of the mutated organism, we employ only one set of entry variables: those of the nonmutated network. A fixed-point of a mutated network is treated as follows. Let f be the wild-type network and x a fixed point in the mutated network $f^{j=b}$ and not in the wild-type network. Observe first that $\forall i \neq j$, $f_i^{j=b}(x) = x_i$ and $x_j = f_j^{j=b}(x) = b$. If it were the case that $f_j(x) = b$, then x would also be a fixed point of f. Hence, making $f_j(x) = b'$ prevents x from being a fixed point in f. Thus, the key idea is that, if $x_j = b$, the formula $\varphi_{\mathrm{IO}}(G, x, x[b'/j])$ captures the fact that x is not a fixed point of f but, since $f_j^{j=b}(x) = b$, x is a fixed point of $f^{j=b}$.

Definition 14. *Let* G *be an interaction graph,* $j \in \mathbb{N}_n^+$, $b \in \mathbb{B}$, *and* $B \subseteq \mathbb{B}^n$. *The* mutation formula *for a mutation by* $j = b$ *with fixed points* B *is defined as:*

$$\varphi_{f^{j=b}}(G, j, b, B) = \bigwedge_{x \in B} \varphi_{\mathrm{IO}}(G, x, x[b'/j]).$$

For example, using the mutation formula with G_1 of Fig. 1, $j = 1$, $b = 1$, and $B = \{(1, 0, 0)\}$, we get $\varphi_{f^{j=b}}(G_1, j, b, B) = \neg f_1 000 \wedge \neg f_2 100 \wedge \neg f_3 100$. That is, $f(1, 0, 0) = (0, 0, 0)$ and hence $f^{1=1}(1, 0, 0) = (1, 0, 0)$.

We will now assume that we are given the following parameters: an interaction graph G with hypothetical interactions H^+ and H^-, a set A of desired fixed points, and a set M of mutations with their corresponding fixed points. Depending on the query of interest, we can use such parameters and combine the formulas below into a formula ψ to obtain Boolean networks f such that $\sigma_f \models \psi$.

Definition 15. *Let $G = \langle \mathbb{N}_n^+, I^+, I^- \rangle$ be an interaction graph, $H^+ \subseteq I^+$, $H^- \subseteq I^-$, $A \subseteq \mathbb{B}^n$, and $M \subseteq \{(j,b,B) \mid j \in \mathbb{N}_n^+, b \in \mathbb{B}, B \subseteq \mathbb{B}^n\}$. We define the following set of Boolean-network formulas.*

1. Known interactions:

$$\varphi_{interactions}(G) = \left[\bigwedge_{(j,i)\in(I^+ - H^+)} \varphi_{R_G}^+(j,i) \right] \wedge \left[\bigwedge_{(j,i)\in(I^- - H^-)} \varphi_{R_G}^-(j,i) \right]$$

2. Nonambiguity:

$$\varphi_{nonambiguity}(G) = \left[\bigwedge_{(j,i)\in I^+} \neg\varphi_{R_G}^-(j,i) \right] \wedge \left[\bigwedge_{(j,i)\in I^-} \neg\varphi_{R_G}^+(j,i) \right]$$

3. Fixed points:
   (a) A is exactly the set of fixed points of f:

$$\varphi_{A=\mathrm{FP}(f)}(G,A) = \left[\bigwedge_{x\in(\mathbb{B}^n - A)} \neg\varphi_{\mathrm{FP}}(G,x) \right] \wedge \left[\bigwedge_{x\in A} \varphi_{\mathrm{FP}}(G,x) \right]$$

   (b) A is a subset of the set of fixed points of f:

$$\varphi_{A\subseteq\mathrm{FP}(f)}(G,A) = \bigwedge_{x\in A} \varphi_{\mathrm{FP}}(G,x)$$

4. Mutations (For all $(j,b,B) \in M$, the states in B are fixed points in $f^{j=b}$):

$$\varphi_{mutation}(G,M,A) = \bigwedge_{(j,b,B)\in M} \varphi_{f^{j=b}}(G,j,b,B-A)$$

3 Experiments

In this section, we report on the effectiveness of our method in a practical biology case, using a prototype called *Griffin* (for "Gene regulatory interaction formulator for inquiring networks"). After encoding the problem, *Griffin* converts the formula to an equisatisfiable conjunctive normal form (CNF) and feeds it to the SAT engine Sat4j [6]. If the SAT engine returns a satisfying assignment, such an assignment is decoded back to a Boolean network. Sat4j operates in an incremental manner, adding a blocking clause for every found satisfying assignment, and iteratively attempting to find as many satisfying networks as time permits. *Griffin* is written in the Java programming language version 7. Times below are those of *Griffin* running on a laptop computer (Intel Core i3 CPU at 2.13 GHz, 6GB RAM), averaged over 100 repetitions.

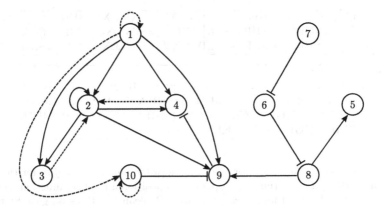

Fig. 3. Interaction graph for *A. thaliana* root stem cell niche coupled GRN model, based on [3] (Fig. 1.b). Node indices map to gene names as follows: 1-SHR, 2-SCR, 3-JKD, 4-MGP, 5-PLT, 6-IAA, 7-AUX, 8-ARF, 9-WOX5, and 10-CLEX. Continuous arcs denote known regulations; dashed arcs denote hypothetical regulations.

3.1 The *Arabidopsis thaliana* Root Stem Cell Niche GRN Model

We analyzed the *Arabidopsis thaliana* root stem cell niche gene regulatory network (GRN) model proposed by Azpeitia et al. [3], which has 10 genes. There are minor differences between the interaction graph in [3, Fig 1.b] and the one in Fig 3. These differences, however, are justifiable for the following reasons. SHR expression was experimentally reported only in a subset of the cell types of the stem cell niche [16] and CLEX was introduced in [3] as a missing component. Nevertheless, by 2010, CLEX had not yet been experimentally described or discovered. Thus, the self-regulations of these genes were hypothetical interactions and were included as such in the present paper. The regulations of MGP and JKD over SCR were grounded on weak experimental information [21] and their existence was discussed in [3]. Finally, in [3] we also added as a hypothetical interaction the SHR regulation over CLEX. CLEX could represent not only a single gene, but a complete signaling pathway, so this regulation was also hypothetical.

In the first experiment, *Griffin* constructed a formula ψ taking as inputs: (i) the interaction graph shown in Fig. 3, and (ii) the set of *five* known fixed points for the network. We chose $\varphi_{A=\text{FP}(f)}$ as the formula for encoding the set of desired fixed points, i.e., no additional fixed points were allowed. Compared to the previous manual analysis [3], where only *two* Boolean networks were found, *Griffin* obtained a total of 74 Boolean networks. *Griffin* took 6.75 ms to build the formula and 84.10 ms to find the networks.

In a second experiment, instead of using $\varphi_{A=\text{FP}(f)}$, the formula $\varphi_{A\subseteq\text{FP}(f)}$ was selected, leaving all other inputs as in the first experiment. A total of 236 Boolean networks were found. In this case, *Griffin* took 0.85 ms to build the formula and 169.96 ms to find the networks.

In a final experiment, we kept the two inputs previously described, and added a third input to *Griffin* consisting of 20 single mutations with fixed points, cor-

ARF$'$ = ¬IAA	Auxin$'$ = true	CLEX$'$ = CLEX ∧ SHR
IAA$'$ = ¬ Auxin	JKD$'$ = SHR ∧ SCR	MGP$'$ = ¬ WOX5 ∧ SHR ∧ SCR
PLT$'$ = ARF	SHR$'$ = SHR	WOX5$'$ = ¬ CLEX ∧ ARF ∧ SHR ∧ SCR

SCR$'$ = (SHR ∧ SCR)

SCR$'$ = (SHR ∧ SCR ∧¬ JKD) ∨ (¬ SHR ∧ SCR ∧ JKD) ∨ (SHR ∧ SCR ∧ JKD)

SCR$'$ = (SHR ∧¬ MGP ∧ SCR ∧¬ JKD) ∨ (SHR ∧ MGP ∧ SCR ∧¬ JKD)∨
(¬ SHR ∧¬ MGP ∧ SCR ∧ JKD) ∨ (SHR ∧¬ MGP ∧ SCR ∧ JKD)∨
(SHR ∧ MGP ∧ SCR ∧ JKD)

Fig. 4. Set of Boolean functions for the three networks found by *Griffin* for the mutations with fixed-points experiments. Top: nine propositional Boolean functions common to all these networks. Bottom: three alternative Boolean functions on gene SCR. The network corresponding to the first alternative for SCR coincides with that found by Azpeitia et al. [3] for the *A. thaliana* root stem cell niche model. We use names of genes for components of the states, e.g., IAA instead of x_6. We use gene names with a prime for the respective components of the Boolean network, e.g., ARF$'$ instead of $f_8(x)$.

responding to 10 *loss-of-function* mutations ($b = 0$) and 10 *gain-of-function* mutations ($b = 1$). We also asked for networks with exactly the same set of fixed points as those of the wild-type network. In this case, only three satisfying assignments were found. The corresponding Boolean networks are shown in Fig. 4. All three are different instantiations of the hypothetical regulations. *Griffin* took 16.23 ms to build the formula and 18.97 ms to find the networks. All three experiments were done with nonambiguous interactions.

4   Concluding Remarks

The problem of inferring a Boolean network from a gene interaction graph augmented with biological constraints is important. The reason is that a number of Boolean networks have been so inferred [2,7,9,11,13,15]. Such inferences have mainly been done manually, however. Thus, an efficient algorithm for this problem is likely to have an impact.

The search space of this problem is vast. Hence, nonsymbolic methods [4,18], explicitly representing each network, are only practical for small networks. Here we have developed a symbolic method, employing a SAT solver, and having the potential of inferring larger Boolean networks.

The key idea of our method is straightforward: There is a variable for each entry of the truth table of the desired networks. Now, the formulas directly derived from the definition of interaction [19] can be unacceptably large, as they encode an existential quantifier ranging over all genes. Hence, we exploit an important observation: such a quantifier can equivalently range over the regulators only. As a gene is typically regulated by only a few genes, this observation has three consequences: (i) the number of entry variables is reduced, (ii) the size of the interaction formulas is also reduced, and (iii) it is not necessary to include explicit constraints preventing the addition of fictitious regulators.

To the interaction formulas we add formulas representing biological constraints. Relevant constraints are networks having: (1) a certain known set of fixed points and (2) a certain known set of fixed points in *mutations* [2] of the network of interest. Formulas for the fixed points of the network of interest are readily expressed, unlike those for the fixed points in a mutated network. A direct approach for representing the fixed points in a mutated network would employ a new set of entry variables, thus creating a source of inefficiency. So as to avoid the representation of the mutated network, we use the following indirect approach. We exploit the fact that each fixed point occurring in a mutated network (and not in the network of interest) corresponds to an input-output pair which is not a fixed point.

We tested a prototype of our method in the analysis of the *A. thaliana* root stem cell niche Boolean GRN. Our prototype produced more precise results than those performed previously (first experiment), as well as results of analyses that had not been done before (second and third experiments).

As future work, we plan test our prototype on larger examples, incorporate canalization, and explore ways of reducing the size of the formula for the prevention of additional fixed points.

Acknowledgments. We should like to thank Álvaro Chaos, Pedro Góngora, Elizabeth Ortiz, and Nathan Weinstein for fruitful discussions. We also gratefully acknowledge the facilities provided by IIMAS, UNAM, as well as financial support from DGAPA grant PAPIIT IN113013.

References

1. Akutsu, T., Miyano, S., Kuhara, S.: Identification of genetic networks from a small number of gene expression patterns under the Boolean network model. In: Pacific Symposium on Biocomputing, vol. 4, pp. 17–28 (1999)
2. Alvarez-Buylla, E.R., Benítez, M., Corvera-Poiré, A., Candor, A.C., de Folter, S., de Buen, A.G., Garay-Arroyo, A., García-Ponce, B., Jaimes-Miranda, F., Pérez-Ruiz, R.V., Pineiro-Nelson, A., Sánchez-Corrales, Y.E.: Flower development. The Arabidopsis Book p. 8:e0999 (2010), doi:10.1199/tab.0999
3. Azpeitia, E., Benítez, M., Vega, I., Villarreal, C., Alvarez-Buylla, E.R.: Single-cell and coupled GRN models of cell patterning in the *Arabidopsis thaliana* root stem cell niche. BMC Syst. Biol. 4(134) (2010)
4. Azpeitia, E., Weinstein, N., Benítez, M., Mendoza, L., Alvarez-Buylla, E.R.: Finding missing interactions of the Arabidopsis thaliana root stem cell niche gene regulatory network. Frontiers in Plant Science 4(10) (2013), doi:10.3389/fpls.2013.00110
5. Berestovsky, N., Nakhleh, L.: An evaluation of methods for inferring Boolean networks from time-series data. PloS One 8(6), e66031 (2013)
6. Berre, D.L., Parrain, A.: The Sat4j library, release 2.2. Journal on Satisfiability, Boolean Modeling and Computation 7, 59–64 (2010)
7. Davidich, M.I., Bornholdt, S.: Boolean network model predicts cell cycle sequence of fission yeast. PLoS One 3(2), e1672 (2008), doi:10.1371/journal.pone.0001672
8. Dubrova, E., Teslenko, M.: A SAT-based algorithm for computing attractors in synchrounous Boolean networks. IEEE/ACM Transactions on Computational Biology and Bioinformatics 8(5), 1393–1399 (2011)

9. Fauré, A., Naldi, A., Chaouiya, C., Thieffry, D.: Dynamical analysis of a genetic Boolean model for the control of the mammalian cell cycle. Bioinformatics 22(14), e124–e131 (2006)
10. Kauffman, S.: Homeostasis and differentiation in random genetic control networks. Nature 224(5215), 177–178 (1969)
11. Klamt, S., Saez-Rodriguez, J., Lindquist, J.A., Simeoni, L., Gilles, E.D.: A methodology for the structural and functional analysis of signaling and regulatory networks. BMC Bioinformatics 7(56) (2006), doi:10.1186/1471-2105-7-56
12. Lähdesmäki, H., Shmulevich, I., Yli-Harja, O.: On learning gene regulatory networks under the Boolean network model. Machine Learning 52(1-2), 147–167 (2003)
13. Li, F., Long, T., Lu, Y., Ouyang, Q., Tang, C.: The yeast cell-cycle network is robustly designed. Proc. Natl. Acad. Sci. U.S.A. 101(14), 4781–4786 (2004)
14. Liang, S., Fuhrman, S., Somogyi, R.: REVEAL, a general reverse engineering algorithm for inference of genetic network architectures. In: Pacific Symposium on Biocomputing, vol. 3, pp. 18–29 (1998)
15. Mendoza, L., Xenarios, I.: A method for the generation of standardized qualitative dynamical systems of regulatory networks. Theoretical Biology and Medical Modelling 3(13) (2006)
16. Nakajima, K., Sena, G., Nawy, T., Benfey, P.: Intercellular movement of the putative transcription factor SHR in root patterning. Nature 413(6853), 307–3011 (2001)
17. Naldi, A., Thieffry, D., Chaouiya, C.: Decision diagrams for the representation and analysis of logical models of genetic networks. In: Calder, M., Gilmore, S. (eds.) CMSB 2007. LNCS (LNBI), vol. 4695, pp. 233–247. Springer, Heidelberg (2007)
18. Pal, R., Ivanov, I., Datta, A., Bittner, M.L., Dougherty, E.R.: Generating Boolean networks with a prescribed attractor structure. Bioinformatics 21(21), 4021–4025 (2005)
19. Richard, A., Rossignol, G., Comet, J.P., Bernot, G., Guespin-Michel, J., Merieau, A.: Boolean models of biosurfactants production in *Pseudomonas fluorescens*. PLoS One 7(1), e24651 (2012), http://dx.doi.org/10.1371/journal.pone.0024651
20. Tamura, T., Akutsu, T.: Detecting a singleton attractor in a Boolean network utilizing SAT algorithms. IEICE Transactions on Fundamentals of Electronics, Communications and Computer Sciences 92(2), 493–501 (2009)
21. Welch, D., Hassan, H., Blilou, I., Immink, R., Heidstra, R., Scheres, B.: Arabidopsis JACKDAW and MAGPIE zinc finger proteins delimit asymmetric cell division and stabilize tissue boundaries by restricting SHORT-ROOT action. Genes Dev. 21(17), 2196–2204 (2007)

RRCA: Ultra-Fast Multiple In-species Genome Alignments

Sebastian Wandelt and Ulf Leser

Knowledge Management in Bioinformatics
Humboldt-University of Berlin, Berlin, Germany
{wandelt,leser}@informatik.hu-berlin.de

Abstract. Multiple sequence alignment is an important method in Bioinformatics, for instance, to reconstruct phylogenetic trees or for identifying functional domains within genes. Finding an optimal MSA is computationally intractable, and therefore many alignment heuristics were proposed. However, computing MSA for sequences at chromosome/genome scale in a reasonable time with good alignment results remains an open challenge.

In this paper we propose RRCA, a very fast method to compute high-quality in-species MSAs at genome scale. RRCA uses referential compression to efficiently find long common subsequences in to-be-aligned sequences. A colinear sub collection of these subsequences is used for an initial alignment and the not yet covered subsequences are aligned following the same approach recursively. Our evaluation shows that RRCA achieves MSAs at similar quality as current state-of-the-art methods, while often being orders of magnitude faster for all our datasets. For instance, RRCA aligns eight human Chromosome 22 (around 50 MB each) within one minute on a consumer computer; a task that takes hours to days with competitors.

Keywords: Multiple sequence alignment, referential compression.

1 Introduction

A multiple sequence alignment (MSA) arranges a set of sequences in a rectangular array such that one obtains the greatest number of similar characters in every column of the alignment with the minimal number of columns. An optimal MSA is usually computed following Carillo-Lipman [6] or generalizations based on dynamic programming. Since computing an optimal MSA is NP-complete under the most common cost models [33], the development of scalable approximate alignment methods, necessary in a context where information is growing by the day, is an open challenge [19]. The main application for MSA is in bioinformatics, where it is used, for instance, to reconstruct a phylogenetic tree [34] or for function/gene prediction [15]. Recently, several research results on multi-genome read mapping were published [14,17,31], most of which are based on the alignment of many long sequences (genomes or chromosomes) [14,17]. Also recently,

A.-H. Dediu, C. Martín-Vide, and B. Truthe (Eds.): AlCoB 2014, LNBI 8542, pp. 247–261, 2014.

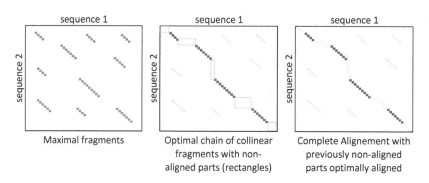

Fig. 1. Dot plot matrices for principal steps in chaining algorithms

it was shown that very high compression ratios are possible when compressing aligned genome data [9]. Therefore, we believe that scalable sequence alignment (in terms of length and number of sequences) will become even more important in the future.

Practical implementations make use of heuristics to guide the assembly of a MSA (see [7,28] for reviews and assessment of existing methods). Progressive alignment [22,27] builds a MSA by combining pairwise sequence alignments (PSA), usually by aligning most similar pairs of sequences first. Hence, progressive alignment methods need a guide tree as an input or compute it on-the-fly. In contrast to progressive alignment, iterative alignment methods [13,4] repeatedly re-align previously aligned sequences and thus often obtain a better score, because alignment errors obtained in the beginning of the alignment process can be repaired at later stages. A third class of heuristics, often used for PSA only, is based on chaining [36,29]. The idea underlying chaining algorithms is visualized in Figure 1: first a set of maximal identical fragments of both sequences is computed (Figure 1, left), then a colinear non-overlapping chain of fragments is identified (Figure 1, middle), and finally all the subsequences in between fragments are aligned following an (usually optimal) MSA algorithm (Figure 1, right); one example in the MSA field is Mugsy [3]. However, many chaining-based methods fail when it comes to long sequences, e.g. whole human chromosomes or MSA problems with more than a few sequences (see our evaluation). There exists further work on mixtures of the three approaches and also on slightly different problems, e.g. alignment-free sequence comparison [35] and alignment of short protein sequences [10].

Our Solution: We propose RRCA, Recursive Referential Compression Alignment. Our technique is based on the idea of chaining and uses the referential compression framework FRESCO [32] to identify identical fragments within the collection of sequences. A non-overlapping chain of fragments (Step 2 of chaining in Figure 1) is identified by a greedy strategy, which always selects the longest non-overlapping fragment next. For Step 3 of Figure 1, i.e. aligning subsequences between chain-aligned fragments, we apply the same approach recursively, therefore the name *Recursive Referential Compression Alignment*. The recursion stops if either 1) all to-be-aligned sequences are equal or 2) all sequences are shorter than a given threshold. In the latter case an optimal MSA algorithm is applied.

Although our method is very intuitive and simple, we show that RRCA finds MSA with comparable quality scores as more sophisticated competitors. At the same time RRCA is often orders of magnitude faster alignment for biological and non-biological datasets. For instance, RRCA can align eight human Chromosome 22 (around 50 MB each) within one minute on a consumer computer; a task that takes hours to days with competitors. Besides, since our technique is based on a compressed representation of all sequences against a reference, RRCA does not have to store the complete collection of raw sequences in main memory during any step of the algorithm. In fact, the more similar the sequences, the more compressed is our representation during the execution of RRCA.

Structure: The structure of the paper is as follows. In Section 2, we define the problem of multiple sequence alignments. We show in Section 3 how RRCA computes a multiple sequence alignment by using referential compression. In Section 4, we evaluate our recursive referential compression algorithm. The paper is concluded in Section 5.

2   Preliminaries

In the following, we present our recursive referential alignment algorithm RRCA. First, the multiple sequence alignment problem is defined.

Definition 1 (Multiple Sequence Alignment). *Given a collection $C = \{s_1, ..., s_n\}$ of sequences over an alphabet Σ, let $-$ be a symbol not in Σ. A* multiple sequence alignment (MSA) *of C is a collection of sequences $\{a_1, ..., a_n\}$, such that $|a_1| = ... = |a_n|$ and each a_i is obtained from s_i by inserting any number of occurrences of symbol $-$. The term* column i *of an alignment $\{a_1, ..., a_n\}$, refers to the symbols $\{a_1(i), ..., a_n(i)\}$. The* length *of an alignment is the number of columns, i.e. the length of any sequence in the MSA. The special case of $n = 2$ is called* pairwise sequence alignment (PSA).

Example 1. A collection $C_{EX} = \{s_1, s_2, s_3, s_4\}$ contains four highly-similar sequences[1]:

> s_1 : JANE*HAD*LOST*HER*JOB*AND*SHE*WAS*UNHAPPY.
>
> s_2 : WHEN*JANE*HAD*LOST*HER*JOB*SHE*FELT*REALLY*UNHAPPY.
>
> s_3 : JANE*HAD*LOST*HER*JOB*AND*SHE*REALLY*WAS*UNHAPPY.
>
> s_4 : WHEN*JANE*HAD*LOST*HER*JOB*SHE*BECAME*UNHAPPY.

One multiple sequence alignment of C_{EX} is $MSA_{EX} = \{a_1, a_2, a_3, a_4\}$, shown in Figure 2: Column 2 of our example alignment is $\{-, H, -, H\}$.

Usually, one is interested in alignments that maximizing a given scoring function. An often used scoring function is sum-of-pairs [16]. A *scoring function*

---

[1] We use * instead of white-spaces for presentation purposes.

```
a_1:  -----JANE*HAD*LOST*HER*JOB*AND*SHE--*-W-AS------*UNHAPPY.
a_2:  WHEN*JANE*HAD*LOST*HER*JOB*SHE*-FELT*RE-ALLY----*UNHAPPY.
a_3:  -----JANE*HAD*LOST*HER*JOB*AND*SHE--*RE-ALLY*WAS*UNHAPPY.
a_4:  WHEN*JANE*HAD*LOST*HER*JOB*S----HE--*BECAME-----*UNHAPPY.
```

Fig. 2. Optimal alignment MSA_{EX} of C_{EX}

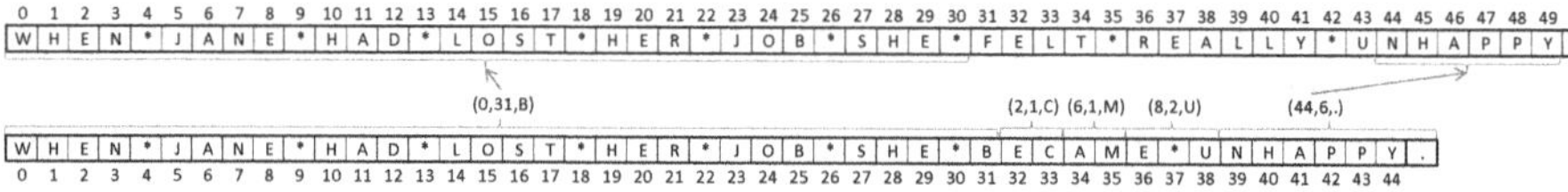

Fig. 3. Example for a referential compression of s_4 (below) against reference s_2 (up)

takes a pair of symbols from $\Sigma \cup \{-\}$ and returns a real number. Given a scoring function *score*, the *sum-of-pairs score for a column* $\{c_1, ..., c_n\}$ is defined as $\sum_{i<j\leq n} score(c_i, c_j)$. Given an MSA $\{a_1, ..., a_n\}$ of length m, the *sum-of-pairs score* is defined as the sum of the sum-of-pairs score for each column. A MSA is optimal for a collection of sequences $C = \{s_1, ..., s_n\}$, if there exists no other MSA for C with a higher score.

Example 2. Let *score* be defined as follows:

$$score(c_1, c_2) = \begin{cases} 1 & \text{if } c_1 \equiv c_2 \\ 0 & \text{if } c_1 \equiv - \wedge c_2 \equiv - \\ -1 & \text{if } else \end{cases}$$

The score of column $\{A, S, A, S\}$ is $1 * score(A, A) + 1 * score(S, S) + 2 * score(A, S) + 1 * score(S, A) = 1 + 1 + 2 * (-1) + 3 * (-1) = -1$.

3   Computing an Alignment with Referential Compression

We present a method for the computation of an initial alignment, usually Step 1 of chaining-based MSA-approaches, i.e. computation of colinear fragments. We use a technique recently emerged in compression of biological sequences: Referential compression. Similar to dictionary-based techniques [37,23], referential compression algorithms replace long subsequences of the to-be-compressed input with references to a distinct sequence, called reference. The reference is not part of the to-be-compressed input data. Furthermore, the reference is usually static, while dictionaries are being extended during compression phase. During the last years several referential compression algorithms emerged [21,20,11,32]. These algorithms work best if the to-be-compressed sequences are similar to the reference sequence. Impressive results are reported when compressing large collections of sequences: referential compression algorithms achieve compression rates of up to 1000:1 for human genomes, i.e. more than 3 TB of raw data for 1092 genomes

---

Algorithm 1. Referential Compression Algorithm

    Input: to-be-compressed sequence s and reference sequence ref
    Output: referential compression rc of s with respect to ref
1: Let rc be an empty list
2: **while** $|s| \neq 0$ **do**
3:     Let pre be the longest prefix of s occurring in ref, and let i be a position of an occurrence of pre in ref
4:     **if** $s \neq pre$ **then**
5:         Add $\langle i, |pre|, s(|pre|) \rangle$ to the end of rc
6:         Remove the first $|pre| + 1$ symbols from s
7:     **else**
8:         Add $\langle i, |pre| - 1, s(|pre| - 1) \rangle$ to the end of rc
9:         Remove the prefix pre from s
10:    **end if**
11: **end while**

---

is compressed down to few GB, at compressions speeds close to maximum read speeds for state-of-the-art hard disks. We proceed with a formal definition of the referential compression algorithm from [32].

Definition 2 (Referential Compression [32]). *A referential match entry (rme) is a triple* $\langle start, length, mismatch \rangle$, *where start is a number indicating the start of a match within a reference sequence, length denotes the match length[2], and mismatch denotes a symbol. The length of a referential match entry rme, denoted* $|rme|$, *is* $length + 1$. *Given sequences s and a reference ref, a referential compression of s with respect to ref, is a list of referential match entries,* $[\langle s_1, l_1, m_1 \rangle, ..., \langle s_n, l_n, m_n \rangle]$, *such that* $(ref(s_1, l_1) \circ m_1) \circ (ref(s_2, l_2) \circ m_2) \circ ... \circ (ref(s_n, l_n) \circ m_n) = s.$, *where $\circ$ denotes the concatenation of two strings.*

The *offset* of a referential match entry rme_i in a referential compression $rc = [rme_1, ..., rme_n]$, denoted $offset(rc, rme_i)$, is defined as $\sum_{j<i} |rme_j|$. Given a rme $\langle start, length, mismatch \rangle$, we write the expression $(start, length, mismatch) \in rc$, if and only if $\langle start, length, mismatch \rangle$ is an element in the referential compression rc.

An algorithm for computing a referential compression is shown in Algorithm 1. To create a referential compression of input sequence s with respect to ref, the algorithm matches prefixes of s with subsequences of ref using a compressed suffix tree on ref. The longest such prefix is removed from s, encoded as a rme and added to rc. The algorithm terminates once s contains no more symbols. Algorithm 1 is a greedy algorithm, i.e. it always takes the longest prefix of the to-be-compressed which can be found in the reference. Any greedy algorithm computes a minimal representation, i.e. the size of the compressed sequence is minimal, if the dictionary for the reference is fixed and the size of a dictionary entry is constant [8].

Example 3. One example referential compression for Sequence s_4 with respect to the reference sequence s_2 is shown in Figure 3. The input is compressed into

---

[2] Match length: Number of symbols for which to-be-compressed sequence and reference coincide.

five referential match entries. The first referential match entry is $\langle 0, 31, B \rangle$ and describes a match for the first 31 characters of sequence s_4 at position 0 of the reference. The mismatch character is B (in the reference an F is found instead of a B). The offset of $\langle 6, 1, M \rangle$ is $|\langle 0, 31, B \rangle| + |\langle 2, 1, C \rangle| = 34$.

3.1 Computing an Initial Alignment

In RRCA, an initial alignment is a chain of colinear fragments, where large overlapping parts of all to-be-aligned the sequences are used as fragments (see Figure 1). These fragments can be obtained, for instance, by computing the longest common subsequences. However, often q-gram-based methods are used: all q-grams of to-be-aligned sequences are computed and then the longest chain of colinear q-grams is extracted. The process of computing and chaining these q-grams is highly time-consuming for long sequences, because a sequence of length n contains $n - q + 1$ q-grams. Even increasing q slightly does not make the problem easier to solve for long sequences. Moreover, if q is chosen too large then similarities between to-be-aligned sequences might be missed by the alignment algorithm.

We use referential compression for the computation of an initial alignment instead. Given a collection of to-be-aligned sequences, we pick one sequence as a reference ref and compress all sequences referentially against ref. Given the referential compressions of all sequences, we extract overlapping parts from the referential match entries, as a base for a chain of colinear fragments. The main advantage of our approach, compared to q-gram-based algorithms, is that referential match entries can represent arbitrary long sequences, and therefore arbitrary long fragments. This allows us to identify fragments with different degrees of similarity using a homogeneous approach, independent from a fixed value q. In our implementation we have always chosen the longest sequence as a reference. Given k sequences of maximum length n, finding the longest sequence takes $O(k)$ and compression of all sequences against the reference takes $O(k * n)$, since the compression is computed in linear time in the length [32].

Another advantage of using referential compression is as follows: We do not need to keep all uncompressed sequence in main memory at any time. For computing a referential compression of a sequence s, we only need s plus the reference sequence and an index over the reference sequence in main memory. After compression of s we proceed with the compression of remaining sequences, and only keep the compressed representations of previously compressed sequences in main memory. This is an important step towards alignment of many very long sequences on consumer computers. In Example 4, we show the referential compression of s_1 to s_4 (from Example 1) with s_2 as a reference.

Example 4. The longest sequence in C_{EX} is s_2. We obtain the following referential compressions $RC = \{rc_1, ..., rc_4\}$ for each sequence against s_2 as a reference:

$$rc_1 = \{(5, 22, A), (7, 1, D), (26, 5, W), (6, 1, S), (42, 8, .)\}$$
$$rc_2 = \{(0, 50, .)\}$$
$$rc_3 = \{(5, 22, A), (7, 1, D), (26, 5, R), (37, 6, W), (6, 1, S), (42, 8, .)\}$$
$$rc_4 = \{(0, 31, B), (2, 1, C), (6, 1, M), (8, 2, U), (44, 6, .)\}$$

Definition 3 (Alignment Fragments). *Given a collection of sequences $C = \{s_1, ..., s_n\}$, we say that $f = ((astart_1, ..., astart_n), alength)$ is a fragment for C, if $s_1[astart_1, alength] = ... = s_n[astart_n, alength]$. Two fragments $f_1 = ((astart_{1,1}, ..., astart_{1,n}), alength_1)$ and $f_2 = ((astart_{2,1}, ..., astart_{2,n}), alength_2)$ are strictly consecutive, if $astart_{1,i} + alength_1 \leq astart_{2,i}$ for all $i \leq n$. An initial alignment for C is a collection of fragments $\{f_1, ..., f_m\}$ for C, such that each pair of fragments f_i and f_{i+1} is strictly consecutive.*

An initial alignment from Definition 3 splits a collection of sequences into different blocks, such that all sequences in C coincide for every second block (with unaligned blocks in between). Below, we describe how to compute an initial alignment by using referential compression. Intuitively, if two referential match entries overlap, i.e. point to the same subsequence of a reference, then the overlapping part is identical in referential match entries, and thus in their uncompressed sequences. We extract all intersections of referential match entries from all sequences in C. In Definition 4, we define an intersection operation on referential match entries in order to identify equal referenced subsequences.

Definition 4. *Given a collection R of referential match entries, $rme_1 = (s_1, l_1, m_1), ..., rme_n = (s_n, l_n, m_n)$, let $s = MAX(s_i)$ and $l = MIN(s_i + l_i) - MAX(s_i)$. The intersection of R, denoted $\bigcap_{i \leq n} rme_i$, is defined as the pair (s, l), if $l \geq 0$, and undefined otherwise.*

The result of intersecting $rme_1 = (42, 8, .)$ with $rme_2 = (44, 6, .)$ is the pair $(44, 6)$. The intersection between $rme_1 = (5, 22, A)$ with $rme_2 = (44, 6, .)$ is not defined, since they refer to different (non-overlapping) parts of the reference.

Definition 5 (Referential Agreement). *Given a set of referentially compressed sequences $RC = \{cs_1, ..., cs_n\}$, the referential agreement of RC is defined as $RefAgree(RC) = \{(s, l) \mid \exists rme_1 \in cs_1, ..., rme_n \in cs_n.(s, l) = \bigcap_{i \leq n} rme_i\}$.*

Informally, the referential agreement of RC defines all the areas of the reference which are referenced by at least one referential match entry of *each* referentially compressed sequence. An upper bound for the time-complexity for computation of the referential agreement is quadratic in the number of referential match entries, since all referential match entries have to be intersected. We reduce the time complexity for this step to $O(k^2 * n * \log n)$ as follows. We create an interval tree [24] for each compressed sequence in $O(k * n)$ (with intervals defined by start and length of each referential match entry), and then find for each referential match entry (there are $O(k * n)$ such entries) its overlapping counterparts by probing k interval trees in $O(k * \log n)$.

Given the set of agreements for RC, we compute partial alignments, which build the basis for an initial alignment. To compute the partial alignments, we need to trace back positions in the sequences which contributed to the referential agreement. The function $TRACE$ computes all such positions for an element of a referential agreement and a referentially compressed sequence.

Definition 6 (TRACE). *Given a referential compression cs and a pair (s, l), we define $TRACE((s, l), cs) = \{s - start_i \mid \exists length_i, mismatch_i.(start_i, length_i, mismatch_i) \in cs \land start_i \leq s \land s + l \leq start_i + length\}$.*

The traces from compressed sequences need to be combined carefully, since several traces of a single compressed sequence might cause an overlap in the same region of the reference. For instance, a compressed sequence such as $\{(5, 10, A), (7, 12, .)\}$ references the same reference subsequence $ref(7, 8)$ two times. For the computation of an alignment, we only want to use one of the two referential match entries, either $(5, 10, A)$ or $(7, 12, .)$, once an overlap with another referential match entry at the same subsequence is found. The function TRACEALL in Definition 7 applies the following heuristic: We pick trace positions from each compressed sequence, such that the difference to the average of all traces is minimized. Thus, local subsequence matches are preferred over more distant matches.

Definition 7 (TRACEALL). *Given a set of referentially compressed sequences $RC = \{cs_1, ..., cs_n\}$ and an element (s, l) of the referential agreement of RC, we let $U = \bigcup_{i \leq n} TRACE((s, l), cs_i)$. $TRACEALL((s, l), RC) = \{p_1, ..., p_n\}$, such that each p_i is the nearest value to $AVG(U)$ in $TRACE((s, l), cs_i)$. Note that there is at least one value in $TRACE((s, l), cs_i)$ for each i.*

Given Definition 7, we compute a set of partial alignments (of different quality). For computation of a complete initial alignment, we need to select a consistent (strictly consecutive) subset of these partial alignments. Our algorithm for computing an initial alignment by referential compression is shown in Algorithm 2. The consistent sub collection of *fragments* (Line 10) is computed by always choosing the longest fragment next, i.e. starting form an empty set of fragments, we repeatedly add the longest not yet used consistent fragment, until no more consistent fragment is left. There exists multiple other heuristics. Finding an optimal chain of colinear non-overlapping fragments is exponential in the number of sequences [2]. Continuing Example 4, we have that $RefAgree(RC) = \{(5, 22), (26, 5), (44, 6)\}$. The initial alignment returned by Algorithm 2, is $\{((0, 5, 0, 5), 22), ((35, 44, 42, 39), 6), ((25, 26, 25, 26), 5)\}$. This initial alignment is shown in Figure 4.

3.2   Completing an Alignment with Recursive Referential Compression

In the previous subsection, we showed how to computed an initial alignment, based on referential compression. Chaining-based approaches usually compute an

---

Algorithm 2. Initial Alignment Algorithm

Input: Collection of sequences $C = \{s_1, ..., s_n\}$
Output: Collection of alignment fragments
1: Let $fragments = \emptyset$
2: Select one $s \in C$ as ref
3: Compress all $s_i \in C$ against ref. The result is $RC = \{cs_1, ..., cs_n\}$
4: Compute $RefAgree(RC)$
5: **for all** $(s, l) \in RefAgree(RC)$ **do**
6:     Compute $TRACEALL((s, l), RC)$
7:     Add fragment $(TRACEALL((s, l), RC), l)$ to $fragments$
8: **end for**
9: Sort $fragments$ by second component (i.e. length of the fragment)
10: Let $consfragments$ be a consistent sub collection of $fragments$
11: Sort $consfragments$ by second component (i.e. start positions of the fragment)
12: return $consfragments$

Algorithm 3. RRCA Algorithm

Input: Collection of sequences $C = \{s_1, ..., s_n\}$
Output: MSA for C
1: **if** $MAX_{i \leq n}(|s_i|) < \delta$ **then**
2:     Return an optimal MSA of C
3: **else**
4:     Let $fragments$ be an initial alignment of C
5:     **if** $fragments$ is empty **then**
6:         Return an optimal MSA of C
7:     **else**
8:         **for all** (non-empty) sequences S not aligned in $fragments$ **do**
9:             Let $res_S = RRCA(S)$
10:        **end for**
11:        Return the (alternating) concatenation of all initially aligned fragments and recursively aligned sequence collections res_S
12:    **end if**
13: **end if**

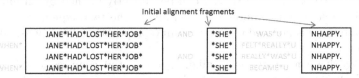

Fig. 4. Initial alignment for C_{EX}

optimal alignment for subsequences not contained in any fragment, for instance the sequences 'WHEN*' and 'WHEN*' in Figure 4. Three reasons can cause these subsequences not to be part of an initial alignment in RRCA:

1. There is a larger insertion/deletion and *not all* sequences contribute a referential match entry.
2. Our greedy referential compression algorithm chose different ways to encode the same subsequences (equality of subsequences cannot be decided by referential agreement).
3. The subsequences are really just not similar.

Therefore, we propose a new strategy as follows: instead computing an optimal MSA of unaligned sequences directly, we repeat the computation of an

			SeqAn		Mugsy		T-Coffee		Mafft		RRCA	
	Dataset	Length	Score	Time (s)	Score	Time (s)	Score	Time (s)	Score	Time (s)	Score	Time (s)
4 sequences	AT-1	500	2,958	0.1	2,958	0.5	2,958	0.1	2,958	0.1	2,958	0.0
	AT-1	16,000	95,712	174.5	95,712	0.8	95,712	109.8	95,712	0.4	95,712	0.0
	AT-1	2,048,000	*	*	12,216,048	23.6	*	*	12,213,022	431.5	12,216,012	0.6
	AT-1	30,000,000	*	*	*	*	*	*	*	*	178,138,742	24.1
	H-22	500	3,000	0.0	3,000	0.5	3,000	0.1	3,000	0.1	3,000	0.0
	H-22	16,000	95,884	86.6	95,884	0.6	95,884	72.5	95,884	0.3	95,884	0.0
	H-22	1,024,000	*	*	6,138,592	28.6	*	*	5,750,316	203.5	6,138,558	0.2
	H-22	30,000,000	*	*	*	*	*	*	*	*	179,407,238	12.8
	Y-wg	500	-1,040	0.2	-7,362	0.5	-2,330	0.1	-3,848	0.1	-1,370	0.0
	Y-wg	16,000	-7,948	290.8	-117,826	0.9	-100,066	148.2	-104,334	24.5	-119,590	0.0
	Y-wg	512,000	*	*	2,643,044	5.8	*	*	2,666,154	298.7	2,623,888	0.3
	Y-wg	8,192,000	*	*	43,999,160	169.1	*	*	*	*	43,086,892	6.6
8 sequences	AT-1	500	13,782	0.3	13,782	1.0	13,782	0.2	13,758	0.1	13,782	0.0
	AT-1	16,000	*	*	446,636	1.8	446,636	509.7	446,612	0.7	446,636	0.0
	AT-1	2,048,000	*	*	57,035,820	94.8	*	*	*	*	57,028,046	1.4
	AT-1	30,000,000	*	*	*	*	*	*	*	*	831,372,614	70.6
	H-22	500	14,000	0.0	14,000	1.0	14,000	0.2	14,000	0.1	14,000	0.0
	H-22	16,000	447,508	592.7	447,508	1.8	447,508	497.6	447,508	0.7	447,508	0.0
	H-22	2,048,000	*	*	56,965,626	265.0	*	*	*	*	56,912,358	0.6
	H-22	30,000,000	*	*	*	*	*	*	*	*	837,014,656	17.3
	Y-wg	500	-12,756	1.0	-63,924	1.0	-18,070	0.9	-18,048	0.2	-12,756	0.0
	Y-wg	16,000	*	*	-639,344	2.2	-603,404	737.5	-575,412	59.3	-416,700	0.1
	Y-wg	2,048,000	*	*	46,053,822	175.0	*	*	*	*	43,192,182	4.4
	Y-wg	8,192,000	*	*	*	*	*	*	*	*	197059310	15.5

Fig. 5. Comparison of MSA-methods for biological datasets (time in seconds). The score is computed as sum-of-pairs of the computed MSA with (match=1, mismatch=-1, gap=-1); larger scores are better. Computations that did not finish on time are marked with *.

initial alignment for unaligned (non-empty) subsequences. For instance, the two sequences 'WHEN*' and 'WHEN*' can be perfectly aligned by an initial alignment using referential compression with one of the two sequences as reference. In this case there is no need to compute an (computationally expensive) optimal alignment.

Our algorithm for recursively aligning referential compressions is shown in Algorithm 3. If the maximum length of a sequence in C is shorter than a fixed δ, then the algorithm computes an optimal MSA, following Needleman-Wunsch [26] (Line 1-2), and returns the result. Otherwise, an initial alignment following Algorithm 2 is computed (Line 4-12). If the initial alignment contains no fragment, i.e. referential compression cannot identify a common subsequence of all sequences in the input, then the algorithm computes and returns an optimal MSA as well (Line 6). If the initial alignment of the input contains at least one fragment, then the algorithm recursively computes a MSA for each set of subsequences not covered by fragments (Line 8-11).

Given the initial alignment from Figure 4, we have three blocks not covered by the initial alignment: one block containing two times 'WHEN*', one block containing two times 'AND' and one block before the fragment containing 'NHAPPY'. The first two blocks are aligned immediately by one recursive call each. The last block will be aligned by computing an optimal alignment, since no initial alignment can be found. The result of RRCA is optimal and shown in Figure 2.

4   Discussion

In the following section, we evaluate our proposed scheme. All experiments were run on a computer with 16 GB RAM and Intel Core i7-2670QM. We evaluate our method on five different datasets with different degrees of similarity. Three biological datasets: a collection of eight human Chromosome 22 (H-22) of the 1000 Genome project [1], a collection of eight Chromosome 1 from Arabidopsis thaliana (AT-1), taken from the 1001 Genomes project [5], release GMINord-borg2010[3], and a collection of eight yeast genomes [25] (Y-wg). We have chosen these species since their sequences have different degrees of inner-species similarity. In our experiments RRCA was set to always choose the longest sequence as a reference. If all sequences have the same length, as initially in our experiment, one sequence is chosen randomly.

We compare RRCA against one optimal MSA algorithm (part of SeqAn [12]) and three approximate solutions (Mugsy [3], T-Coffee [27], and Mafft [18]), in Figure 5. We ran tests on the biological datasets with different lengths. If a program took longer than 15 minutes to complete a test, it was stopped (indicated by a * in Figure 5). It can be seen that the optimal algorithm can only compute a MSA for rather short sequences within 15 minutes. The score obtained by all algorithms is quite similar, with the exception of the least self-similar dataset Y-wg. Overall, RRCA is the fastest MSA algorithm for each single test case, usually orders of magnitude faster than all three approximate competitors.

We performed experiments regarding the exact alignment time of random sequences with an extension of Needleman-Wunsch to MSA, as implemented in Seqan. We generated 500 collections of k random sequences with a fixed length. The result for the alignment of the sequences with $k = 4$ and $k = 8$ is shown in Figure 6. We have used the symbolic regression solver Eureqa [30] to estimate a formula for the alignment time in ms, given input length and the number of sequences k. The best solution with a size (number of terms) smaller than 10 is $time = 0.0000598 * length^2 * k^2$. This formula helps to estimate the alignment time, and thus, can be used to set the constant δ (the maximum length for exact alignment) from Algorithm 3. In our experiments with RRCA, we have set δ such that computing an optimal alignment in recursive call should not take longer than 100 ms, e.g. for $k = 8$, we obtained $\delta = 161.6$.

We analyzed how much time RRCA spends on different parts of the algorithm for aligning 10 human Chromosome 1 (total runtime was three minutes). Creating the index structure for references, i.e. initial reference and references in recursive calls, dominates the runtime (45.8%). The exact alignment of small fragments has the second highest share of the runtime(16.7%). Decompression of sequences (13.4%), compression of sequences (11.7%), and other parts of RRCA (12.2%) follow, respectively.

We have performed additional experiments for the alignment of protein sequences using benchmark BaliBase 3. Even for the most similar set of sequences (around 40% identity), RRCA cannot find good initial alignments and falls back

---

[3] http://1001genomes.org/data/GMI/GMINordborg2010/releases/current/

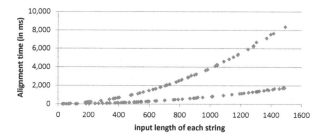

Fig. 6. Alignment of four (lower curve) and eight (upper curve) random sequences of different lengths with SeqAn

to computing an optimal alignment. Similar substrings are not long enough in these short protein sequences and often only shared by small subsets of the whole collection. RRCA will not work for the alignment of sequences from different species, e.g. derived from human and mouse: Similar substrings are not long enough to exploit the benefit of referential compression. In addition, computing the referential compression of a mouse chromosome against a human chromosome is very time consuming. To sum up, alignment technique implemented RRCA cannot easily deal with large rearrangements and synteny. RRCA is tailored towards alignment of long sequences from the same species.

5   Conclusion

RRCA recursively computes a MSA using referential compression for fast identification of chaining fragments. We show that RRCA computes nearly optimal alignments for shorter sequences and for long sequences results with a similar score as competitors. RRCA is orders of magnitude faster than competitors and allows to align sequences within few seconds that take hours with other programs.

We see two major directions for future work. First, it should be investigated how to further improve MSA for very long sequences, in terms of alignment time and alignment quality. The key for improvement is 1) to extend our simple greedy strategy for selection of colinear fragments and 2) to find a heuristic for selecting a reference during the recursive compression step. Our results show that the selection of the longest reference already produces good results, but more sophisticated strategies might yield alignments with higher scores. On the other hand, the run time of sophisticated techniques, which analyze (parts of) each sequence, will undoubtedly increase alignment time. Thus, selecting an efficient strategy for improving alignment times and alignment quality is challenging problem.

In addition, we think that it will be helpful to run an iterative alignment on top of RRCA to improve the quality (scores) of alignments. Second, running times of RRCA can be reduced by investigating different index structures for referential compression. It is important to note that the indexing time of the reference sequences is dominating the runtime (and not the lookup of matches alone). Thus we believe that a lightweight index structure, in terms of indexing time, can further decrease alignment times.

References

1. 1000 Genomes Project Consortium. A map of human genome variation from population-scale sequencing 467(7319), 1061–1073 (October 2010), http://dx.doi.org/10.1038/nature09534
2. Abouelhoda, M.I., Ohlebusch, E.: Multiple genome alignment: Chaining algorithms revisited. In: Baeza-Yates, R., Chávez, E., Crochemore, M. (eds.) CPM 2003. LNCS, vol. 2676, pp. 1–16. Springer, Heidelberg (2003), http://dx.doi.org/10.1007/3-540-44888-8_1
3. Angiuoli, S.V., Salzberg, S.L.: Mugsy: fast multiple alignment of closely related whole genomes. Bioinformatics 27(3), 334–342 (2011)
4. Brudno, M., Chapman, M., Göttgens, B., Batzoglou, S., Morgenstern, B.: Fast and sensitive multiple alignment of large genomic sequences. BMC Bioinformatics 4, 66 (2003)
5. Cao, J., Schneeberger, K., Ossowski, S., Günther, T., Bender, S., Fitz, J., Koenig, D., Lanz, C., Stegle, O., Lippert, C., Wang, X., Ott, F., Müller, J., Alonso-Blanco, C., Borgwardt, K., Schmid, K.J., Weigel, D.: Whole-genome sequencing of multiple Arabidopsis thaliana populations. Nature Genetics 43(10), 956–963 (2011), http://dx.doi.org/10.1038/ng.911
6. Carillo, H., Lipman, D.: The multiple sequence alignment problem in biology. SIAM Journal of Applied Math 48, 1073–1082 (1988)
7. Chen, X., Tompa, M.: Comparative assessment of methods for aligning multiple genome sequences. Nat. Biotech. 28(6), 567–572 (2010), http://dx.doi.org/10.1038/nbt.1637
8. Cohn, M., Khazan, R.: Parsing with prefix and suffix dictionaries. In: Data Compression Conference, pp. 180–189 (1996)
9. Deorowicz, S., Danek, A., Grabowski, S.: Genome compression: a novel approach for large collections. Bioinformatics 29(20), 2572–2578 (2013)
10. Deorowicz, S., Debudaj-Grabysz, A., Gudyś, A.: Kalign-LCS — A more accurate and faster variant of kalign2 algorithm for the multiple sequence alignment problem. In: Gruca, A., Czachórski, T., Kozielski, S. (eds.) Man-Machine Interactions 3. AISC, vol. 242, pp. 499–506. Springer, Heidelberg (2014), http://dx.doi.org/10.1007/978-3-319-02309-0_54
11. Deorowicz, S., Grabowski, S.: Robust Relative Compression of Genomes with Random Access. Bioinformatics, Oxford, England (September 2011), http://dx.doi.org/10.1093/bioinformatics/btr505
12. Döring, A., Weese, D., Rausch, T., Reinert, K.: Seqan an efficient, generic C++ library for sequence analysis. BMC Bioinformatics 9 (2008)
13. Edgar, R.C.: Muscle: a multiple sequence alignment method with reduced time and space complexity. BMC Bioinformatics 5(1) (August 2004), http://dx.doi.org/10.1186/1471-2105-5-113
14. Ferrada, H., Gagie, T., Hirvola, T., Puglisi, S.J.: AliBI: An Alignment-Based Index for Genomic Datasets. ArXiv e-prints (July 2013)

15. Gross, S.S., Brent, M.R.: Using multiple alignments to improve gene prediction. J. Comput. Biol., 379–393 (2005)
16. Gusfield, D.: Algorithms on strings, trees, and sequences: computer science and computational biology. Cambridge University Press, New York (1997)
17. Huang, L., Popic, V., Batzoglou, S.: Short read alignment with populations of genomes. Bioinformatics 29(13), i361–i370 (2013),
    http://dx.doi.org/10.1093/bioinformatics/btt215
18. Katoh, K., Standley, D.M.: MAFFT Multiple Sequence Alignment Software Version 7: Improvements in Performance and Usability. Molecular Biology and Evolution 30(4), 772–780 (2013), http://dx.doi.org/10.1093/molbev/mst010
19. Kemena, C., Notredame, C.: Upcoming challenges for multiple sequence alignment methods in the high-throughput era. Bioinformatics 25(19), 2455–2465 (2009)
20. Kreft, S., Navarro, G.: Lz77-like compression with fast random access. In: Proceedings of the 2010 Data Compression Conference, pp. 239–248. IEEE Computer Society Press, Washington, DC (2010), http://dx.doi.org/10.1109/DCC.2010.29
21. Kuruppu, S., Puglisi, S., Zobel, J.: Optimized relative lempel-ziv compression of genomes. In: Australasian Computer Science Conference (2011)
22. Larkin, M., Blackshields, G.: Brown: Clustal w and clustal x version 2.0. Bioinformatics 23(21), 2947–2948 (2007),
    http://dx.doi.org/10.1093/bioinformatics/btm404
23. Larsson, J., Moffat, A.: Offline dictionary-based compression. In: Proceedings of the IEEE Data Compression Conference, pp. 296–305 (March 1999)
24. McCreight, E.: Efficient algorithms for enumerating intersection intervals and rectangles. Tech. rep., Xerox Paolo Alte Research Center (1980)
25. Mewes, H., Albermann, K., Bähr, M., Frishman, D., Gleissner, A., Hani, J., Heumann, K., Kleine, K., Maierl, A., Oliver, S., Pfeiffer, F., Zollner, A.: Overview of the yeast genome. Nature 387(6632 Suppl.), 7–65 (1997),
    http://www.nature.com/doifinder/10.1038/42755
26. Needleman, S.B., Wunsch, C.D.: A general method applicable to the search for similarities in the amino acid sequence of two proteins. Journal of molecular biology 48(3), 443–453 (1970), http://view.ncbi.nlm.nih.gov/pubmed/5420325
27. Notredame, C., Higgins, D.G., Heringa, J.: T-Coffee: A novel method for fast and accurate multiple sequence alignment.. Journal of molecular biology 302(1), 205–217 (2000), http://dx.doi.org/10.1006/jmbi.2000.4042, doi:10.1006/jmbi.2000.4042
28. Notredame, C.: Recent Evolutions of Multiple Sequence Alignment Algorithms. PLoS Computational Biology 3(8), e123 (2007),
    http://dx.doi.org/10.1371/journal.pcbi.0030123
29. Roytberg, M., Gambin, A., Noe, L., Lasota, S., Furletova, E., Szczurek, E., Kucherov, G.: On subset seeds for protein alignment. IEEE/ACM Transactions on Computational Biology and Bioinformatics 6(3), 483–494 (2009),
    http://dx.doi.org/10.1109/TCBB.2009.4
30. Schmidt, M., Lipson, H.: Distilling free-form natural laws from experimental data. Science 324(5923), 81–85 (2009)
31. Schneeberger, K., Hagmann, J., Ossowski, S., Warthmann, N., Gesing, S., Kohlbacher, O., Weigel, D.: Simultaneous alignment of short reads against multiple genomes. Genome biology 10(9), R98+ (2009),
    http://dx.doi.org/10.1186/gb-2009-10-9-r98
32. Wandelt, S., Leser, U.: FRESCO: Referential compression of highly-similar sequences. IEEE/ACM Transactions on Computational Biology and Bioinformatics 99(PrePrints), 1 (2013)

33. Wang, L., Jiang, T.: On the complexity of multiple sequence alignment. J. Comput. Biol. 1(4), 337–348 (1994), http://view.ncbi.nlm.nih.gov/pubmed/8790475

34. Wong, K.M., Suchard, M.A., Huelsenbeck, J.P.: Alignment Uncertainty and Genomic Analysis. Science 319(5862), 473–476 (2008), http://dx.doi.org/10.1126/science.1151532

35. Yu, H.J., Huang, D.S.: Normalized feature vectors: A novel alignment-free sequence comparison method based on the numbers of adjacent amino acids. IEEE/ACM Transactions on Computational Biology and Bioinformatics 10(2), 457–467 (2013), http://dx.doi.org/10.1109/TCBB.2013.10

36. Zhang, Z., Raghavachari, B., Hardison, R.C., Miller, W.: Chaining multiple-alignment blocks. Journal of Computational Biology 1(3), 217–226 (1994)

37. Ziv, J., Lempel, A.: A universal algorithm for sequential data compression. IEEE Transactions on Information Theory 23(3), 337–343 (1977)

Exact Protein Structure Classification Using the Maximum Contact Map Overlap Metric

Inken Wohlers[1], Mathilde Le Boudic-Jamin[2], Hristo Djidjev[3],
Gunnar W. Klau[4], and Rumen Andonov[2]

[1] Genome Informatics, University of Duisburg-Essen, Germany
inken.wohlers@uni-due.de
[2] INRIA Rennes - Bretagne Atlantique and University of Rennes 1, France
{rumen.andonov,mathilde.le_boudic-jamin}@irisa.fr
[3] Los Alamos National Laboratory, Los Alamos, NM, USA
djidjev@lanl.gov
[4] Life Sciences, CWI, Science Park 123, 1098 XG Amsterdam, The Netherlands
gunnar.klau@cwi.nl

Abstract. In this work we propose a new distance measure for comparing two protein structures based on their contact map representations. We show that our novel measure, which we refer to as the *maximum contact map overlap (max-CMO) metric*, satisfies all properties of a metric on the space of protein representations. Having a metric in that space allows to avoid pairwise comparisons on the entire database and thus to significantly accelerate exploring the protein space compared to non-metric spaces. We show on a gold-standard classification benchmark set of $6,759$ and $67,609$ proteins, resp., that our exact k-nearest neighbor scheme classifies up to 95% and 99% of queries correctly. Our k-NN classification thus provides a promising approach for the automatic classification of protein structures based on contact map overlap.

Keywords: k-nearest neighbours, metric spaces, maximum contact map overlap, automatic classification of proteins.

1 Introduction

Understanding the functional role and evolutionary relationships of proteins is key to answering many important biological and biomedical questions. Because the function of a protein is determined by its structure and because structural properties are usually conserved throughout evolution, such problems can be better approached if proteins are compared based on their representations as three-dimensional structures rather than as sequences. Databases such as SCOP [14] and CATH [15] have been built to organize the space of protein structures. Both SCOP and CATH, however, are constructed partly based on manual curation, and many of the currently over $98,000$ protein structures in the protein data bank (PDB) [3] are still unclassified. Moreover, classifying a newly found structure manually is both expensive in terms of human labor and slow. Therefore,

A.-H. Dediu, C. Martín-Vide, and B. Truthe (Eds.): AlCoB 2014, LNBI 8542, pp. 262–273, 2014.
© Springer International Publishing Switzerland 2014

computational methods that can accurately and efficiently complete such classifications will be highly beneficial. Basically, given a query protein structure, the problem is to find its place in a classification hierarchy of structures, for example, to predict its family or superfamily in the SCOP database.

One approach to solving that problem is based on having introduced a meaningful distance measure between any two protein structures. Then the family of a query protein q can be determined by comparing the distances between q and members of candidate families and choosing a family whose members are "closer" to q than members of the other families, where the precise criteria for deciding which family is closer depend on the specific implementation. The key condition and a crucial factor for the quality of the classification result is having an appropriate distance measure between proteins.

Several such distances have been proposed, each having its own advantages. Recently, a number of approaches based on a graph-based measure of closeness called *contact map overlap (CMO)* [7] have been shown to perform well [2,5,11, 12,16,19,20]. Informally, CMO corresponds to the maximum size of a common subgraph of the two contact map graphs, see the next section for the formal definition. Although CMO is a widely used measure, none of the CMO-based distance methods suggested so far satisfies the triangle inequality and, hence, introduces a metric on the space of protein representations. Having a metric in that space establishes a structure that allows much faster exploration of the space compared to non-metric spaces. For instance, all previous CMO-based algorithms require pairwise comparisons of the query with the entire database. With the rapid increase of the protein databases, such a strategy will unavoidably create performance problems even if the individual comparisons are fast.

In this work we propose a new distance measure for comparing two protein structures based on their contact map representations. We show that our novel measure, which we refer to as the *maximum contact map overlap (max-CMO) metric*, satisfies all properties of a metric. This enables us to describe a given protein database as a metric space where we model each protein family as a ball with a specially chosen protein from the family as center. We exploit this representation to accurately and efficiently classify a query protein according to its k nearest neighbors. We demonstrate that using polynomial-time approximations of max-CMO in terms of lower-bound upper-bound intervals speeds up the classification process significantly, without sacrificing its accuracy. We point out that our approach is not heuristic and guarantees solving the classification problem to *provable optimality* with respect to our max-CMO metric and that we do so without having to compute all query-target alignments to optimality.

We show on a gold-standard classification benchmark set of 6,759 proteins that our exact k-nearest neighbor scheme classifies up to 224 out of 236 queries correctly, and on a large, extended version of the data set that contains 67,609 proteins even up to 1361 out of 1369 queries. Our k-NN classification thus provides a promising approach for the automatic classification of protein structures based on flexible contact map overlap alignments.

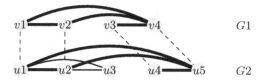

Fig. 1. The alignment visualized with dashed lines $((v_1 \leftrightarrow u_1)(v_2 \leftrightarrow u_2)(v_3 \leftrightarrow u_4)(v_4 \leftrightarrow u_5))$ maximizes the number of the common edges between the graphs G_1 and G_2. The alignment activates four common edges that are emphasized in bold (i.e., $\mathrm{CMO}(G_1, G_2) = 4$).

Amongst the other existing (non-CMO) protein structure comparison methods we are aware of only one satisfying the triangle inequality. This so called scaled Gauss metric (SGM) introduced in [17] and further developed in [8] is shown to be very successful in automatic classification. In our work, however, we focus on contact map overlap and a comparison to classification algorithms based on different concepts is outside the scope of this paper. There have also been works on graph distances motivated by applications unrelated to protein structure comparison. Closest to our work are [4, 9], where a graph distance is defined based on the maximum common subgraph of two graphs. But unlike max-CMO, which takes into account the number of edges in contact map overlap, the authors in [4, 9] use the number of nodes for the size of a graph.

2   The Maximum Contact Map Overlap Metric

We introduce here the notions of contact map overlap (CMO) and the related max-CMO distance between protein structures. A *contact map* describes the structure of a protein P in terms of a simple, undirected graph $G = (V, E)$ with vertex set V and edge set E. The vertices of V are linearly ordered and correspond to the sequence of residues of P. Edges denote residue *contacts*, that is, pairs of residues that are close to each other. More precisely, there is an edge (i, j) between residues i and j iff the Euclidean distance in the protein fold is smaller than a given threshold. The *size* $|G| := |E|$ of a contact map is the number of its contacts. Given two contact maps $G_1(V, E_1)$ and $G_2(U, E_2)$ for two protein structures, let $I = (i_1, i_2, \ldots, i_m)$ and $J = (j_1, j_2, \ldots, j_m)$ be subsets of V and U, respectively, respecting the linear order. Vertex sets I and J encode an *alignment* of G_1 and G_2 in the sense that vertex i_1 is aligned to j_1, i_2 to j_2 and so on. In other words, the alignment (I, J), is a one-to-one mapping between the sets V and U. Given an alignment (I, J), a shared contact (or common edge) occurs if both $(i_k, i_l) \in E_1$ and $(j_k, j_l) \in E_2$ exist. We say in this case that the shared contact (i_k, i_l) is *activated* by the alignment (I, J). The maximum contact overlap problem consists in finding an alignment (I^*, J^*) that maximizes the number of shared contacts and $\mathrm{CMO}(G_1, G_2)$ denotes then this maximum number of shared contacts between the contact maps G_1 and G_2, see Figure 1.

Computing $CMO(G_1, G_2)$ is NP-hard following from [10]. Nevertheless, maximum contact map overlap has been shown to be a meaningful way for comparing two protein structures [2,5,11,12,19,20]. Previously, several distances have been proposed based on the maximum contact map overlap, for example, $D_{\min}$ [5,16] and D_{sum} [2,11,20] with

$$D_{\min}(G_1, G_2) = 1 - \frac{CMO(G_1, G_2)}{\min\{|E_1|, |E_2|\}} \text{ and } D_{\text{sum}}(G_1, G_2) = 1 - \frac{2CMO(G_1, G_2)}{|E_1| + |E_2|} .$$

These distances have the disadvantage that they are no metrics as the following lemma shows (see the extended version [18] for a proof).

Lemma 1. *Distances $D_{\min}$ and D_{sum} do not satisfy the triangle inequality.*

Let $G_1(V, E_1), G_2(U, E_2)$ be two contact maps graphs. We propose a new distance

$$D_{\max}(G_1, G_2) = 1 - \frac{CMO(G_1, G_2)}{\max\{|E_1|, |E_2|\}} . \tag{1}$$

The following claim states that $D_{\max}$ is indeed a distance metric on the space of contact maps and we refer to it as the max-CMO metric.

Lemma 2. *$D_{\max}$ is a metric on the space of contact maps.*

Proof. To prove the triangle inequality for the function $D_{\max}$, we consider three contact maps $G_1(V, E_1), G_2(U, E_2), G_3(W, E_3)$, and we want to prove that $D_{\max}(G_1, G_2) + D_{\max}(G_2, G_3) \geq D_{\max}(G_1, G_3)$. We will use the fact that a similar function d_{max} on sets is a metric, which is defined as

$$d_{\max}(A, B) = 1 - \frac{|A \cap B|}{\max\{|A|, |B|\}} . \tag{2}$$

The mapping $\mathcal{M}$ corresponding to $CMO(G_1, G_2)$ generates an alignment (V', U'), where $V' \subseteq V$ and $U' \subseteq U$ are ordered sets of vertices preserving the order of V and U, correspondingly. Since $\mathcal{M}$ is a one-to-one mapping, we can rename the vertices of U' to the names of the corresponding vertices of V' and keep the old names of the vertices of $U \setminus U'$. Denote the resulting ordered vertex set by $\overline{U}$ and denote by $\overline{E_2}$ the corresponding set of edges. Define the graph $\overline{G_2} = (\overline{U}, \overline{E_2})$. Note that $|\overline{E_2}| = |E_2|$ and any common edge discovered by $CMO(G_1, G_2)$ has the same endpoints (after renaming) in $\overline{E_2}$ as in E_1; hence $CMO(G_1, G_2) = CMO(G_1, \overline{G_2}) = |E_1 \cap \overline{E_2}|$. Then from (2)

$$D_{\max}(G_1, G_2) = 1 - \frac{CMO(G_1, G_2)}{\max\{|E_1|, |E_2|\}} = 1 - \frac{|E_1 \cap \overline{E_2}|}{\max\{|E_1|, |\overline{E_2}|\}} = d_{\max}(E_1, \overline{E_2}) .$$

Similarly, we compute the mapping corresponding to $CMO(\overline{G_2}, G_3)$ and generate an optimal alignment $(\overline{U'}, W')$. As before, we use the mapping to rename the vertices of W' to the corresponding vertices of $\overline{U'}$ and denote the resulting sets of vertices and edges by $\overline{W}$ and $\overline{E_3}$. Similarly to the above case, it follows that $D_{\max}(G_2, G_3) = d_{\max}(\overline{E_2}, \overline{E_3})$. Combining the last two equalities, we get

$$D_{\max}(G_1, G_2) + D_{\max}(G_2, G_3) = d_{\max}(E_1, \overline{E_2}) + d_{\max}(\overline{E_2}, \overline{E_3})$$
$$\geq d_{\max}(E_1, \overline{E_3}). \tag{3}$$

On the other hand, $E_1 \cap \overline{E_3}$ contains only edges jointly activated by the alignments (V', U') and $(\overline{U'}, W')$ and its cardinality is not larger than $CMO(G_1, G_3)$, which corresponds to the optimal alignment between G_1 and G_3. Hence $|E_1 \cap \overline{E_3}| \leq CMO(G_1, G_3)$ and, since $|\overline{E_3}| = |E_3|$,

$$d_{\max}(E_1, \overline{E_3}) = 1 - \frac{|E_1 \cap \overline{E_3}|}{\max\{|E_1|, |\overline{E_3}|\}} \geq 1 - \frac{CMO(G_1, G_3)}{\max\{|E_1|, |E_3|\}} = D_{\max}(G_1, G_3).$$

Combining the last inequality with (3) proves the triangle inequality for $D_{\max}$. The other two properties of a metric, that $D_{\max}(G_1, G_2) \geq 0$ with equality if and only if $G_1 = G_2$ and $D_{\max}(G_1, G_2) = D_{\max}(G_2, G_1)$, are obviously also true. □

If instead of $CMO(G_1, G_2)$ one computes lower or upper bounds for its value, replacing those values in (1) produces an upper or lower bound for $D_{\max}$, respectively.

3   Nearest Neighbor Classification of Protein Structures

We suggest to approach the problem of classifying a given query protein structure with respect to a database of target structures based on a majority vote of the k nearest neighbors in the database. Nearest neighbor classification is a simple and popular machine learning strategy with strong consistency results, see for example [1]. An important feature of our approach is that it is based on a metric and we fully profit from all usual benefits when exploring a metric space [13].

3.1   Finding Family Representatives

In order to minimize the number of targets with which a query has to be compared directly, i.e., via computing an alignment, we designate a representative central structure for each family. Let d denote any metric. Each family $\mathcal{F} \in \mathcal{C}$ can then be characterized by a representative structure $R_{\mathcal{F}}$ and a family radius $r_{\mathcal{F}}$ determined by

$$R_{\mathcal{F}} = \arg \min_{A \in \mathcal{F}} \max_{B \in \mathcal{F}} d(A, B), \quad r_{\mathcal{F}} = \min_{A \in \mathcal{F}} \max_{B \in \mathcal{F}} d(A, B). \tag{4}$$

In order to find $R_{\mathcal{F}}$ and $r_{\mathcal{F}}$, we compute, during a preprocessing step, all pairwise distances within $\mathcal{F}$. We aim to compute these distances as precise as possible, using a sufficiently long run time for each pairwise comparison. Since proteins from the same family are structurally similar, the alignment algorithm performs favorably and we can usually compute intra-family distances optimally. These distances obtained during preprocessing are later re-used during k-NN classification for computing triangle bounds.

3.2   Dominance between Target Protein Structures

In order to find the target structures which are closest to a query q, we have to decide for a pair of targets A and B which one is closer. We call such a relationship between two target structures *dominance*:

Definition 3 (dominance). *Protein A dominates protein B with respect to a query q if and only if $d(q, A) < d(q, B)$.*

In order to conclude that A is closer to q than B, it may not be necessary to know $d(q, A)$ and $d(q, B)$ exactly. It is sufficient that A directly dominates B according to the following rule.

Lemma 4 (direct dominance). *Protein A dominates protein B with respect to a query q if $\overline{d}(q, A) < \underline{d}(q, B)$, where $\overline{d}(q, A)$ and $\underline{d}(q, B)$ are an upper and lower bound on $d(q, A)$ and $d(q, B)$, respectively.*

Proof. Follows from the inequalities $d(q, A) \leq \overline{d}(q, A) < \underline{d}(q, B) \leq d(q, B)$.   □

The idea of dominance is crucial for reducing the number of computations in our approach. Based on the relationship of polynomial-time computed lower and upper bounds a dominated protein is discarded from further consideration. Although the precise distance between proteins and the associated alignment are not computed, which is an NP-hard problem, the accuracy of the classification is not sacrificed. In its simplest form this idea has been first proposed in [12]. Here we extend it by exploiting the properties of a metric space as shown below.

Given a query q, a representative r and a target A, the triangle inequality provides an upper bound, while the reverse triangle inequality provides respectively a lower bound on the distance from query q to target A

$$d(q, A) \leq d(q, r) + d(r, A) \text{ and } d(q, A) \geq |d(q, r) - d(r, A)| \ .$$

We define the *triangle upper (resp. lower) bound* as

$$\overline{d}^{\triangle}(q, A) = \min_{r \in R}\{\overline{d}(q, r) + \overline{d}(r, A)\} \ ,$$

$$\underline{d}_{\triangledown}(q, A) = \max_{r \in R}\max\{\underline{d}(q, r) - \overline{d}(r, A), \underline{d}(r, A) - \overline{d}(q, r)\} \ .$$

Lemma 5. $\underline{d}_{\triangledown}(q, A) \leq d(q, A) \leq \overline{d}^{\triangle}(q, A)$

Proof. $\underline{d}_{\triangledown}(q, A) = \max_{r \in R}\max\{\overline{d}(q, r) - \underline{d}(r, A), \overline{d}(r, A) - \underline{d}(q, r)\} \leq \max_{r \in R}|d(q, r) - d(r, A)| \leq d(q, A) \leq \min_{r \in R} d(q, r) + d(r, A) \leq \min_{r \in R} \overline{d}(q, r) + \overline{d}(r, A) = \overline{d}^{\triangle}(q, A)$.   □

Using Lemma 5 we derive supplementary sufficient conditions for dominance, which we call *indirect dominances*.

Lemma 6 (indirect dominance). *Protein A dominates protein B with respect to query q if $\overline{d}^{\triangle}(q, A) < \underline{d}_{\triangledown}(q, B)$.*

Proof. $d(q, A) \overset{\text{Lemma 5}}{\leq} \overline{d}^{\triangle}(q, A) < \underline{d}_{\triangledown}(q, B) \overset{\text{Lemma 5}}{\leq} d(q, B)$.   □

3.3   Classification Algorithm

K-nearest neighbor classification is a scheme which assigns the query to the class to which most of the k targets belong which are closest to the query. In order to classify, we therefore need to determine the k structures with minimum distance to the query and assign the super-family to which the majority of the neighbors belong. As seen in the previous section, we can use bounds to decide whether a structure is closer to the query than another structure. This can be generalized to deciding whether or not is it possible for a structure to be among the k closest structures in the following way. We construct two priority queues LB and UB whose elements are $(t, lb(q, t)))$ and $(t, ub(q, t))$, respectively, where q is the query and t the target. The algorithm works with any lower bound $lb(q, t)$ on the distance between q and t, for example $\underline{d}(q, t)$ or $\underline{d}_{\bigtriangledown}(q, t)$ and with any upper bound $ub(q, t)$ on $d(q, t)$, for example $\overline{d}(q, t)$ or $\overline{d}^{\triangle}(q, t)$. In our current implementation we use D_{max} as a distance while lower and upper bounds are polynomially computed based on Lagrangian relaxation as explained in [2]. The quality of these bounds for the purpose of protein classification has been already demonstrated in [11,12]. LB and UB are sorted in the order of increasing distance. The k-th element in queue UB is denoted by t_k^{UB}. Its distance to the query, $d(q, t_k^{\mathrm{UB}})$, is the distance for which at least k target elements are closer to the query. Therefore we can safely discard all those targets which have a lower bound distance of more than $d(q, t_k^{\mathrm{UB}})$ to query q. That is, t_k^{UB} dominates all targets t for which $lb(q, t) > ub(q, t_k^{\mathrm{UB}})$.

4   Validation Setup

We evaluated the classification performance and efficiency of different types of dominance of our algorithm on domains from SCOPCath [6], a benchmark that consists of a consensus of the two major structural classifications SCOP [14] (version 1.75) and Cath [15] (version 3.2.0). We use this consensus benchmark in order to obtain a gold-standard classification that very likely reflects structural similarities that are detectable automatically, since two classifications, each using a mix of expert knowledge and automatic methods, agree in their super-family assignments. SCOPCath has been filtered such that it only contains proteins with less than 50% sequence identity. Since this results in a rather small benchmark with only 6, 759 structures, we added these filtered structures for our evaluation in order to have a benchmark representative of the merged databases SCOP and Cath. There were 264 domains in extended SCOPCath which share more than 50% sequence similarity with a domain in SCOPCath, but do not both belong to the same SCOP family; since their families are perhaps not in SCOPCath and their classification in SCOP and Cath may not agree, we removed them. This way we obtained 60, 850 additional structures. These belong to 1, 348 super-families and 2, 480 families of which 2, 093 families have more than one member. For SCOPCath, there are 1, 156 multi-member families. Structures and families are divided into classes according to Table 1. For super-family assignment, we compared a structure only to structures of the corresponding class

Table 1. For every protein class, the table lists the number of structures in SCOPCath (str) and extended SCOPCath (ext), the corresponding number of families (fam) and superfamilies (sup)

class	a	b	c	d	e	f	g	h	i	j	k
# str	1195	1593	1774	1591	30	103	342	72	11	38	10
# ext	10796	19215	17497	15679	349	1006	2398	520	43	81	25
# fam	524	516	548	632	6	59	121	32	5	29	8
# sup	303	266	191	375	6	52	82	31	5	29	8

since class membership can in most cases be determined automatically, for example by a program that computes secondary structure content. We then computed all-versus-all distances (2) or distance bounds within each family using optimal maximum contact map overlap or the upper Lagrangian bound on it and determined the family representative according to Equation (4). For every pairwise distance computation, we used a maximum time limit of 10 s. Since most comparisons were computed optimally, the average run time is approximately 2 s.

For classification, we randomly selected one query from every family with at least six members. This resulted in 236 queries for SCOPCath and 1,369 queries for the extended SCOPCath benchmark. For every query, the $k = 10$ nearest neighbor structures from SCOPCath and extended SCOPCath, respectively, were computed using our k-NN Algorithm. The algorithm is a two-step procedure. First it improves distance bounds by applying several rounds of triangle dominance, in which the maximum contact map overlap bounds from query to representatives are updated, and second it switches to pairwise dominance, for which the distance to any remaining target is computed. In the first step, query representative distances are computed using an initial time limit of $\tau = 1$ s, then triangle dominance is applied to all targets and the algorithm iterates with time limit doubled until a termination criterion is met. This way, bounds on query target distances are improved successively. The computation of triangle dominance terminates if any of the following holds (i) k targets are left (ii) all query-representative distances have been computed optimally or with a time limit of 32 CPU seconds (iii) the number of targets did not reduce from one round to the next. Pairwise dominance terminates if any of the following holds (i) k targets are left or all remaining targets belong to the same super-family (ii) all query-target distances have been computed with a time limit of 32 CPU seconds. The query is then assigned to the super-family to which the majority of the k nearest neighbors belongs. In cases in which the pairwise dominance terminates with more than k targets or more than one super-family remains, the exact k nearest neighbors are not known. In that case we order the targets based on the upper bound distance to the query and assign the super-family using the top ten queries. In the case that there is a tie among the superfamilies to which the top ten targets belong, we report this situation. In order to investigate the impact of k on classification accuracy, we additionally decreased k from 9 to 1, using each time the $k + 1$ nearest neighbors from the classification result for

$k + 1$. In the case that for a query more than $k + 1$ queries remained in this classification, we used all of them for searching for the k nearest neighbors, but put an additional termination criterion which prevents extremely long run times for a few queries. Due to the large number of computations, classifications were run on different architectures on clusters with various load and are therefore only used for order of magnitude comparison.

5   Computational Results

5.1   Characterizing the Distance Measure

In a first, preprocessing step we evaluate how well our distance metric captures known similarities and differences between protein structures by computing intra-family and inter-family distances. A good distance for structure comparison should pool similar structures, i.e., from the same family, whereas it should locate dissimilar structures from different families far apart from each other. In order to quantify such characteristics, we compute for each family with at least two members a central, representative structure according to Equation (4). Therefore, we compute the distance between any two structures that belong to the same family. Such intra-family distances should ideally be small. We observe that the distribution of intra-family distances differ between classes and are usually smaller than 0.5, except for class c. For the four major protein classes, there is a distance peak close to 0 and another one around 0.2.

We then compute a radius around the representative structure that encompasses all structures of the corresponding family. The number of families with a given radius decreases nearly linearly from 0 to 0.6, with most families having a radius close to zero, and almost no families having a radius greater than 0.6.

Considering that the distance metric is bound to be within 0 and 1, inter-family distances and radii show that our distance overall captures well the similarity between structures. Further, we investigate the distance between protein families by computing their overlap value as defined by $d(R_{\mathcal{F}_1}, R_{\mathcal{F}_2}) - r_{\mathcal{F}_1} - r_{\mathcal{F}_2}$. Most families are not close to each other according to our distance metric. Families of the four most populated classes which belong to different superfamilies overlap in 23-25% of cases for class a, 11-18% for class b, 10-22% for class c and 11-18% for class d. These bounds on the number of overlapping families can be obtained by using the lower and upper bounds on the distances between representatives and the distances between family members appropriately.

5.2   Results for the SCOPCath Benchmark

When classifying the 236 queries of SCOPCath, we achieve between 89% and 95% correct super-family assignments, see Table 2. Remarkably, the highest accuracy is reached for $k=1$, so here just classifying the query as belonging to the super-family of the nearest neighbor is the best choice. Our k-NN classification resulted for any k in a large number of ties, especially for $k=2$, see Table 2. These currently

Table 2. Classification results showing the number of queries out of overall 236 queries for SCOPCath and 1369 queries for extended SCOPCath that have been assigned to a super-family, the number of assignments to the correct superfamily (cor), the number of assignments computed exactly, i.e. queries which terminate with the provable k nearest neighbors (exc), thereof the number of correct classifications (e&c) and the number of ties which do not allow a superfamily assignment based on majority vote

	SCOPCath				ext. SCOPCath			
k	cor	exc	e&c	ties	cor	exc	e&c	ties
10	210	117	110	10	1303	1120	1104	35
9	211	143	134	9	1331	1182	1166	5
8	213	156	149	11	1334	1228	1215	12
7	213	165	155	8	1341	1271	1257	6
6	214	188	178	10	1341	1286	1276	11
5	217	206	198	10	1346	1339	1329	7
4	217	204	195	10	1344	1341	1330	9
3	219	211	205	10	1351	1352	1341	3
2	213	209	206	20	1348	1347	1343	17
1	224	234	224	0	1361	1368	1360	0

unresolved ties also decrease assignment accuracy compared to $k = 1$, for which a tie is not possible. Table 2 further lists the number of queries which have been assigned, where exact denotes that the provable k nearest neighbors have been computed. The percentage of exactly computed nearest neighbors varies between 50% and 99% and increases with decreasing k. A likely reason for this is that the larger k, the weaker is the k-th distance upper bound that is used for domination, especially if the target on rank k is dissimilar to the query. Since SCOPCath domains have low sequence similarity, this is likely to happen. It is also interesting to note that there are for any k quite a few queries which have been assigned exact, but which are nonetheless wrongly assigned, see Table 2. These are cases in which our distance metric fails in ranking the targets correctly with respect to gold standard.

5.3   Results for the Extended SCOPCath Benchmark

Our exact k-NN classification can also be successfully applied to larger benchmarks like extended SCOPCath, which are more representative of databases such as SCOP. Here, the benefit of using a metric distance, triangle inequality and k-NN classification is more pronounced. Remarkably, our classification run time on this benchmark that is about an order of magnitude larger than SCOPCath is for most queries of the same order of magnitude as run times on SCOPCath (except for some queries which need an extremely long run time and finally cannot be assigned exactly). Also here, run time varies extremely between queries, between 0.15 and 85.63 hours for queries of the four major classes which could be assigned exactly. The median run time for all 1120 exactly assigned extended SCOPCath queries is 3.8 hours.

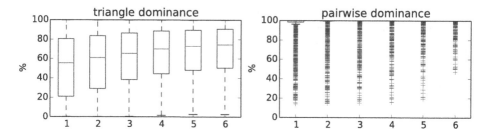

Fig. 2. Boxplots of the percentage of removed targets at each iteration during triangle and pairwise dominance for the 1369 queries of the extended SCOPCath benchmark

The classification results for extended SCOPCath are shown in Table 2. Slightly more queries have been assigned correctly compared to SCOPCath, between 95% and 99%, and significantly more queries have been assigned exactly. Both may reflect that there are now more similar structures within the targets. Further, the number of ties is decreased. Figure 2 displays the progress of the computation. Here, many more target structures are removed by triangle dominance and within the very first iteration of pairwise dominance compared with the SCOPCath benchmark. For example, for most queries, more than 60% of targets are removed by triangle dominance alone. Only very few queries need to explicitly compute the distance to a large percentage of the targets, and most can be assigned after only one round of pairwise dominance.

6   Conclusion

In this work we introduced a new distance based on the CMO measure and proved that it is a true metric, which we call the max-CMO metric. We analyzed the potential of max-CMO for solving the k-NN problem efficiently and *exactly* and built on that basis a protein superfamily classification algorithm. Depending on the value of k, our accuracy varies between 89% for $k = 10$ and 95% for $k = 1$ for SCOPCath and between 95% and 99% for extended SCOPCath. The fact that the accuracy is highest for $k = 1$ indicates that using more sophisticated rules than k-NN may produce even better results.

In summary, our approach provides a general solution to k-NN classification based on a computationally intractable metric for which polynomial upper and lower bounds are available that can successfully be applied for exact large-scale protein superfamily classification.

References

1. Altman, N.S.: An introduction to kernel and nearest-neighbor nonparametric regression. The American Statistician (1992)
2. Andonov, R., Malod-Dognin, N., Yanev, N.: Maximum contact map overlap revisited. J. Comput. Biol. 18(1), 27–41 (2011)

3. Bernstein, F., Koetzle, T., Williams, G., Meyer Jr., E., Brice, M., Rodgers, J., Kennard, O., Shimanouchi, T., Tasumi, M.: The protein data bank: A computer-based archival file for macromolecular structures. J. of Mol. Biol. 112, 535 (1977)
4. Bunke, H., Shearer, K.: A graph distance metric based on the maximal common subgraph. Pattern Recognition Letters 19, 255–259 (1998)
5. Caprara, A., Carr, R., Istrail, S., Lancia, G., Walenz, B.: 1001 optimal PDB structure alignments: integer programming methods for finding the maximum contact map overlap. J. Comput. Biol. 11(1), 27–52 (2004)
6. Csaba, G., Birzele, F., Zimmer, R.: Systematic comparison of SCOP and CATH: a new gold standard for protein structure analysis. BMC Struct. Biol. 9, 23–23 (2009)
7. Godzik, A., Skolnick, J., Kolinski, A.: Regularities in interaction patterns of globular proteins. Protein Eng. 6(8), 801–810 (1993)
8. Harder, T., Borg, M., Boomsma, W., Røgen, P., Hamelryck, T.: Fast large-scale clustering of protein structures using Gauss integrals. Bioinformatics 28(4), 510–515 (2012)
9. Hidovic, D., Pelillo, M.: Metrics for attributed graphs based on the maximal similarity common subgraph. IJPRAI 18(3), 299–313 (2004)
10. Lathrop, R.H.: The protein threading problem with sequence amino acid interaction preferences is NP-complete. Protein Eng. 7(9), 1059–1068 (1994)
11. Malod-Dognin, N., Przulj, N.: Gr-align: fast and flexible alignment of protein 3d structures using graphlet degree similarity. Bioinformatics (2014)
12. Malod-Dognin, N., Le Boudic-Jamin, M., Kamath, P., Andonov, R.: Using dominances for solving the protein family identification problem. In: Przytycka, T.M., Sagot, M.-F. (eds.) WABI 2011. LNCS, vol. 6833, pp. 201–212. Springer, Heidelberg (2011)
13. Moreno-Seco, F., Mico, L., Oncina, J.: A modification of the laesa algorithm for approximated k-nn classification. Pattern Recognition Letters 24, 47–53 (2003)
14. Murzin, A.G., Brenner, S.E., Hubbard, T., Chothia, C.: SCOP: a structural classification of proteins database for the investigation of sequences and structures. J. Mol. Biol. 247(4), 536–540 (1995)
15. Orengo, C.A., Michie, A.D., Jones, S., Jones, D.T., Swindells, M.B., Thornton, J.M.: CATH–a hierarchic classification of protein domain structures. Structure 5(8), 1093–1108 (1997)
16. Pelta, D.A., González, J.R., Moreno Vega, M.: A simple and fast heuristic for protein structure comparison. BMC Bioinformatics 9, 161–161 (2008)
17. Rogen, P., Fain, B.: Automatic classification of protein structure by using gauss integrals. Proceedings of the National Academy of Sciences of the United States of America 100(1), 119–124 (2003)
18. Wohlers, I., Boudic-Jamin, M.L., Djidjev, H., Klau, G.W., Andonov, R.: Exact protein structure classification using the maximum contact map overlap metric. Tech. Rep. LA-UR-14-20815, Los Alamos National Laboratory (2014)
19. Wohlers, I., Malod-Dognin, N., Andonov, R., Klau, G.W.: CSA: comprehensive comparison of pairwise protein structure alignments. Nucleic Acids Research 40(W1), W303–W309 (2012)
20. Xie, W., Sahinidis, N.V.: A reduction-based exact algorithm for the contact map overlap problem. J. Comput. Biol. 14(5), 637–654 (2007)

Author Index

Ahmad, Aitzaz   196
Almirantis, Yannis   220
Althaus, Ernst   25
Amaya Moreno, Liana   35
Andonov, Rumen   262
Arruda, Thiago da Silva   59
Azimi, Sepinoud   95
Azpeitia, Eugenio   235

Carrillo, Miguel   235
Chance, Mark R.   171
Chateau, Annie   47
Ciortuz, Liviu   119

Defterli, Ozlem   35
Dias, Ulisses   59
Dias, Zanoni   59, 146, 158
Djidjev, Hristo   262

Euler, Reinhardt   131

Fügenschuh, Armin   35

Galperin, Michael Y.   1
Giannakopoulos, George   220
Giroudeau, Rodolphe   47
Gratie, Diana-Elena   95
Grigoriev, Alexander   71
Grosse, Ivo   83

Hedtke, Ivo   83
Hildebrandt, Andreas   25
Hildebrandt, Anna Katharina   25

Iancu, Bogdan   95

Kawai, Yosuke   107, 208
Kelk, Steven   71
Klau, Gunnar W.   262
Kojima, Kaname   107, 208
Koonin, Eugene V.   1
Koyutürk, Mehmet   171
Krithara, Anastasia   220

Le Boudic-Jamin, Mathilde   262
Lekić, Nela   71
Lemaitre, Claire   119
Lemarchand, Laurent   131
Lemnian, Ioana   83
Leser, Ulf   247
Lin, Congping   131
Lintzmayer, Carla Negri   146, 158

Maxwell, Sean   171
Mimori, Takahiro   107, 208
Mishra, Bud   183
Müller-Hannemann, Matthias   83
Muñoz, Stalin   235

Nagasaki, Masao   107, 208
Nariai, Naoki   107, 208
Narzisi, Giuseppe   183
Nikolaou, Christoforos   220
Noor, Amina   196
Nounou, Hazem   196
Nounou, Mohamed   196

Ohtsuki, Tomohiko   208

Paliouras, Giorgos   220
Peterlongo, Pierre   119
Petre, Ion   95
Polychronopoulos, Dimitris   220

Rosenblueth, David A.   235

Sato, Yukuto   107, 208
Schatz, Michael C.   183
Serpedin, Erchin   196
Shibuya, Testuo   208
Sparkes, Imogen   131

Wajid, Bilal   196
Wandelt, Sebastian   247
Weber, Gerhard-Wilhelm   35
Wohlers, Inken   262

Yamaguchi-Kabata, Yumi   107, 208